Qualitative Theory of Volterra Difference Equations

Youssef N. Raffoul

Qualitative Theory of Volterra Difference Equations

 Springer

Youssef N. Raffoul
Department of Mathematics
University of Dayton
Dayton, OH, USA

ISBN 978-3-030-07318-3   ISBN 978-3-319-97190-2   (eBook)
https://doi.org/10.1007/978-3-319-97190-2

Mathematics Subject Classification: 369A05, 39A06, 39A10, 39A12, 39A22, 39A23, 39A30, 39A60, 39A70, 39B05, 45Bxx, 45Dxx, 45Jxx

This Springer imprint is published by the registered company Springer Nature Switzerland AG.
The registered company address is: Gewerbestrasse 11, 6330 Cham, Switzerland

Dedicated to the memory of my parents, Naim and Nazha Raffoul, who have worked tirelessly and relentlessly to provide their children with better lives.

Dedicated to my wonderful wife Nancy, who is the wind beneath my wings, and to our awesome children Hannah, Paul, Joseph, and Daniel, and to our adopted daughter Venicia Al Hawach.

Preface

I hope you will enjoy my book. Inside you will find advanced and contemporary topics and techniques on Volterra difference equations. It is my intention to make this book accessible to postgraduate students with basic knowledge in differential equations, difference equations, and real analysis. The subject of Volterra difference equations is a vast and widely open subject with plenty of opportunities for new research. Volterra difference equations provide more realistic models for broad range of phenomena in the natural and engineering sciences. In particular, they are used as mathematical representations of systems for which we know, based on physical principles, that they satisfy conditions of stability and boundedness.

Motivated by recent increased activity in research, this book will provide a systematic approach to the study of the qualitative theory of boundedness, periodicity, and stability of Volterra difference equations. Stability theory is an area that will continue to be of great interest to researchers for a long time because of the usefulness it demonstrates in real life applications. This book attempts to connect two different aspects: the theoretical part of functional difference equations and its applications to Volterra difference equations and to population dynamics. The author will mainly use fixed point theory and Lyapunov functionals to arrive at major results. By comparing both methods to certain equations, some interesting conclusions will be drawn regarding the particular application of each method. Since no existing book or monograph that is solely devoted to Volterra difference equations which encompasses recent results, this book attempts to fill this void.

Researchers and graduate students who are interested in the method of Lyapunov functions or functionals, in the study of boundedness of solutions, in the stability of the zero solution, or in the existence of periodic solutions should be able to use this book as a primary reference and as a resource of current findings. This book contains many open problems and should be of great benefit to those who are pursuing research in functional difference equations and in Volterra difference equations. Great efforts were made to present detailed proofs of theorems so that the book would be self-contained.

The work done by Volterra and Lyapunov over 100 years ago is still attractive to researchers today, but first we provide some background on these mathematicians. Vito Volterra (3 May 1860–11 October 1940) was an Italian mathematician and physicist, known for his contributions to mathematical biology and integral equations. Born in Ancona, then part of the Papal States, into a very poor family, Volterra showed early promise in mathematics before attending the University of Pisa, where he fell under the influence of Enrico Betti, and where he became professor of rational mechanics in 1883. He immediately started work developing his theory of functionals which led to his interest and later contributions in integral and integro-differential equations. His work is summarized in his book *Theory of Functionals and of Integral and Integro-Differential Equations* (1930). Volterra emphasized consistently that differential equations are, at best, only rough approximations of actual ecological systems. They would apply only to animals without age or memory, which eat all the food they encounter and immediately convert it into offspring. Anything more realistic would yield integro-differential rather than differential equations. This phenomenon will be discussed in Chapter 5.

Lyapunov functions are named after Alexander Lyapunov, a Russian mathematician who in 1892 published his book *The General Problem of Stability of Motion*. In [110], Lyapunov was the first to consider the modifications necessary in nonlinear systems to the linear theory of stability based on linearizing near a point of equilibrium. His work, initially published in Russian and then translated to French, received little attention for many years. Interest in Lyapunov stability started suddenly during the Cold War period when his method was found to be applicable to the stability of aerospace guidance systems, which typically contain strong nonlinearities not treatable by other methods. More recently the concept of the Lyapunov exponent related to Lyapunov's First Method of discussing stability has received wide interest in connection with chaos theory.

In this book, Chapter 1 offers an introduction to basic difference calculus including variation of parameters and the concept of fundamental matrix. In addition, we provide basic properties of the z-transform and its usefulness in obtaining stability results concerning Volterra difference equations of convolution type. Later in Chapter 1 we utilize the concept of total stability, which has never been developed and successfully used in Volterra difference equations. In our case, total stability relies on using the notion of the resolvent to express the solution of a given Volterra difference equation. We will derive one of the resolvent equations that we make use of along with Lyapunov functionals to verify our conditions that explicitly depend on the resolvent. Then we discuss necessary and sufficient conditions for the uniform asymptotic stability of the zero solution which we apply to perturbed Volterra difference equations. The materials of Chapter 1 are relatively new and it should serve as an important source for new research ideas which should be of interest for a long time. We end the chapter by offering new research problems that combine the concept of resolvent with Lyapunov functionals.

Chapter 2 is devoted to functional difference equations. Functional equations are general in nature and encompass results concerning existence of periodic solutions, boundedness, and stability for autonomous and nonautonomous difference equations, finite and infinite delay difference equations, and Volterra difference equations. We consider a functional difference system and prove general theorems regarding boundedness of solutions and stability of the zero solution. Our theorems are based on the assumption that there is a Lyapunov functional that satisfies certain inequalities in terms of wedges. The general theorems are applied to finite and scalar and vector Volterra difference equations. In particular, we discuss boundedness of solutions and the stability of the zero solution by the method of Lyapunov functionals. As for the application part to Volterra difference equations, we will have to construct suitable Lyapunov functionals that meet all the requirements of our general theorems. Such a task is difficult but possible with some guidance. We offer two general theorems that guide us through the process of constructing such a Lyapunov functional. It is the author's view that Lyapunov's direct method is the leading technique for dealing with stability and boundedness in many areas of differential and difference equations. Finally, we provide open problems regarding Volterra difference systems of advanced type and offer a discussion that makes it possible for the interested reader to carry on with the proposed research.

Chapter 3 is entirely devoted to the study of fixed point theory in analyzing boundedness and stability in Volterra difference equations or functional difference equations that we write in the form of Volterra difference equations. The chapter serves as an introduction to the theory of fixed point and its use in functional difference equations. Most of the work depends on the three principles: complete metric spaces, the contraction mapping principle and other type of known fixed point theorems, and an elementary variation of parameters formula. In the past one hundred and fifty years, Lyapunov functions/functionals have been exclusively and successfully used in the study of stability and existence of bounded solutions. However, with the method of Lyapunov functionals, we are continually faced with unrelenting difficulties. Those persisting obstacles have forced us to look for other ways to relax stringent conditions, and in particular, for the use of fixed point theory. Lyapunov method requires pointwise conditions, while many real life applications ask for averages. Difference equations have solutions expressed as summation equations which justifies conditions on averages. Another major difficulty is the construction of a suitable Lyapunov function or functional that yields meaningful results. Deep knowledge of the theory of difference equations and real analysis is usually needed when using Lyapunov direct method, unlike the use of fixed point theory. We ease into an exposition on the use of Lyapunov functionals and fixed point theory in Volterra summation equations. We provide a variety of results on existence theory and resolvent. We include a collection of Lyapunov functionals and techniques, which enable researchers to obtain fruitful results concerning the qualitative analysis of solutions. Toward the end of the chapter we compare the method of fixed point theory and Lyapunov functionals and include open problems.

Chapter 4 is entirely devoted to the study of periodic solutions and asymptotically periodic solutions. We begin by introducing known results regarding Volterra difference equations and then proceed to considering functional difference equations with finite and infinite delays and appeal to Schaefer fixed point theorem [159] for the existence of periodic solutions. Then we apply the results to difference equations and Volterra difference equations. The application in Volterra difference equations heavily depends on Lyapunov type functionals for obtaining the a priori bound. We define a homotopy and then appeal to Schaefer's fixed point theorem [159] to show the existence of periodic solutions. The results are applied to scalar nonlinear Volterra difference equations via the use of Lyapunov type functionals. We devote the next section to the study of the existence of periodic and asymptotically periodic solutions in coupled infinite delay Volterra systems by using Schauder fixed point theorem (Theorem 4.7.1). We end the chapter by considering functional difference systems that have constants as their solutions. We use the contraction mapping principle to prove theorems concerning the determination of the unique constants to which each solution converges. It turns out that those constants serve as global attractors. The chapter concludes by proposing open problems regarding the existence of periodic and asymptotically periodic solutions of a Volterra type infinite delay difference equation with the absence of a linear term.

General theorems occupy a central place in the first four chapters. One must ultimately face the task of applications. Abstract ideas can be better understood by analyzing the dynamics of specific equations, especially with applications. With this philosophy in mind, Chapter 5 is solely dedicated to the applications of functional difference equations and particularly, the applications of Volterra difference equations to population dynamics. We begin the chapter by introducing different types of population models. Cone theory is introduced and utilized to prove the existence of positive periodic solutions for functional difference equations. Then, we introduce an infinite delay population model that governs the growth of population $N(n)$ of a single species whose members compete among themselves for the limited amount of food that is available to sustain the population. We use the results of the previous section to obtain the existence of a positive periodic solution. Moreover, from a biologist's point of view, the idea of permanence plays a central role in any competing species. This phenomenon is studied by considering an $(l+m)$-species Lotka-Volterra competition-predation system with several delays.

Chapter 6 pertains to results on the exponential stability, l_p-stability, and instability of the zero solution, using Lyapunov functionals. In our quest for the perfect Lyapunov functional, we discover different types of boundedness and stabilities. We begin the chapter by considering the use of Lyapunov direct method to obtain exponential stability for scalar Volterra difference equations and compare our results to existing literature. It turns out that our Lyapunov direct method for scalar equations cannnot be extended to systems, nor to Volterra equations with infinite delay. We next proceed to the study of nonlinear Volterra difference systems and use a combination of Lyapunov functionals of convolution type coupled with the z-transform to obtain boundedness and stability. In addition, we look into the l_p-stability of the zero solution and its connection with exponential stability. Toward the end of the chapter, we present relatively unexplored area. That is, we utilize a nonstandard discretization scheme and apply it to continuous Volterra integro-differential equations. We will show that under our discretization scheme the stability of the zero solution of the continuous dynamical system is preserved. Also, under the same discretization, using a combination of Lyapunov functionals, Laplace transforms, and z-transforms, we show that the boundedness of solutions of the continuous dynamical system is preserved. We end the chapter with a brief section introducing semigroup, which should give rise to increased research in the application of semigroup to Volterra difference equations. The chapter concludes with multiple open problems.

I would like to thank current and previous chairs of the Department of Mathematics at the University of Dayton for their unwavering support. Also, I would like to thank my colleagues Dr. Paul Eloe for his encouragement and reading the manuscript, Dr. Muhammad Islam, Dr. Aparna Higgins for her careful reading part of the book, and Dr. Arthur Busch for his technical support.

A special thank you goes to my brothers Melhem, Tony, Khalil, and Hanna, and to my sisters Mona, Samira, and Lola.

Heartfelt appreciation goes to my two favorite teachers from my hometown Miziara: Father Hanna Al Bacha and Mr. Fayez Karam.

Dayton, OH, USA   Youssef N. Raffoul

Contents

Chapter 1
Stability and Boundedness

In this chapter we provide a brief introduction to difference calculus including basic material on Volterra difference equations. Using the z-transform we state some known theorems regarding stability of the zero solution of Volterra difference equations of convolution types. We move on to introducing Lyapunov functions for autonomous difference equations and state some known results concerning stability and boundedness. In Section 1.3 we introduce the concept of total stability and its correlation with uniform asymptotic stability for perturbed Volterra difference equations. In addition, we introduce the corresponding resolvent equations and utilize them to express the solution of the considered Volterra difference equation. Once the solution is explicitly given, we prove theorems concerning the stability of the zero solution. In Section 1.4, we obtain necessary and sufficient conditions of the stability of the zero solution of Volterra vector difference equations using the notion of resolvent. The resolvent is an abstract term which makes it difficult, if not impossible, to make efficient use of it. However, by the help of Lyapunov functionals, we will be able to verify all the conditions that are related to the resolvent. In Section 1.4.1 we apply the results of Section 1.4 to perturbed nonlinear scalar Volterra difference equations. Conditions will be verified using a combination of Lyapunov functional and the resolvent. In addition, we review some known results that have incorporated the concept of resolvent and point out some of the difficulties in verifying some of the conditions on the resolvent. We end the chapter with some interesting open problems that should be somewhat challenging.

Most of the work in this chapter can be found in [36, 52, 61, 65, 83, 115, 117, 143] and the references therein.

© Springer Nature Switzerland AG 2018
Y. N. Raffoul, *Qualitative Theory of Volterra Difference Equations*,
https://doi.org/10.1007/978-3-319-97190-2_1

1.1 Introduction

In this book, the sets $\mathbb{Z}^-$, $\mathbb{Z}^+$, $\mathbb{N}$, $\mathbb{R}$, and $\mathbb{R}^+$ denote the set of nonpositive integers, all integers, the set of nonnegative integers, the set of natural numbers, the set of real numbers, and the set of nonnegative real numbers, respectively. Let $x(t)$ be a sequence such that $x : \mathbb{Z} \to \mathbb{R}$. We define the *delta difference operator* $\triangle y(t)$ by

$$\triangle y(t) = y(t+1) - y(t).$$

Also, the *shift operator* E where $E : \mathbb{Z} \to \mathbb{R}$ is defined by

$$Ey(t) = y(t+1).$$

The second order difference is given by

$$\triangle^2 y(t) = \triangle(\triangle y(t)) = y(t+2) - 2y(t+1) + y(t).$$

The anti-difference of $y(t)$ denoted by $\sum y(t)$ is any function so that $\triangle(\sum y(t)) = y(t)$ for all t in the domain of y. It is easy to see that $\triangle\left(\sum_{s=p}^{t-1} y(s)\right) = y(t)$ for fixed p with $t > p$, and $\triangle\left(\sum_{s=t-1}^{p} y(s)\right) = -y(t-1)$ for fixed p with $p > t$. Corresponding to the fundamental theorem of calculus we have $\sum_{s=a}^{t-1} \triangle y(s) = y(t) - y(a)$. We will assume that $\sum_{s=a}^{b} y(s) = 0$ for all $a > b$. We denote the product of $y(t)$ from $t = a$ to b by $\prod_{t=a}^{b} y(t)$ with the understanding that $\prod_{t=a}^{b} y(t) = 1$ for all $a > b$.

A linear Volterra difference equation is of the form

$$x(n+1) = a(n)x(n) + \sum_{s=0}^{n-1} C(n,s)x(s) \tag{1.1.1}$$

where a is a given sequence and x is an unknown function to be found. Equation (1.1.1) is scalar if all sequences are scalars, and vector if a and x are $k \times 1$ sequences and C is an $k \times k$ matrix. Equation (1.1.1) is of convolution type if $C(n,s) = C(n-s)$. To see the usefulness of Volterra difference equations, one may not look any further than the completely delayed difference equation

$$x(t+1) = b(t)x(t-h) \tag{1.1.2}$$

where $h \in \mathbb{Z}^+$ and $b : \mathbb{Z}^+ \to \mathbb{R}$. Due to the absence of a linear term in $x(t)$ in (1.1.2), it is hard to obtain any useful information regarding the solutions. To overcome such difficulties, we rewrite (1.1.2) in the form of Volterra difference equation, see [138].

That is,

$$\triangle x(t) = \big(b(t+h)-1\big)x(t) - \triangle_t \sum_{s=t-h}^{t-1} b(s+h)x(s).$$

As we shall see later in the book that a Lyapunov functional of the form

$$V(t) = \Big[x(t) + \sum_{s=t-h}^{t-1} b(s+h)x(s)\Big]^2$$

$$+ \delta \sum_{s=-h}^{-1} \sum_{z=t+s}^{t-1} b^2(z+h)x^2(z)$$

leads to the exponential stability of the zero solution of (1.1.2) under the assumption that

$$-\frac{\delta}{(\delta+1)h} \le Q(t) \le -\delta h b^2(t+h) - Q^2(t),$$

hold for some $\delta > 0$ and $Q(t) = b(t+h) - 1$.

Also, it is well known, [12, 20], and [109], that Volterra difference equations play a major role in numerical methods applied to Volterra integro-differential equations. Discrete Volterra systems arise in studying numerical solutions of Volterra integro-differential equations and from modeling systems that are digital, such as digital filters and computer controlled systems and traffic control [160]. Also, Volterra discrete systems are accurately used to model nonlinear systems (such as aircraft flight in high angle-of-attack/sideslip flight) see [124]. Hence, studying the qualitative behavior of solutions of Volterra discrete systems is essential. Throughout this book, we loosely interchange "sequences" and "functions."

The study of difference equations in general is necessitated by the fact that the passage from the continuous case to the discrete case is not trivial, as we show next. First, we have the following. Consider the nonautonomous nonlinear system

$$x'(t) = f(t,x(t)), \quad t \ge 0, \tag{1.1.3}$$

$$\tag{1.1.4}$$

where $x(t) \in \mathbb{R}^k$, $f : \mathbb{Z}^+ \times D \to \mathbb{R}^k$ where $D \subset \mathbb{R}^k$ and open. If $V(t,x)$ is a scalar function and $x(t)$ is an unknown solution of (1.1.3), we may compute

$$\frac{dV(t,x(t))}{dt} = gradV(t,x) \cdot f(t,x) + \frac{\partial V}{\partial t}.$$

We denote this derivative by $V'(t,x)$.

Consider the nonlinear delay differential equation

$$x'(t) = -a(t)x^3(t) + b(t)x^3(t-\tau), \quad \tau \in \mathbb{R}^+$$

where the functions $a(t)$ and $b(t)$ are continuous. Assume that

$$a(t) \geq |b(t+\tau)| + k,$$

for some $k > 0$. Define the Lyapunov functional

$$V(t,x_t) = |x(t)| + \int_{t-\tau}^{t} |b(s+\tau)||x^3(s)|ds.$$

Then along the solutions we have that

$$V'(t,x_t) \leq (-a(t) + |b(t+\tau)|)x^3(t) \leq -k|x^3(t)|.$$

Then an argument in a result of Burton [21] yields uniform asymptotic stability. As for the discrete case, we consider

$$x(n+1) = -a(n)x^3(n) + b(n)x^3(n-\tau), \ \tau \in \mathbb{Z}^+.$$

Define the Lyapunov functional

$$V(n,x_n) = |x(n)| + \sum_{s=n-\tau}^{n-1} \left(|b(s+\tau)||x^3(s)| \right).$$

Then along the solutions we have

$$\begin{aligned}
\triangle V(n,x_n) \leq\ & |a(n)||x^3(n)| + |b(n)||x^3(n-\tau)| - |x(n)| \\
& + |b(n+\tau)||x^3(n)| - |b(n)||x^3(n-\tau)| \\
& - (|a(n)| + |b(n+\tau)|)|x^3(n)| - |x(n)|.
\end{aligned} \tag{1.1.5}$$

It is clear that nothing can be concluded from inequality (1.1.5).

Definition 1.1.1. A sequence $x(t)$ is said to be of exponential order if there exists a number $r > 0$, and an integer $N > 0$ such that

$$|x(t)| \leq r\rho^t \ \ for\ all\ t > N,$$

where $\rho \geq 0$ is some suitable constant.

Definition 1.1.2. The z-transform $Z[x(t)]$ of a sequence $x(t)$ of exponential order is defined by

$$\tilde{x}(z) = Z[x(t)] = \sum_{t=0}^{\infty} x(t)z^{-t}, \ \ |z| > \rho,$$

where z is a complex number, and ρ is the radius of convergence of $Z[x(t)]$.

Next we consider the scalar difference equation

$$y(n+1) = a(n)y(n) + g(n,x(n)), \ y(n_0) = y_0. \tag{1.1.6}$$

We have the following theorem, which introduces the variation of parameters formula of (1.1.6).

Theorem 1.1.1. *Suppose $a(n) \neq 0$ for all $n \in \mathbb{N}$. Then, $y(n)$ is a solution of (1.1.6) if and only if*

$$y(n) = \left(\prod_{i=n_0}^{n-1} a(i) \right) y_0 + \sum_{r=n_0}^{n-1} \prod_{i=r+1}^{n-1} a(i) g(r, x(r)). \tag{1.1.7}$$

Note that if $a(n) = 0$, then (1.1.7) is not a solution of (1.1.6). We have a parallel variation of parameters formula for vector difference equations. To be precise, we consider the vector difference equation

$$y(n+1) = D(n)y(n) + g(n), \ y(n_0) = y_0, \tag{1.1.8}$$

where D is a $k \times k$ matrix function, g is a $k \times 1$ vector function, and y is a $k \times 1$ unknown vector.

Definition 1.1.3. If $\Phi(n)$ is a matrix that is nonsingular for $n = n_0$ and satisfies

$$x(n+1) = D(n)x(n), \tag{1.1.9}$$

then it is said to be a *fundamental matrix* for system (1.1.9).

In the autonomous case when D is a constant matrix, $\Phi(n) = D^{n-n_0}$, and if $n_0 = 0$, then $\Phi(n) = D^n$.

Theorem 1.1.2. *Let $\Phi(n)$ be the fundamental matrix of (1.1.9) and I be the $k \times k$ identity matrix. Then, $y(n)$ is a solution of (1.1.8) if and only if*

$$y(n) = \Phi(n)\Phi^{-1}(n_0)y_0 + \sum_{s=n_0}^{n-1} \Phi(n)\Phi^{-1}(s+1)g(s). \tag{1.1.10}$$

For the autonomous case when D is a constant matrix, the solution of (1.1.8) is given by

$$y(n) = D^{n-n_0}y_0 + \sum_{s=n_0}^{n-1} D^{n-s-1}g(s).$$

For more on the fundamental matrix solution, we ask the reader to consult [57] and [92].

In the next few pages we review some of the literature that exists on the scalar Volterra difference equation

$$x(n+1) = a(n)x(n) + \sum_{j=0}^{n} b(n-j)x(j), \tag{1.1.11}$$

where $n \in \mathbb{Z}^+, a(n) \in \mathbb{R}$, and $b: \mathbb{Z}^+ \to \mathbb{R}$ are given sequences. Equation (1.1.11) may be considered as the discrete analogue of the famous Volterra integro-differential equation

$$x'(t) = a(t)x(t) + \int_0^t b(t-s)x(s)ds.$$

Equation (1.1.11) represents a system in which the future state $x(n+1)$ depends not only on the present state $x(n)$ but also on all past states $x(n-1), x(n-2), \ldots, x(0)$. Equation (1.1.11) plays a major role in modeling as we shall see in the next example.

Example 1.1 ([57, 100]). Let $x(n)$ denote the fraction of susceptible individuals in a certain population during the nth day of an epidemic, and let $a(k) > 0$ be the measure of how infectious the infected individuals are during the kth day. Then the spread of an epidemic may be modeled by the equation

$$Ln\frac{1}{x(n+1)} = \sum_{j=0}^n (1+\varepsilon - x(n-j))a(j),$$

where $\varepsilon > 0$, and small and $n \in \mathbb{Z}^+$. Let $x(n) = e^{-y(n)}$, then the above model transforms into

$$y(n+1) = \sum_{j=0}^n (a(n-j)(1+\varepsilon - e^{-y(j)})). \tag{1.1.12}$$

Since $x(n) \in [0,1)$, we have $y(n) \geq 0$ for all solutions of (1.1.12). During the early stages of the epidemic $x(n)$ is close to 1, and consequently $y(n)$ is close to zero. Hence it is reasonable to linearize (1.1.12) around 0. So if we replace $e^{-y(j)}$ by $1 - y(j)$, (1.1.12) becomes

$$y(n+1) = \sum_{j=0}^n (a(n-j)(\varepsilon + y(j))), \ y(0) = 0. \tag{1.1.13}$$

We will return to (1.1.13). Next we give a brief introduction on how discrete Volterra systems arise in studying numerical solutions of Volterra integro-differential equations (see [161]). Consider the Volterra integro-differential equation that models the growth of a single specie (see [43, 44])

$$x'(t) = -\lambda(1+cx(t))\int_{-\infty}^t (t-s)e^{-(t-s)}x(s)ds, \ t \geq 0 \quad x(t) = \phi(t), \ t \leq 0 \tag{1.1.14}$$

where $\phi(t)$ is any continuous known initial function. Set

$$\psi(t) = \int_{-\infty}^0 (t-s)e^{-(t-s)}\phi(s)ds$$

and rewrite (1.1.14)

$$x'(t) = -\lambda(1+cx(t))\left(\psi(t) + \int_0^t (t-s)e^{-(t-s)}x(s)ds\right), \ t \geq 0. \tag{1.1.15}$$

For simplicity we set $c = 1$ and the kernel

$$k(t - s) := (t - s)e^{-(t-s)}.$$

Divide the interval $[0, t]$ into n intervals of equal length h with $t = t_n$, and $t_j = jh$, $j = 0, \ldots n$. Let $\phi_j = \phi(jh)$; $\psi_j = \Psi(jh)$, $k_j = k(jh)$, where h is fixed. We use the θ-rule to approximate the integral. Thus we have

$$\int_0^{t_n} k(t_n - s)x(s)ds = \sum_{j=0}^{n-1} \int_{t_j}^{t_{j+1}} k(t_n - s)x(s)ds \approx h \sum_{j=0}^{n} w_j^{(n)} k(t_n - j)x(j),$$

where $x(j)$ denotes a numerical approximation to $x(t_j)$ and

$$\{w_0^{(n)}, w_1^{(n)}, \ldots, w_{n-1}^{(n)}, w_n^{(n)}\} = \{\theta, \ldots, 1, 1 - \theta\}, \ 0 \le \theta \le 1$$

and $\sum_{j=0}^{n} w_j^{(n)} = n$, $n \ge 0$. Letting $\theta = 1$, and $x' = \dfrac{x(n+1) - x(n)}{h}$, we get

$$x(n+1) = x(n) - \lambda h(1 + x(n))\left(\psi_n + h \sum_{j=0}^{n} k(n - j)x(j)\right), \ n \ge 0 \ \ x(0) = \phi(0)$$

Letting $b(n) = -\lambda h^2 k(n)$, $g(n, x(n)) = -\lambda h \psi_n x(n)$, $f(n) = -\lambda h \psi_n$ and $q(x(n)) = x(n) \sum_{j=0}^{n} b(n - j)x(j)$, we arrive at the Volterra difference equation of convolution type

$$x(n+1) = x(n) + \sum_{j=0}^{n} b(n - j)x(j) + f(n) + g(n, x(n)) + q(x(n)),$$

$$x(0) = \phi(0).$$

The above Volterra difference equation of convolution type will be studied in Section 5.3 of Chapter 5. We will revisit the subject of discretization scheme in Chapter 6.

Let $C(n)$ denote the set of functions $\phi : [0, n] \cap \mathbb{Z}^+ \to \mathbb{R}$ and $\|\phi\| = \sup\{|\phi(s)| : 0 \le s \le n\}$. For each $n_0 \in \mathbb{Z}^+$ and $\phi \in C(n_0)$, there is a unique function $x : \mathbb{Z}^+ \to \mathbb{R}$ which satisfies (1.1.11) on $[n_0, +\infty)$ with $x(s) = \phi(s)$ for $0 \le s \le n_0$. Such a function $x(n)$ is called a solution of (1.1.11) through (n_0, ϕ) and is denoted by $x(n, n_0, \phi)$.

Definition 1.1.4. The zero solution of (1.1.11) is

1. stable (S) if for each $\varepsilon > 0$, there is a $\delta = \delta(n_0, \varepsilon) > 0$ such that $[n_0 \ge 0, \phi \in C(n_0), \|\phi\| < \delta]$ imply $|x(n, n_0, \phi)| < \varepsilon$ for all $n \ge n_0$.
2. It is uniformly stable (US) if it is stable and δ is independent of n_0.
3. It is asymptotically stable (AS) if it is (S) and $|x(n, n_0, \phi)| \to 0$, as $n \to \infty$.
4. It is uniformly asymptotically stable (UAS) if it is uniformly stable and if there is an $\eta > 0$ such that, for each $\gamma > 0$, there is a $T > 0$ such that

---

8 1 Stability and Boundedness

$$|x_0| < \eta, \ n_0 \geq 0, \ \text{and} \ n \geq n_0 + T$$

imply $|x(n,n_0,x_0)| < \gamma$.

The convolution of two sequences $x(n)$ and $y(n)$ is defined as

$$x(n) * y(n) = \sum_{j=0}^{n} x(n-j)y(j) = \sum_{j=0}^{n} x(n)y(n-j),$$

and hence it is known that

$$Z(x(n) * y(n)) = \tilde{x}(z) \cdot \tilde{y}(z).$$

With this in mind, (1.1.11) maybe written as

$$x(n+1) = a(n)x(n) + b(n) * x(n).$$

By taking the z-transform on both sides, one arrives at

$$\tilde{x}(z) = zx(0)g^{-1}(z), \tag{1.1.16}$$

where

$$g(z) = z - a - \tilde{b}(z). \tag{1.1.17}$$

Theorem 1.1.3 ([56, 57, 59]). *The zeros of $g(z)$ all lie in the region $|z| < c$ for some real positive constant c. Moreover, $g(z)$ has finitely many z with $|z| \geq 1$, provided that $x(n) \in l^1$ (summable $\sum_{i=0}^{\infty} |x(i)| = ||x||_1 < \infty$.)*

Theorem 1.1.4 ([56, 57, 58, 59]). *Suppose that $b(n)$ does not change sign for $n \in \mathbb{Z}^+$. Then the zero solution of (1.1.11) is (AS) if*

$$|a(n)| + \left| \sum_{n=0}^{\infty} b(n) \right| < 1.$$

Theorem 1.1.5 ([56, 57, 59]). *Suppose that $b(n)$ does not change sign for $n \in \mathbb{Z}^+$. Then the zero solution of (1.1.11) is **not** (AS) if one of the following conditions holds.*

(i) $a(n) + \sum_{n=0}^{\infty} b(n) \geq 1$,

(ii) $a(n) + \sum_{n=0}^{\infty} b(n) \geq -1$, for some $n \in \mathbb{Z}^+$,

(iii) $a(n) + \sum_{n=0}^{\infty} b(n) \geq -1$, and $b(n) < 0$, for some $n \in \mathbb{Z}^+$, and $\sum_{n=0}^{\infty} b(n)$ is sufficiently small.

In the next theorem, we use z-transform to asymptotically analyze the solutions of (1.1.13) when $a(n) = ca^n$, for positive constants a and c.

Theorem 1.1.6 ([57, 100]). *Suppose $a(n) = ca^n$, such that $0 < a + c < 1$. Then any solution $y(n)$ of (1.1.13) satisfies*

$$\lim_{n \to \infty} y(n) = \frac{\varepsilon c}{1 - (a + c)}$$

which implies the spread of the disease will not reach an epidemic proportion.

Proof. Taking the z-transform on both sides of (1.1.13) yields,

$$z\tilde{y}(z) = \tilde{a}(z)\frac{\varepsilon z}{z - 1} + \tilde{a}(z)\tilde{y}(z),$$

or

$$\tilde{y}(z) = \frac{\varepsilon z \tilde{a}(z)}{(z + 1)(z - \tilde{a}(z))}.$$

By setting $a(n) = ca^n$, and performing partial fraction we arrive at

$$\tilde{y}(z) = \frac{\varepsilon cz)}{(z - 1)(z - (a + c))} = \frac{\varepsilon c}{1 - a - c}\Big[\frac{1}{z - 1} - \frac{a + c}{z - (a + c)}\Big].$$

Taking the inverse z-transform gives

$$y(n) = \frac{\varepsilon c)}{1 - (a + c))}[1 - (a + c)^n]$$

and hence the results.

In Chapter 5, we will use Lyapunov functional and obtain uniform asymptotic stability for nonlinear systems that are similar to (1.1.12).

Next we discuss the role that finite delay Volterra difference equations play in modeling neural networks with delay. Suppose an artificial neural network consisting of electronic neurons (amplifiers) interconnected through a matrix of resistors. Here an electronic neuron, the building block of the network, consists of a nonlinear amplifier that transforms an input signal u_i into the output signal v_i, and the input impedance of the amplifier unit is described by the combination of a resistor ρ_i and a capacitor C_i. We assume that the input-output relation is completely characterized by a voltage amplification function $v_i = f_i(u_i)$. The synaptic connections of the network are represented by resistors R_{ij} that connect the output terminal of the amplifier j with the input part of the neuron i. In order for the network to function properly, the resistances R_{ij} must be able to take on negative values. This can be realized by supplying each amplifier with an inverting output line that produces the signal $-v_j$. Applying Kirchhoff's law and input-output relation replaced by $v_i = f_i(u_i(t - \tau_i))$

with a positive constant $\tau_i \in \mathbb{Z}^+$, we obtain after some calculations and simplifications the Volterra discrete equation

$$u(n+1) = C_i R_i [-u(n) + \sum_{j=1}^{k} \frac{R_i}{R_{ij}} f_j(u_j(n - \tau_j))], \quad 1 \le i \le k.$$

For more on such construction we refer to [73].

1.2 Introduction to Lyapunov Functions

In this section, we briefly go over some definitions and theorems regarding Lyapunov functions/functionals. Let $G \subset \mathbb{R}^k$ be an open set and consider the autonomous difference equation

$$x(n+1) = f(x(n)), \tag{1.2.1}$$

where $f : G \to \mathbb{R}^k$, is continuous . We assume that x^* is an equilibrium solution of (1.2.1), that is $f(x^*) = x^*$.

Definition 1.2.1. Let the function $V : \mathbb{R}^k \to \mathbb{R}$ be continuous.

1. The variation of V with respect to (1.2.1) is defined as

$$\triangle V(x(n)) = V(f(x(n))) - V(x(n)) = V(x(n+1)) - V(x(n)).$$

2. The function V is said to be a Lyapunov function/functional on a subset H of $\mathbb{R}^k$ if
   i) $V(x^*) = 0$, and $V(x) > 0$, for $x \ne x^*$ and
   ii) $\triangle V(x) \le 0$, whenever x and $f(x)$ belong to the set H.
3. The function V is said to be a strict Lyapunov function/functional on a subset H of $\mathbb{R}^k$ if $\triangle V(x) < 0$.
4. Let $B(x, \gamma)$ denote the open ball in $\mathbb{R}^k$ of radius γ and center x defined by $B(x, \gamma) = \{y \in \mathbb{R}^k \mid ||y - x|| < \gamma\}$. We say V is positive definite at x^* if $V(x^*) = 0$, and $V(x^*) > 0$, for all $x \in B(x^*, \gamma), x \ne x^*$, for some $\gamma > 0$.

We have the following standard stability and boundedness theorems.

Theorem 1.2.1 (Lasalle Stability Theorem). *Suppose V is a Lyapunov function for (1.2.1) in a neighborhood of H of the equilibrium solution x^*, then x^* is stable. Moreover, x^* is asymptotically stable if V is a strict Lyapunov function.*

Theorem 1.2.2. *Suppose V is a Lyapunov function for (1.2.1) on the set $H = \{x \in \mathbb{R}^k : ||x|| > \alpha\}$ for some $\alpha > 0$, and if*

$$V(x) \to \infty \text{ as } ||x|| \to \infty,$$

then all solutions of (1.2.1) are bounded.

The theory for boundedness stability and existence of periodic solutions for non-functional difference equations have been fully developed, unlike functional difference equations which include Volterra difference equations. In the next theorems we provide some results taken from [85] and [106] concerning the nonautonomous nonlinear discrete system

$$x(n+1) = f(n,x(n)), \quad n \geq 0, \tag{1.2.2}$$
$$x(n_0) = x_0, \quad n_0 \geq 0$$

where $x(n) \in \mathbb{R}^k$, $f : \mathbb{Z}^+ \times D \to \mathbb{R}^k$ where $D \subset \mathbb{R}^k$ is an open set containing the origin, is a given nonlinear function satisfying $f(n,0) = 0$ for all $n \in \mathbb{Z}^+$. For $x(n) \in \mathbb{R}^k$, $||x||$ denotes the Euclidean norm of x. For any $k \times k$ matrix A, define the norm of A by $|A| = \sup\{|Ax| : ||x|| \leq 1\}$.

Definition 1.2.2. A solution $x(n)$ of (1.2.2) is said to be bounded if for any $n_0 \in \mathbb{Z}^+$ and number r there exists a number $\alpha(n_0,r)$ depending on n_0 and r such that $||x(n,n_0,x_0)|| \leq \alpha(n_0,r)$ for all $n \geq n_0$ and x_0, $|x_0| < r$. It is uniformly bounded if α is independent of the initial time n_0.

Definition 1.2.3. Let $x(n)$ be solution of (1.2.2) with respect to initial condition x_0 and $y(n)$ be solution of (1.2.2) with respect to initial condition y_0. The solution $x(n)$ is then said to be sable, if, whenever $\varepsilon > 0$ is given, there exists $\delta(\varepsilon)$ for which

$$||x(n) - y(n)|| < \varepsilon, \text{ whenever } ||x_0 - y_0|| < \delta.$$

Consider the linear difference equation

$$x(n+1) = x(n) + 1, \, x(0) = x_0. \tag{1.2.3}$$

It is easy to check that $x(n) = x_0 + (n - n_0)$ is the solution of (1.2.3). If $y(n)$ is another solution with $y(n_0) = y_0$, then we have $y(n) = y_0 + (n - n_0)$. For any $\varepsilon > 0$, let $\delta = \varepsilon$. Then

$$||x(n) - y(n)|| = ||x_0 + (n - n_0) - y_0 - (n - n_0)|| = ||x_0 - y_0)|| < \varepsilon$$

whenever, $||x_0 - y_0|| < \delta$. Hence, the solution $x(n)$ is stable, but unbounded. This simple example shows that the properties of boundedness of all solutions and stability of a solution do not coincide. On the other hand, the difference equation

$$x(n+1) = x^{1/3}(n),$$

has its solution $x(n)$ satisfies $|x(n)| \to 1$ for all initial values $x_0 \neq 0$. Therefore, the solution $x(n) = 0$ is unstable.

Definition 1.2.4. The zero solution of system (1.2.2) is said to be exponentially stable if any solution $x(n,n_0,x_0)$ of (1.2.2) satisfies

$$||x(n,n_0,x_0)|| \leq C\Big(||x_0||,n_0\Big)a^{-\delta(n-n_0)}, \quad \text{for all } n \geq n_0,$$

where a is constant with $a > 1$, $C : \mathbb{R}^+ \times \mathbb{Z}^+ \to \mathbb{R}^+$, and δ is a positive constant. The zero solution of (1.2.2) is said to be uniformly exponentially stable if C is independent of n_0.

Theorem 1.2.3 ([85]). *Let a be a constant with $a > 1$. Let $D \subset \mathbb{R}^k$ be an open set containing the origin, and let $V(n,x) : \mathbb{Z}^+ \times D \to \mathbb{R}^+$ be a given function satisfying*

$$\lambda_1 ||x||^p \leq V(n,x) \leq \lambda_2 ||x||^q, \tag{1.2.4}$$

and

$$\triangle V(n,x) \leq -\lambda_3 ||x||^r + ka^{-\delta n}, \tag{1.2.5}$$

for some positive constants $\lambda_1, \lambda_2, \lambda_3, p, q, r, k$ and δ. Moreover, if for some positive constants α and γ,

$$0 < \frac{\lambda_3}{\lambda_2^{r/q}} \leq \alpha < 1 \tag{1.2.6}$$

such that

$$V(n,x) - V^{r/q}(n,x) \leq \gamma a^{-\delta n} \tag{1.2.7}$$

with

$$\delta > -\frac{\ln(1 - \lambda_3/\lambda_2^{r/q})}{\ln(a)},$$

then the zero solution of (1.2.2) is uniformly exponentially stable.

The next theorem does not require an upper bound on the Lyapunov function.

Theorem 1.2.4 ([85]). *Let a be a constant with $a > 1$. Let $D \subset \mathbb{R}^k$ be an open set containing the origin, and let $V(n,x) : \mathbb{Z}^+ \times D \to \mathbb{R}^+$ be a given function satisfying*

$$\lambda_1 ||x||^p \leq V(n,x), \tag{1.2.8}$$

and

$$\triangle V(n,x) \leq -\lambda_2 V(n,x) + ka^{-\delta n}, 0 < \lambda_2 < 1 \tag{1.2.9}$$

for some positive constants $\lambda_1, \lambda_2, p, k$ and δ.
Then, the zero solution of (1.2.2) is exponentially stable.

Theorems 1.2.3 and 1.2.4 were applied to the following nonlinear difference equations

$$x(n+1) = \sigma x(n) + Rx^{1/3}(n)a^{-ln},$$

and

$$x(n+1) = \sigma x + Rx^{1/3} + a^{\gamma_1 n}\sin(x)$$

respectively, under appropriate conditions. However, the above theorems cannot be applied to Volterra difference equations, and hence the need for new and general theory that deal with the issues that lie ahead of us. Next we briefly discuss some of the results in [36] about boundedness of solutions of (1.2.2). We will use notations that are more suitable and flow better with the book. The purpose of the next example is to emphasize that there is a difference between uniform boundedness and boundedness.

Example 1.2 ([36]). Consider the two-dimensional system

$$x(n+1) = \begin{pmatrix} 1 & (n+2)^2(n+3)^{-1} \\ 0 & (n+2)^3(n+3)^{-3} \end{pmatrix} x(n) + \begin{pmatrix} 0 & (n+1)^{-2} \\ 0 & 0 \end{pmatrix} x_0 \qquad (1.2.10)$$

Then the solution of (1.2.10) is given by

$$x(n) = R(n,n_0)x_0 + \sum_{s=n_0}^{n} R(n,s)g(s-1), \; n \geq n_0 \qquad (1.2.11)$$

where the matrix $R = (r_{lj}), l, j = 1, 2$ and vector g are

$$R(n,n_0) = \begin{pmatrix} 1 & -(n_0+1)(n_0+2)^2 + (n_0+2)^3(n+1)(n+2)^{-1} \\ 0 & (n_0+2)^3(n+2)^{-3} \end{pmatrix}$$

$$g(n) = \begin{pmatrix} 0 & (n+2)^{-2} \\ 0 & 0 \end{pmatrix}.$$

Note that at n_0 the matrix $R(n_0,n_0)$ is the identity matrix. Let $x_0 = \begin{pmatrix} x_{01} \\ x_{02} \end{pmatrix}$. The term

$$\sum_{s=n_0}^{n} R(n,s)g(s-1) = x_{02} \sum_{s=n_0+1}^{n} \begin{pmatrix} s^{-2} \\ 0 \end{pmatrix} \leq \pi^2 x_{02} \begin{pmatrix} 1 \\ 0 \end{pmatrix},$$

which is uniformly bounded. On the other hand, for any positive integer K, we obtain

$$r_{12}(Kn_0,n_0) = \frac{(n_0+2)^2}{Kn_0+2} \left[-(n_0+1)(Kn_0+2)^2 + (n_0+2)(Kn_0+1) \right]$$

$$= (n_0+1)^2(Kn_0-1)^2 + (Kn_0+2) \to \infty, \text{ as } n_0 \to \infty.$$

Then it is clear from the solution given by (1.2.11) that boundedness of solutions depends on the initial value n_0 and hence solutions are bounded but not uniformly bounded.

Theorem 1.2.5 ([36]). *Suppose there exists a function $V(n,x)$ and W_1 such that*

$$W_1(\|x\|) \leq V(n,x), \; n \geq n_0 \qquad (1.2.12)$$

$$\triangle V(x(n)) = V(f(n,x(n))) - V(x(n)) = V(x(n+1)) - V(x(n)) \leq 0, \qquad (1.2.13)$$

and

$$W_1(\|x\|) \to \infty, \text{ as } \|x\| \to \infty. \qquad (1.2.14)$$

Assume for any initial time n_0 with $x(n_0) = x_0$, $V(n_0, x_0)$ is bounded, then solutions of (1.2.2) *are bounded.*

Proof. Let r_0 be any positive constant such that $||x_0|| \leq r_0$. By (1.2.14) and since $V(n_0, x_0)$ is bounded, there exists a function $\alpha(n_0, r_0)$ such that

$$V(n_0, x_0) \leq W_1\big(\alpha(n_0, r_0)\big).$$

Utilizing conditions (1.2.12) and (1.2.13) we have

$$W_1(||x||) \leq V(n, x) \leq V(n_0, x_0) \leq W_1\big(\alpha(n_0, r_0)\big). \qquad (1.2.15)$$

Taking inverse in (1.2.15) we arrive at $||x(n, n_0, x_0)|| \leq \alpha(n_0, r_0)$. This completes the proof.

Theorem 1.2.6 ([163]). *Let $B_\alpha := B(\alpha, x)$ denote the ball of radius α centered at x. Suppose there exist functions $V(n, x) : \mathbb{Z}^+ \times B_\alpha \to \mathbb{R}$ and $W_1 : \mathbb{Z}^+ \times B_\alpha \to \mathbb{R}^+$, $\alpha > 0$ where $V(n, x), W(n, x)$ are continuous in x. Suppose that*

$$\triangle V(x(n)) = -p(n)a\big(W(n, x)\big) + g(n) \qquad (1.2.16)$$

where $g, p : \mathbb{Z}^+ \to \mathbb{R}^+$, $a(r) > 0, a(0) = 0$ is continuous monotone and nondecreasing, $\sum_{s=1}^{\infty} p(m_s) = \infty$, for any subsequence $\{m_s\}$ with $m_s \to \infty$ and $s \to \infty$ and there exists an $E > 0$ such that $\sum_{n=1}^{\infty} p(n) = E$. If there exists a number $M > 0$ such that $M \leq V(n, x(n))$ for all $(n, x) \in \mathbb{Z}^+ \times B_\alpha$, then for every solution $x(n)$ of (1.2.2) with $x(n) \in B_\alpha$ we have $W(n, x(n)) \to 0$ as $n \to \infty$.

Proof. Suppose that $W(n, x(n)) \nrightarrow 0$ as $n \to \infty$ for some $\{x(n)\} \subset B_\alpha$. Then there exists an $\varepsilon > 0$ and a subsequence $\{n_j\}$ with $n_j \to \infty$ as $j \to \infty$ such that $W(n_j, x(n_j)) \geq \varepsilon$. Summing (1.2.16) from n_1 to n_j yield

$$V(n_j + 1, x(n_j + 1)) \leq V(n_1, x(n_1)) - a(\varepsilon) \sum_{s=1}^{n_j} p(s) + E, \ s \geq 1,$$

which yields a contradiction. Thus, $W(n, x(n)) \nrightarrow 0$ as $n \to \infty$ for every $\{x(n)\} \subset B_\alpha$. This completes the proof.

It is evident that Theorem 1.2.6 gives sufficient conditions for the asymptotic stability of the zero solution of (1.2.2) in the case $f(n, 0) = 0$. Thus, if we replace condition (1.2.16) by

$$\triangle V(x(n)) = -W(n, |x|)$$

then the zero solution of (1.2.2) is asymptotically stable. As an application we have the following example.

Example 1.3. Consider the scalar nonlinear Volterra difference equation

$$x(n+1) = a(n)x(n) + b(n)\frac{x(n)}{1 + \sum_{s=0}^{n-1} x^2(s)}, \ n \geq 0. \tag{1.2.17}$$

If

$$|a(n)| + |b(n)| \leq 1,$$

then all solutions of (1.2.17) are bounded and the zero solution is asymptotically stable. To see this, consider the Lyapunov function $V(n,x) = |x(n)|$. Then along the solutions of (1.2.17) we have

$$\Delta V(x(n)) = |x(n+1)| - |x(n)| \leq (-1 + |a(n)|)|x(n)| + |b(n)|\frac{|x(n)|}{1 + \sum_{s=0}^{n-1} x^2(s)}$$

$$\leq (-1 + |a(n)| + |b(n)|)|x(n)|.$$

The results follow from Theorems 1.2.5 and 1.2.6.

Theorems 1.2.5 and 1.2.6 have limitations as they cannot be applied to Volterra difference equations of the form

$$x(n+1) = a(n)x(n) + \sum_{s=0}^{n-1} b(n,s)x(s), \ n \geq 0.$$

These type of equations will be handled in Chapter 2 once we develop the appropriate theorems. We refer to [36] for the proofs of the next two theorems.

Theorem 1.2.7 ([36]). *Suppose*

$$\|f(n,x(n))\| \leq \sum_{s=0}^{n-n_0} a(n,s)\|x(n-s)\| + b(n), n \geq n_0,$$

and

$$1 - \sup_{n \geq n_0} \sum_{s=n_0}^{\infty} a(n+s,s) > 0, \ \sum_{s=n_0}^{\infty} |b(s)| < \infty.$$

Then solutions of (1.2.2) *are uniformly bounded.*

Definition 1.2.5. A continuous function $W : [0,\infty) \to [0,\infty)$ with $W(0) = 0, W(s) > 0$ if $s > 0$, and W is strictly increasing is called a wedge. (In this book wedges are always denoted by W or W_i, where i is a positive integer.)

One might asks if boundedness implies the existence of such a Lyapunov function that satisfies (1.2.12)–(1.2.14). The answer is positive and is provided in the next theorem.

Theorem 1.2.8 ([36]). *Suppose solutions of* (1.2.2) *are bounded. Then there exists a Lyapunov function $V(n,x(n))$ and a wedge W_1 satisfying* (1.2.12)–(1.2.14).

Theorem 1.2.7 will be generalized in Chapter 2. The rest of the materials in this section is unpublished and belongs to the author, and hence the labelings.

Definition 1.2.6 (Raffoul). A function $U : \mathbb{Z}^+ \times D \to [0, \infty)$ is called

1. positive definite if $U(n, 0) = 0$ and there is a wedge W_1 with $U(n, x) \geq W_1(|x|)$,
2. decrescent if there is a wedge W_2 with $U(n, x) \leq W_2(|x|)$,
3. negative definite if $-U(n, x)$ is positive definite,
4. radially unbounded if $D = \mathbb{R}^k$ and there is a wedge $W_3(|x|) \leq U(n, x)$ and $W_3(r) \to \infty$ as $r \to \infty$.

Theorem 1.2.9 (Raffoul). *Suppose there is a Lyapunov function V for (1.2.2) (see Definition 1.2.1.)*

1. *If V is positive definite, then $x = 0$ is stable.*
2. *If V is positive definite and decrescent, then $x = 0$ is uniformly stable.*
3. *If V is positive definite and decrescent, and $\triangle V(n, x)$ is negative definite, then $x = 0$ is uniformly asymptotically stable .*
4. *If $D = \mathbb{R}^k$ and if V is radially unbounded, then all solutions of (1.2.2) are bounded.*

Proof.
1. We have $\triangle V(n, x) \leq 0$, V is continuous in x, $V(n, 0) = 0$, and $W_1(|x|) \leq V(n, x)$. Let $\varepsilon > 0$ and $n_0 \geq 0$ be given. We must find δ such that $|x_0| < \delta$ and $n \geq n_0$ imply $|x(n, n_0, x_0)| < \varepsilon$. (Throughout these proofs we assume ε is small enough so that $|x(n, n_0, x_0)| < \varepsilon$ implies that $x \in D$.) As V is continuous in x and $V(n, 0) = 0$ there is a $\delta > 0$ such that $|x_0| < \delta$ implies $V(n_0, x_0) < W_1(\varepsilon)$. Thus, if $n \geq n_0$ and $|x_0| < \delta$ and $x = x(n, n_0, x_0)$, we have

$$W_1(|x(n)|) \leq V(n, x) \leq V(n_0, x_0) < W_1(\varepsilon),$$

or $|x(n)| < \varepsilon$ as required.
2. For a given ε we select a $\delta > 0$ such that $W_2(\delta) < W_1(\varepsilon)$ where $W_1(|x|) \leq V(n, x) \leq W_2(|x|)$. If $n_0 \geq 0$, we have

$$\begin{aligned} W_1(|x(n)|) \leq V(n, x) &\leq V(n_0, x_0) \\ &\leq W_2(|x_0|) < W_2(\delta) < W_1(\varepsilon), \end{aligned}$$

or $|x(n)| < \varepsilon$ as required.
3. Let $\varepsilon = 1$, and find δ of uniform stability and call it η. Let γ be given. We must find $T > 0$ such that

$$|x_0| < \eta, \ \ n_0 \geq 0, \ \text{and } n \geq n_0 + T$$

imply $|x(n, n_0, x_0)| < \gamma$. Pick $\mu > 0$ with $W_2(\mu) < W_1(\gamma)$, so that there is $n_1 \geq n_0$ with $|x(n_1)| < \mu$, then, for $n \geq n_1$, we have

$$\begin{aligned} W_1(|x(n)|) \leq V(n, x) &\leq V(n_1, x_1) \\ &\leq W_2(|x_1|) < W_2(\delta) < W_1(\gamma), \end{aligned}$$

or $|x(n_1)| < \gamma$. Since $\triangle V(n,x) \le -W_3(|x|)$, so as long as $|x(n)| > \mu$, then $\triangle V(n,x) \le -W_3(\mu)$; thus

$$V(n,x(n)) \le V(n_0,x_0) - \sum_{s=n_0}^{n-1} W_3(|x(s)|)$$

$$\le W_2(|x_0|) - W_3(\mu)(n-n_0)$$

$$\le W_2(\eta) - W_3(\mu)(n-n_0),$$

which vanishes at

$$n = n_0 + \frac{W_2(\eta)}{W_3(\mu)} \ge n_0 + T,$$

where $T \ge \frac{W_2(\eta)}{W_3(\mu)}$. Hence, if $T > \frac{W_2(\eta)}{W_3(\mu)}$, then $|x(n)| > \mu$ fails, and we have $|x(n)| < \gamma$ for all $n \ge n_0 + T$. This proves (UAS).

4. Since V is radially unbounded, we have $V(n,x) \ge W_1(|x|) \to \infty$ as $|x| \to \infty$. Thus, given $n_0 \ge 0$, and x_0, there is an $r > 0$ with $W_1(r) > V(n_0,x_0)$. Hence, if $n \ge n_0$ and $x(n) = x(n,n_0,x_0)$, then

$$W_1(|x(n)|) \le V(n,x(n)) \le V(n_0,x_0) < W_1(r),$$

or $|x(n)| < r$. The proof of Theorem 1.2.9 is complete.

According to Theorem 1.2.9, all solutions of (1.2.17) are bounded and its zero solution is (UAS).

The next example will show that $\triangle V(n,x) \le 0$ is not enough to drive solutions to zero.

Example 1.4 (Raffoul). Let $g : [0,\infty) \to (0,\beta]$ with $g(0) = 1$, and $0 < \beta \le 1$. Consider the nonautonomous difference equation

$$x(n+1) = [g(n+1)/g(n)]x(n). \qquad (1.2.18)$$

It is clear that $x(n) = g(n)$ is a solution of (1.2.18). Our goal is to construct a function

$$V(n,x) = a(n)x^2(n)$$

such that $\triangle V(n,x) = -\alpha(n)x^2(n)$, where $a(n), \alpha(n) > 0$ for $n \in \mathbb{Z}^+$, and $\sum_{s=0}^{\infty} \alpha(s) < \infty$. That is $\triangle V(n,x) \le 0$. Along the solutions of (1.2.18) we have

$$\triangle V(n,x) = \left[a(n+1)\frac{g^2(n+1)}{g^2(n)} - a(n)\right]x^2(n).$$

By setting

$$a(n+1)\frac{g^2(n+1)}{g^2(n)} - a(n) = -\alpha(n),$$

we get the difference equation

$$a(n+1) - \frac{g^2(n)}{g^2(n+1)}a(n) = -\frac{g^2(n)}{g^2(n+1)}\alpha(n). \qquad (1.2.19)$$

Using the variation of parameters formula given by (1.1.7), Equation (1.2.19) has the solution, after some simplification,

$$a(n) = \left(\prod_{i=0}^{n-1} \frac{g^2(i)}{g^2(i+1)}\right)a(0) - \sum_{r=0}^{n-1}\prod_{i=r+1}^{n-1} \frac{g^2(i)}{g^2(i+1)}\frac{g^2(r)}{g^2(r+1)}\alpha(r)$$

$$= \left[a(0)g^2(0) - \sum_{r=0}^{n-1} g^2(r)\alpha(r)\right]/g^2(n)$$

$$\geq \left[a(0) - \beta^2\sum_{r=0}^{n-1}\alpha(r)\right]/g^2(n).$$

Since $0 < g \leq 1$, and $\sum_{s=0}^{\infty}\alpha(s) < \infty$, we may chose $a(0)$ so large to imply that $a(n) > 1$ for all $n \geq 1$. Thus we have shown that $V \geq 0$ and $\triangle V$ is negative definite do not imply that solutions tend to zero. Notice that V is not decrescent. That is there is no wedge W_2 with $V(n,x) \leq W_2(|x|)$.

In the investigation of stability and boundedness for nonlinear nonautonomous systems, the effective approach is to study equations that are related in some way to the linear equations whose behavior is known to be covered by the known theory. For example, the perturbed nonlinear system

$$x(n+1) = Ax(n) + g(n,x(n)), \; n \geq 0 \qquad (1.2.20)$$

where A is an $k \times k$ constant matrix and $g : \mathbb{Z}^+ \rightarrow \mathbb{R}^k$ is continuous in x. The perturbation g is assumed small in some sense. The next theorem pertains to the stability of the perturbed nonlinear difference equation given by (1.2.20). But first for reference, a symmetric matrix C is positive definite if $x^T C x > 0$ for $x \neq 0$.

Theorem 1.2.10 (Raffoul). *Suppose there exists a positive definite symmetric matrix B such that*

$$A^T B A - B = -I, \qquad (1.2.21)$$

$$\lim_{x \to 0} \frac{|g(n,x)|}{|x|} = 0, \; uniformly \; in \; n \qquad (1.2.22)$$

then the zero solution of (1.2.20) is stable.

Proof. We write x for $x(n)$ and we define the Lyapunov function V by $V = x^T Bx$. Using (1.2.21)–(1.2.22), we have along the solutions of (1.2.20) that

$$\triangle V = (Ax + g(n,x))^T B(Ax + g(n,x)) - x^T Bx$$

$$= x^T (A^T BA - B)x + 2x^T A^T Bg(n,x) + g^T(n,x)Bg(n,x)$$

$$= |x|^2 \left(-1 + 2\frac{x^T}{|x|} A^T B \frac{g(n,x)}{|x|} + \frac{g^T(n,x)}{|x|} B \frac{g(n,x)}{|x|} \right)$$

$$\leq |x|^2 \left(-1 + 2|A^T B| \frac{|g(n,x)|}{|x|} + \frac{|g^T(n,x)|}{|x|} B \frac{|g(n,x)|}{|x|} \right)$$

$$\leq 0,$$

for sufficiently small $|x|$ and $x \neq 0$. The result follows from *1.* of Theorem 1.2.9. This completes the proof.

We have the following theorem concerning the existence of positive definite symmetric matrix B. Consider the autonomous linear system of difference equations

$$x(n+1) = Ax(n) \tag{1.2.23}$$

where A is an $k \times k$ constant matrix.

Theorem 1.2.11 (Raffoul). *For a given positive definite matrix C, the equation*

$$A^T BA - B = -C \tag{1.2.24}$$

can be solved for a positive definite symmetric matrix B if and only if all eigenvalues of A lie inside the unit circle.

Proof. Assume (1.2.24) and let $V = x^T Bx$. Then $\triangle V < 0$ for $x \neq 0$. Hence the zero solution of (1.2.23) is asymptotically stable by Theorem 1.2.1 and hence all the eigenvalues lie inside the unit circle.

Suppose all the eigenvalues of A lie inside the unit circle and define the matrix B by

$$B = \sum_{n=0}^{\infty} (A^T)^n CA^n.$$

Since all the eigenvalues of A reside inside the unit circle, we have $|A^n| \leq Ka^{-\delta n}$ for positive constants K, a and δ, such that $0 < a < 1$. Hence, the infinite sum converges (geometric series). It is clear from the definition of B, the matrix B is symmetric. Next we show it is positive definite.

$$x^T (A^T)^n CA^n x = (A^n x)^T C(A^n x) = y^T Cy,$$

and if $x \neq 0$, then $y \neq 0$; as C is positive definite $y^T Cy > 0$ for $x \neq 0$. Thus B is positive definite. Finally,

$$A^T BA - B = A^T \sum_{n=0}^{\infty} (A^T)^n CA^n A - \sum_{n=0}^{\infty} (A^T)^n CA^n$$

$$= \sum_{n=0}^{\infty} (A^T)^{n+1} CA^{n+1} - \sum_{n=0}^{\infty} (A^T)^n CA^n$$

$$= \sum_{n=0}^{\infty} \triangle_n \left((A^T)^n CA^n \right)$$

$$= (A^T)^n CA^n \big|_{n=0}^{\infty} = -C.$$

This completes the proof.

When Lyapunov functionals are used to study the behavior of solutions of functional difference equations, we often end up with a pair of inequalities of the form

$$V(n,x(\cdot)) = W_1(x(n)) + \sum_{s=0}^{n-1} K(n,s) W_2(x(s)),$$

$$\triangle V(n,x(\cdot)) \leq -W_3(x(n)) + F(n).$$

The above two inequalities are rich in information regarding the qualitative behavior of the solutions. However, getting such information will require deep knowledge of Lyapunov functionals and analysis. In Chapter 2, we will develop general theorems to deal with such inequalities that arise from the assumption of the existence of a Lyapunov functional. To be specific, we consider the nonlinear Volterra difference equation

$$x(n+1) = \sum_{s=0}^{n-1} c(n,s) f(x(s)) + g(n,x(n)) \tag{1.2.25}$$

for all integers $n \geq 0$ and for integers, $0 \leq s \leq n$. The functions g and f are continuous in x and satisfy $|f(x)| \leq \delta|x|$, and $|g(n,x)| \leq \lambda(n)(|x|+1)$, where $\lambda : \mathbb{Z}^+ \to (0,1)$. Define the functional V by

$$V(n,x) = |x(n)| + \delta \sum_{s=0}^{n-1} \sum_{u=0}^{\infty} |c(u,s)||x(s)|.$$

After some calculations, by evaluating $\triangle V$ along the solutions of (1.2.25), we arrive at

$$\triangle V(n,x) \leq \left[\lambda(n) + \delta \sum_{u=n+1}^{\infty} |c(u,n)| - 1 \right] |x(n)| + \lambda(n)$$

$$\leq -\alpha|x(n)| + \lambda(n),$$

where we have assumed that $\lambda(n) + \delta \sum_{u=n+1}^{\infty} |c(u,n)| - 1 \leq -\alpha$, for positive constant α. It will be shown in Chapter 2, Theorem 2.1.1 that all solutions of (1.2.25) are uniformly bounded.

Next we address instability of (1.2.2).

Definition 1.2.7. The zero solution of (1.2.2) is unstable if there is an $\varepsilon > 0$, and $n_0 \geq 0$, such that for any $\delta > 0$ there is an x_0 with $|x_0| < \delta$ and there is an $n_1 > n_0$ such that $|x(n_1, n_0, x_0)| \geq \varepsilon$.

Theorem 1.2.12 (Raffoul). *Suppose there exists a continuous Lyapunov function* $V : \mathbb{Z}^+ \times D \to [0, \infty)$ *which is locally Lipschitz in x such that*

$$W_1(|x|) \leq V(n,x) \leq W_2(|x|), \tag{1.2.26}$$

and along the solutions of (1.2.2) we have

$$\triangle V(n,x) \geq W_3(|x|). \tag{1.2.27}$$

Then the zero solution of (1.2.2) is unstable.

Proof. Suppose not, then for $\varepsilon = \min\{1, d(0, \partial D)\}$ we can find a $\delta > 0$ such that $|x_0| < \delta$ and $n \geq 0$ imply that $|x(n, 0, x_0)| < \varepsilon$. We may pick x_0 in such a way so that $|x_0| = \delta/2$ and find $\gamma > 0$ with $W_2(\gamma) = W_1 \delta/2)$. Then for $x(n) = x(n, 0, x_0)$ we have $\triangle V(n,x) \geq 0$ so that

$$W_2(|x(n)|) \geq V(n, x(n)) \geq V(0, x_0) \geq W_1(\delta/2) = W_2(\gamma)$$

from which we conclude that $\gamma \leq |x(n)|$ for $n > 0$. Thus

$$\triangle V(n,x) \geq W_3(|x(n)|) \geq W_3(\gamma).$$

Thus,

$$W_2(|x(n)|) \geq V(n, x(n)) \geq V(0, x_0) + nW_3(\gamma),$$

from which we conclude that $|x(n)| \to \infty$, which is a contradiction. This completes the proof.

We have the following example. Consider the two-dimensional system

$$x(n+1) = \begin{pmatrix} 2 & 1 \\ 0 & -2 \end{pmatrix} x(n) := Ax(n). \tag{1.2.28}$$

Let

$$V(n,x) = x^T(n) \begin{pmatrix} 1/3 & 2/15 \\ 2/15 & 2/5 \end{pmatrix} x(n) := x_1^2(n)/3 + 2x_2^2(n)/5 + 4x_1(n)x_2(n)/15.$$

Then along the solutions of (1.2.28) we have

$$\triangle V(n,x) = x_1^2(n) + x_2^2(n).$$

It is clear that V is positive definite since the matrix $B = \begin{pmatrix} 1/3 & 2/15 \\ 2/15 & 2/5 \end{pmatrix}$ is positive definite. Moreover, we have

$$(x_1(n)/3 + 2x_2(n)/5)^2 \leq V(n,x) \leq x_1^2(n) + x_2^2(n)$$

and hence an application of Theorem 1.2.12 shows the zero solution of (1.2.28) is unstable. Next we explain how we constructed the Lyapunov function V. It is a known fact that if A is an $k \times k$ constant matrix all of whose eigenvalues lie outside the unit circle, then there is a positive definite matrix C ($x^T C x \geq 0$) that is symmetric such that the equation

$$A^T BA - B = C$$

can be solved for $B = B^T$. Thus in the above example, we took $V = x^T B x$ with $C = \begin{pmatrix} 1 & 0 \\ 0 & 1 \end{pmatrix}$.

We state and prove variant forms of discrete Gronwall's Inequality.

Theorem 1.2.13 (Discrete Gronwall's Inequality). *Let $\mathbb{N}_{n_0} = \{n_0, n_0 + 1, n_0 + 2, \cdots\}$ where n_0 is a fixed nonnegative integer. Assume $u(n), \alpha(n), \beta(n)$, and $\gamma(n)$ be nonnegative scalar sequences for all $n \geq n_0$. Let $n \in \mathbb{N}_{n_0}$ and assume, for all $n \geq n_0$, the inequality*

$$u(n) \leq \alpha(n) + \beta(n) \sum_{s=n_0}^{n-1} \gamma(s)u(s) \tag{1.2.29}$$

holds. Then,

$$u(n) \leq \alpha(n) + \beta(n) \sum_{s=n_0}^{n-1} \alpha(s)\gamma(s) \prod_{r=s+1}^{n-1} \big(1 + \beta(r)\big)\gamma(r) \tag{1.2.30}$$

holds for all $n \geq n_0$.

Proof. Define $\varphi : \mathbb{N}_{n_0} \to \mathbb{R}$ by $\varphi(n) = \sum_{s=n_0}^{n-1} \gamma(s)u(s)$. Then,

$$\triangle\varphi(n) = \gamma(n)u(n), \quad \varphi(n_0) = 0.$$

Substituting $u(n) \leq \alpha(n) + \beta(n)\varphi(n)$, in the above equality yields,

$$\triangle\varphi(n) = \gamma(n)u(n) \leq \gamma(n)\big(\alpha(n) + \beta(n)\varphi(n)\big),$$

from which we conclude

$$\varphi(n+1) - (1 + \beta(n)\gamma(n))\varphi(n) \leq \alpha(n)\gamma(n). \tag{1.2.31}$$

Since all sequences are nonnegative, we have that $1 + \beta(n)\gamma(n) > 0$, for all $n \geq n_0$. Thus, (1.2.31) is equivalent to

$$\triangle\Big[\prod_{s=n_0}^{n-1} \big(1 + \beta(s)\gamma(s)\big)^{-1} \varphi(s) \Big] \leq \alpha(n)\gamma(n) \prod_{s=n_0}^{n} \big(1 + \beta(s)\gamma(s)\big)^{-1}.$$

Summing the above inequality from n_0 to $n-1$ yields

$$\prod_{s=n_0}^{n-1} \left(1+\beta(s)\gamma(s)\right)^{-1} \varphi(n) \leq \sum_{s=n_0}^{n-1} \alpha(s)\gamma(s) \prod_{r=n_0}^{n} \left(1+\beta(r)\gamma(r)\right)^{-1}$$

or

$$\varphi(n) \leq \sum_{s=n_0}^{n-1} \alpha(s)\gamma(s) \prod_{r=s+1}^{n-1} \left(1+\beta(r)\gamma(r)\right).$$

As a consequence

$$\sum_{s=n_0}^{n-1} \gamma(s)u(s) \leq \sum_{s=n_0}^{n-1} \alpha(s)\gamma(s) \prod_{r=s+1}^{n-1} \left(1+\beta(r)\gamma(r)\right).$$

Hence

$$u(n) \leq \alpha(n)+\beta(n) \sum_{s=n_0}^{n-1} \gamma(s)u(s)$$

$$\leq \alpha(n)+\beta(n) \sum_{s=n_0}^{n-1} \alpha(s)\gamma(s) \prod_{r=s+1}^{n-1} \left(1+\beta(r)\gamma(r)\right).$$

This completes the proof.

We have the following special cases of Gronwall's inequality. Also by noting that $1+L\beta\gamma(s) \leq e^{L\beta\gamma(s)}$, where $L \geq 1$ and constant, yields the followings.

Corollary 1.1. *If $\alpha(n) = \alpha, \beta(n) = \beta$, for all $n \in \mathbb{N}_{n_0}$, then we have*

$$u(n) \leq \alpha \prod_{s=n_0}^{n-1} \left(1+\beta\gamma(s)\right),$$

or

$$u(n) \leq \alpha e^{\sum_{s=n_0}^{n-1} \beta\gamma(s)}.$$

Corollary 1.2. *Assume the hypothesis of Corollary 1.1. If*

$$u(n) \leq m\left[\alpha+\beta \sum_{s=n_0}^{n-1} \gamma(s)u(s)\right], \text{ for all } n \in \mathbb{N}_{n_0}, n \geq n_0 \text{ and } m \geq 1$$

then we have

$$u(n) \leq \alpha \prod_{s=n_0}^{n-1} \left(1+m\beta(s)\gamma(s)\right),$$

or

$$u(n) \leq \alpha e^{\sum_{s=n_0}^{n-1} m\beta\gamma(s)}.$$

Proof. Define $\varphi : \mathbb{N}_{n_0} \to \mathbb{R}$ by $\varphi(n) = m\beta \sum_{s=n_0}^{n-1} \gamma(s)u(s)$. The rest of the proof follows along the lines of the proof of Theorem 1.2.13. For more on Gronwall's inequality we refer to [57].

Now is the time to introduce functional delay difference equations. It has been observed in modeling physical or biological situations that the rate of change of the systems' current status is more likely to depend on the history of the system and not only on the current status. This usually leads to the delay functional difference system

$$x(n+1) = f(x(n), x(n-\tau)),\qquad(1.2.32)$$

where $x(n)$ is the system's state at time n, $f : \mathbb{R}^k \times \mathbb{R}^k \to \mathbb{R}^k$ is a given function, and the time lag $\tau \in \mathbb{Z}^+$. Equation (1.2.32) arises naturally in modeling population dynamics of a single-species structured population. For example, in such situation, if $x(n)$ denotes the population density of the mature and reproductive population, and if the maturation period is assumed to be constant, then we have

$$f(x(n), x(n-\tau)) = d_m x(n) + e^{-d_i \tau} b(x(n-\tau))$$

where d_m and d_i are the death rates of the mature and immature populations, respectively, and $b : \mathbb{R} \to \mathbb{R}$ is the birth rate. Assuming death is instantaneous, and so the term $d_m x(n)$ is without delay. However, the rate into the mature population is the maturation rate (not the birth rate), that is, the birth rate at time τ, multiplied by the survival probability $e^{-d_i \tau}$ during the maturation process.

To specify a solution $x(n)$ of (1.2.32) for $n \in \mathbb{Z}^+$, we must prescribe the history of x on $[-\tau, 0]$, say $x(s) = \phi(s)$, $s \in [-\tau, 0]$ where

$$\phi : [-\tau, 0] \to \mathbb{Z}^k,$$

which we refer to it throughout the book, as a given initial function. That is

$$f(x(n), x(n-\tau)) = f(x(n), \phi(n-\tau)), \ n \in [0, \tau].$$

Functional delay difference equations will be studied in detail in Chapters 2 and 3.

1.3 Total Stability via Resolvent

We shall obtain asymptotic stability criteria for the discrete Volterra equation

$$x(n+1) = A(n)x(n) + \sum_{s=0}^{n} B(n,s)x(s) + g(n, x(n))\qquad(1.3.1)$$

for all integers $n \geq 0$ and for integers, $0 \leq s \leq n$, where A, C are $k \times k$ matrix functions, and x is a $k \times 1$ unknown vector. We also assume that $|g(n,x)| \leq \lambda(n)|x|$ for some function λ whose properties are given below. Hino and Murakami [82] defined total stability for Volterra integro-differential equations and applied the concept to obtain asymptotic stability criteria. Elaydi and Murakami [61] have extended these definitions and methods to (1.3.4) in the linear case with $g = 0$. Recently, Zhang [177] extended the original work of Hino and Murakami [82], by defining a concept, ψ-total stability. In the case, $\psi = 1$, Zhang's definition reduces to that of Hino and Murakami. Zhang exhibits examples with $\psi \neq 1$ and obtains some new asymptotic stability criteria in the linear case.

For $x \in \mathbb{R}^k$, $|x|$ denotes the Euclidean norm of x. For any $k \times k$ matrix A, define the norm of A by $|A| = \sup\{|Ax| : |x| \leq 1\}$. Let $C(n)$ denote the set of functions $\phi : [0, n] \rightarrow \mathbb{R}$ and $\|\phi\| = \sup\{|\phi(s)| : 0 \leq s \leq n\}$. For each $\psi : \mathbb{Z}^+ \rightarrow (0, \infty)$, we denote by $C_\psi(\tau)$ the space of all functions $p : [\tau, \infty] \rightarrow \mathbb{R}$ such that $\sup_{s \geq \tau} |p(s)/\psi(s)| < \infty$. We set

$$|p|_\psi = \sup\{|p(s)/\psi(s)| : s \geq \tau\}.$$

For each $n_0 \in \mathbb{Z}^+$ and $\phi \in C(n_0)$, there is a unique function $x : \mathbb{Z}^+ \rightarrow \mathbb{R}$ which satisfies (1.3.1) on $[n_0, \infty)$ with $x(s) = \phi(s)$ for $0 \leq s \leq n_0$. Such a function $x(n)$ is called a solution of (1.3.1) through (n_0, ϕ) and is denoted by $x(n, n_0, \phi)$.

Most of the materials here are taken from [65, 128] and the references therein. In this section, we adopt the stability definitions given by Definition 1.1.4.

Definition 1.3.1. The zero solution of (1.3.1) is ψ-totally stable ($\psi - TS$) if for any $\varepsilon > 0$, there exists a $\delta = \delta(\varepsilon) > 0$ such that $[n_0 \geq 0, \phi \in C(n_0), p \in C_\psi(n_0), \|\phi\| < \delta, |p|_\psi < \delta]$ imply $|y(n, n_0, \phi, p)| < \varepsilon$, where $y(n) = y(n, n_0, \phi, p)$ is a solution of

$$y(n+1) = A(n)y(n) + \sum_{s=0}^{n} B(n, s)y(s) + g(n, y(n)) + p(n), \ n \geq n_0 \qquad (1.3.2)$$

such that $y(s) = \phi(s)$ for $s \in [0, n_0]$. It follows from the above definitions that the zero solution of (1.3.1) is $\psi - TS$ implies it is (US).

Next we consider a simple difference equation to illustrate that a zero solution can be uniformly stable and asymptotically stable but not uniformly asymptotically stable. Consider the difference equation

$$x(n+1) = \frac{n}{n+1}x(n), \ x(n_0) = x_0 \neq 0, \ n \geq n_0 \geq 1. \qquad (1.3.3)$$

Let $z(n) = nx(n)$. Then $x(n+1) = z(n+1)/n+1$, and hence (1.3.3) becomes

$$z(n+1) = z(n), \ z(n_0) = x_0 n_0,$$

which has the solution $z(n) = x_0 n_0$ and the solution to (1.3.3) is then found to be

$$x(n) := x(n, n_0, x_0) = \frac{x_0 n_0}{n}.$$

Clearly the zero solution is (US) and (AS). However, for $n_0 = n$, we have

$$x(2n, n, x_0) = \frac{x_0 n}{2n} \rightarrow \frac{x_0}{2} \neq 0$$

which implies that the zero solution is not (UAS).

We begin by considering the Volterra difference equation

$$x(n+1) = A(n)x(n) + \sum_{s=0}^{n} B(n,s)x(s) \tag{1.3.4}$$

for all integers $n \geq 0$ and for integers, $0 \leq s \leq n$, where A, C are $k \times k$ matrix functions, and x is a $k \times 1$ unknown vector. Equation (1.3.4) is of convolution type when $B(n,s) = B(n-s)$. We will develop one of the resolvent equations associated with (1.3.4). In particular, if $R(n,s)$ denote the resolvent of (1.3.4), then it was shown in [52] that $R(n,s)$ must satisfy

$$R(n+1,s) = A(n)R(n,s) + \sum_{u=s}^{n} B(n,u)R(u,s), \tag{1.3.5}$$

if $s \leq n$, $R(s,s) = I$ and $R(n,s) = 0$ if $n < s$. We shall also show that

$$R(n,s+1)(A(s)-I) + \sum_{u=s}^{n-1} R(n,u+1)B(u,s) + \triangle_s R(n,s) = 0, \tag{1.3.6}$$

if $s \leq n$, $R(n,n) = I$ and $R(n,s) = 0$ if $n < s$, where $\triangle_s R(n,s) = R(n,s+1) - R(n,s)$. In the case (1.3.4) is of convolution type; that is $B(n,s) = B(n-s)$, then the resolvent matrix equation (1.3.5) takes the form

$$R(n+1) = A(n)R(n) + \sum_{u=0}^{n} B(n-u)R(u), \, R(0) = I, \, n \in \mathbb{Z}^+. \tag{1.3.7}$$

Note that, if we consider the matrix $A(n) = A$, A constant and nonsingular, then it can be easily shown that the fundamental matrix $F(n)$ of (1.3.4), which is nonsingular with $F(0) = I$ and by the uniqueness of solutions, one can verify that $R(n,0) = F(n)$ and $R(n,s) = F(n-s)$, $n \geq s$. For the rest of this chapter we utilize the resolvent equations coupled with suitable Lyapunov functionals and apply Gronwall's inequality to obtain total stability and hence (UAS) of the zero solution of different forms of (1.3.4). Let $n_0 \geq 0$, $\phi \in C(n_0)$, and $x(n) = x(n,n_0,\phi)$ be a solution of (1.3.4) Let

$$x(n) = R(n,n_0)\phi(n_0) + \sum_{s=n_0}^{n-1} R(n,s+1) \sum_{u=0}^{n_0-1} B(s,u)\phi(u) \tag{1.3.8}$$

where $R(n,s)$ satisfies (1.3.6).

Lemma 1.1 (Variation of Parameters). *Suppose $x(n)$ is a solution of* (1.3.4). *Then $x(n)$ satisfies* (1.3.8), *if and only if $R(n,s)$ satisfies* (1.3.6).

Proof. Let $D(s) = A(s) - I$. Summing

$$\triangle(R(n,s)x(s)) = R(n,s+1)\triangle(x(s)) + (\triangle_s R(n,s))x(s)$$

from $s = n_0$ to $n-1$ we obtain,

$$x(n) - R(n,n_0)\phi(n_0) = \sum_{s=n_0}^{n-1} R(n,s+1)\Delta(x(s)) + (\Delta_s R(n,s))x(s).$$

Thus,

$$x(n) - R(n,n_0)\phi(n_0) = \sum_{s=n_0}^{n-1} R(n,s+1)\left[D(s)x(s) + \sum_{u=0}^{s} B(s,u)x(u)\right]$$

$$+ \sum_{s=n_0}^{n-1} (\Delta_s R(n,s))x(s)$$

$$= \sum_{s=n_0}^{n-1} R(n,s+1)\left[D(s)x(s) + \sum_{u=0}^{n_0-1} B(s,u)\phi(u)\right.$$

$$\left. + \sum_{u=n_0}^{s} B(s,u)x(u)\right] + \sum_{s=n_0}^{n-1} (\Delta_s R(n,s))x(s)$$

$$= \sum_{s=n_0}^{n-1} R(n,s+1) \sum_{u=0}^{n_0-1} B(s,u)\phi(u) + \sum_{s=n_0}^{n-1} R(n,s+1)D(s)x(s)$$

$$+ \sum_{s=n_0}^{n-1} \sum_{u=n_0}^{s} R(n,s+1)B(s,u)x(u) + \sum_{s=n_0}^{n-1} (\Delta_s R(n,s))x(s).$$

Interchange the order of summation to obtain

$$\sum_{s=n_0}^{n-1} \sum_{u=n_0}^{s} R(n,s+1)B(s,u)x(u) = \sum_{u=n_0}^{n-1} \sum_{s=u}^{n-1} R(n,s+1)B(s,u)x(u)$$

$$= \sum_{s=n_0}^{n-1} \sum_{u=s}^{n-1} R(n,u+1)B(u,s)x(s).$$

Hence,

$$x(n) - R(n,n_0)\phi(n_0) - \sum_{s=n_0}^{n-1} R(n,s+1) \sum_{u=0}^{n_0-1} B(s,u)\phi(u)$$

$$= \sum_{s=n_0}^{n-1} \left[R(n,s+1)D(s) + \sum_{u=s}^{n-1} R(n,u+1)B(u,s) + \Delta_s R(n,s)\right]x(s).$$

Since $x(n)$ satisfies (1.3.8), we have the left side of the above equality is zero and hence the summation on the right is equal to zero. This results in $R(n,s)$ satisfying (1.3.6). On the other hand, if $R(n,s)$ satisfies (1.3.6), we have the summation on the right is equal to zero and hence $x(n)$ satisfies (1.3.8).

We note that (1.3.8) is a variation of parameters formula. For more on the subject, we refer to [52] and [135]. The next lemma offers a new variation of parameters formula that we need throughout the book.

Lemma 1.2 (New Variation of Parameters). *[Raffoul] Consider the perturbed Volterra difference equation that is given by (1.3.1) with such that $x(s) = \phi(s)$ for $s \in [0, n_0]$. If $R(n,s)$ satisfies (1.3.6), then any solution $x(n)$ of (1.3.1) is given by*

$$x(n) = R(n,n_0)\phi(n_0) + \sum_{s=n_0}^{n-1} R(n,s+1) \sum_{u=0}^{n_0-1} B(s,u)\phi(u)$$

$$+ \sum_{s=n_0}^{n-1} R(n,s+1)g(s,x(s)). \tag{1.3.9}$$

Proof. The proof follows along the line of the proof of Lemma 1.1 with minor modifications. To see this, we let $D(s) = A(s) - I$. Then from the proof of Lemma 1.1 we have that

$$x(n) - R(n,n_0)\phi(n_0) = \sum_{s=n_0}^{n-1} R(n,s+1)\left[D(s)x(s) + \sum_{u=0}^{s} B(s,u)x(u) + g(s,x(s)) \right]$$

$$+ \sum_{s=n_0}^{n-1} (\triangle_s R(n,s))x(s)$$

$$= \sum_{s=n_0}^{n-1} R(n,s+1)\left[D(s)x(s) + \sum_{u=0}^{n_0-1} (B(s,u)\phi(u) \right.$$

$$\left. + \sum_{u=n_0}^{s} B(s,u)x(u) + g(s,x(s)) \right] + \sum_{s=n_0}^{n-1} (\triangle_s R(n,s))x(s)$$

$$= \sum_{s=n_0}^{n-1} R(n,s+1) \sum_{u=0}^{n_0-1} B(s,u)\phi(u) + \sum_{s=n_0}^{n-1} R(n,s+1)D(s)x(s)$$

$$+ \sum_{s=n_0}^{n-1}\sum_{u=n_0}^{s} R(n,s+1)B(s,u)x(u) + \sum_{s=n_0}^{n-1} R(n,s+1)g(s,x(s))$$

$$+ \sum_{s=n_0}^{n-1} (\triangle_s R(n,s))x(s).$$

By interchanging the order of summation on the same term as in Lemma 1.1 we arrive at

$$x(n) - R(n,n_0)\phi(n_0) - \sum_{s=n_0}^{n-1} R(n,s+1) \sum_{u=0}^{n_0-1} B(s,u)\phi(u) - \sum_{s=n_0}^{n-1} R(n,s+1)g(s,x(s))$$

$$= \sum_{s=n_0}^{n-1} \left[R(n,s+1)D(s) + \sum_{u=s}^{n-1} R(n,u+1)B(u,s) + \triangle_s R(n,s) \right] x(s).$$

As $R(n,s)$ satisfies (1.3.6) for all n and s, we have the summation on the right is equal to zero and hence $x(n)$ satisfies (1.3.9). This completes the proof.

We have the following lemma that provides sufficient conditions for the boundedness of the resolvent.

Lemma 1.3. *If*

$$\sup_{n\geq 0}\frac{1}{1+|A(n)|}\sum_{s=0}^{n}|B(n,s)|\leq L^* \tag{1.3.10}$$

and

$$\sup_{n\geq 1}\sum_{s=0}^{n-1}|R(n,s+1)|\left(1+|A(s)|\right)\leq M^* \tag{1.3.11}$$

for some positive constants L^* and M^*, then there exists a positive constant Q such that $|R(n,s)|\leq Q$ for all $0\leq n_0\leq s\leq n<\infty$.

Proof. By solving equation (1.3.6) for $\triangle_s R(n,s)$, and summing it from s to $n-1$ and then changing the order of summation we arrive at

$$R(n,s) = I - \sum_{u=s}^{n-1} R(n,u+1))(I-A(u)) + \sum_{v=s}^{n-1}\sum_{u=v}^{n-1} R(n,u+1)B(u,v)$$

$$= I - \sum_{u=s}^{n-1} R(n,u+1)(I-A(u)) + \sum_{u=s}^{n-1}\sum_{v=s}^{u} R(n,u+1)B(u,v).$$

Thus

$$|R(n,s)| \leq 1 + \sum_{u=s}^{n-1}|R(n,u+1)|\left(1+|A(u)|\right)$$

$$+ \sum_{u=s+1}^{n-1}|R(n,u+1)|\left(1+|A(u)|\right)\sup_{\tau\geq 0}\frac{1}{1+|A(\tau)|}\sum_{v=0}^{\tau}|B(\tau,v)|$$

$$\leq 1+M^*+M^*L^* =: Q.$$

The resolvent is an abstract term and hence verifying conditions that are associated with it is challenging. With the aid of (1.3.6) and Lyapunov functionals we shall furnish an example in which we verify condition (1.3.11).

Lemma 1.4. *If $y(n) = u(n-n_0)x(n)$ where $x(n)$ is a solution of (1.3.1), then $y(n)$ satisfies (1.3.2) with*

$$p(n) = (\triangle_n u(n-n_0))x(n+1)$$

$$+ \sum_{s=0}^{n} B(n,s)\left[u(n-n_0)-u(s-n_0)\right]x(s)$$

$$+u(n-n_0)g(n,x(n)) - g(n,u(n-n_0)x(n)). \tag{1.3.12}$$

Proof.

$$y(n+1) = u(n+1-n_0)x(n+1)$$
$$= (\triangle_n u(n-n_0))x(n+1) + u(n-n_0)x(n+1).$$

The rest of the proof follows easily by substituting the right side of (1.3.1) in the second term of the above equality and then by adding and subtracting the necessary terms.

Theorem 1.3.1. *Suppose* (1.3.10),

$$\frac{1}{1+|A(n)|} \sum_{s=0}^{n_0} |B(n,s)| \to 0 \qquad (1.3.13)$$

as $n - n_0 \to \infty$ uniformly, and for any $\zeta = \zeta(\varepsilon)$, $0 < \varepsilon < 1$, there exists an $N > 0$ such that

$$\frac{\lambda(n)}{1+|A(n)|} < \zeta(\varepsilon) \qquad (1.3.14)$$

for all $n \geq N$ where ε is the one given below. If the zero solution of (1.3.1) is $\psi - TS$ with $\psi = 1 + |A(n)|$, then it is (UAS).

Proof. By definition the zero solution of (1.3.1) is $\psi - TS$ implies it is (US). Let $n_0 \in Z^+$ and $||\varphi|| < \delta(1)$, where $\delta(\cdot)$ is the one given for the $(\psi - TS)$ of (1.3.1) with $\psi(n) = 1 + |A(n)|$ for all $n \in Z^+$. Then $|x(n,n_0,\varphi)| < 1$ for all $n \geq n_0$. Now for any $\varepsilon > 0, 0 < \varepsilon < 1, \alpha > 0$, we set

$$u(t) = u(t,\alpha,\varepsilon) = \begin{cases} \frac{1+2\alpha t}{1+\alpha \varepsilon t} & \text{for } t \geq 0 \\ 1 & \text{for } t < 0 \end{cases}$$

Set $y(n) = u(n-n_0)x(n)$; then $y(n)$ solves (1.3.8) where $p(n)$ is given by (1.3.2). It follows from (1.3.10) that for any $\eta > 0$, there exists an $S = S(\eta) > 0$ such that

$$\frac{1}{1+|A(n)|} \sum_{s=0}^{n-S(\eta)} |B(n,s)| < \eta$$

for all $n \geq S(\eta)$. Also, $|u(n)| \leq \frac{2}{\varepsilon}$ (see [177]) and $|\triangle_n u(n-n_0)| \leq 2\alpha$. By (1.3.2) we have

$$|p(n)| \leq |\triangle_n u(n-n_0)|$$
$$+ (1+|A(n)|)\frac{1}{(1+|A(n)|)} \sum_{s=n-S(\eta)+1}^{n} |B(n,s)||u(n-n_0) - u(s-n_0)|$$
$$+ (1+|A(n)|)\frac{1}{(1+|A(n)|)} \sum_{s=0}^{n-S(\eta)} |B(n,s)||u(n-n_0) - u(s-n_0)|$$
$$+ \left(1+|A(n)|\right)\left[|u(n-n_0)||g(n,x(n))|\right]$$

$$+|g(n,u(n-n_0)x(n))|\Big]\frac{1}{1+|A(n)|}$$

$$\leq 2\alpha+2\alpha L^*S(\eta)(1+|A(n)|)+4\frac{\eta}{\varepsilon}(1+|A(n)|)$$

$$+\frac{4}{\varepsilon}\frac{\lambda(n)}{1+|A(n)|}(1+|A(n)|)$$

$$\leq \Big[2\alpha+2\alpha L^*S(\eta)+4\frac{\eta}{\varepsilon}+\frac{4\zeta}{\varepsilon}\Big][1+|A(n)|].$$

Take small numbers, $\eta=\eta(\varepsilon)$, $\alpha=\alpha(\varepsilon)$ and $\zeta=\zeta(\varepsilon)$ so that $\alpha(1+L^*S(\eta))<\frac{\delta(1)}{6}$, $\eta<\frac{\delta(1)\varepsilon}{12}$ and $\zeta<\frac{\delta(1)\varepsilon}{12}$. Then, $|p(n)|_\psi<\delta(1)$. Consequently, it follows from the $\psi-TS$ of the zero solution of (1.3.1) that $|y(n)|<1$ for all $n\geq n_0\geq 0$. Hence, if $n\geq n_0+\frac{(1-\varepsilon)}{\alpha\varepsilon}$ we have

$$|x(n,n_0,\varphi)|=\frac{|y(n)|}{|u(n-n_0)|}$$

$$<\frac{1}{|u(n-n_0)|}$$

$$<\frac{1+\varepsilon\alpha(n-n_0)}{1+2\alpha(n-n_0)}<\varepsilon.$$

This completes the proof.

Theorem 1.3.2. *If* (1.3.10), (1.3.11), (1.3.13) *and*

$$\sup_{n\geq 0}\sum_{s=0}^{n}\lambda(n)<\infty \tag{1.3.15}$$

hold, then the zero solution of (1.3.1) *is (UAS).*

Proof. We first show that the zero solution of (1.3.1) is $\psi-TS$ with $\psi(n)=1+|A(n)|$. Let $p\in C_\psi(n_0)$ and $y(n)=y(n,n_0,\varphi,p)$ be a solution of (1.3.2). By replacing $g(n,x(n))$ with $g(n,x(n))+p(n)$ in (1.3.9) we have that

$$y(n)=R(n,n_0)\phi(n_0)+\sum_{s=n_0}^{n-1}R(n,s+1)\sum_{u=0}^{n_0-1}B(s,u)\phi(u)$$

$$+\sum_{s=n_0}^{n-1}R(n,s+1)g(s,y(s))+\sum_{s=n_0}^{n-1}R(n,s+1)p(s)$$

and

$$|y(n)|\leq\|\phi\|\Big[R(n,n_0)$$

$$+\sum_{s=n_0}^{n-1}|R(n,s+1)|(1+|A(n)|)\sup_{\tau\geq 0}\frac{1}{1+|A(\tau)|}\sum_{u=0}^{\tau}|B(\tau,u)|\Big]$$

$$+|p|_\psi \sum_{s=n_0}^{n-1} R(n,s+1)(1+|A(s)|) + \sum_{s=n_0}^{n-1} |R(n,s+1)|\lambda(s)|y(s)|$$

$$\leq ||\phi||[Q+M^*L^*] + |p|_\psi M^* + Q\sum_{s=n_0}^{n-1} \lambda(s)|y(s)|).$$

Applying Gronwall's inequality (Corollary 1.1) we obtain

$$|y(n)| \leq \left[||\phi||(Q+M^*L^*) + |p|_\psi M^*\right] e^{Q\sum_{s=n_0}^{n-1}\lambda(s)}.$$

This implies that the zero solution of (1.3.1) is $\psi - TS$ and therefore by Theorem 1.3.1 it is (UAS).

In the next example we use Lyapunov's method to directly verify condition (1.3.11). This is of special interest to us because, in difference equations, this produces a new summability criteria of the resolvent .

1.3.1 Application to Perturbed Volterra Difference Equations

In this section, we use Lyapunov functional in terms of the resolvent and obtain stability results concerning nonlinear scalar Volterra difference equations.

Example 1.5. Consider the perturbed and scalar Volterra difference equation

$$x(n+1) = a(n)x(n) + \sum_{s=0}^{n} b(n,s)x(s) + g(n,x(n)) \tag{1.3.16}$$

with the assumption that $|g(n,x)| \leq \lambda(n)|x|$. Assume (1.3.10) and (1.3.13) hold for the scalar equation (1.3.16). Also, suppose there is a sequence $\varphi : Z^+ \times Z^+ \to (0,\infty)$ such that

$$\triangle_s \varphi(n,s) \geq |b(n,s)| \tag{1.3.17}$$

and

$$-|a(n)| + K\left(1 - |b(n,n)| - \varphi(n,n)\right) \geq \beta \tag{1.3.18}$$

where β and K are positive constants with $0 < K < 1$. If (1.3.15) holds, then the zero solution of (1.3.16) is (UAS).

Proof. Define the Lyapunov functional $V(s)$ on $[0, n-1]$ by

$$V(s) = |R(n,s)| + \sum_{u=s}^{n-1} \varphi(u,s)|R(n,u+1)| \tag{1.3.19}$$

where $R(n,s)$ is the resolvent of (1.3.6) with $g=0$, satisfying

$$R(n,s+1)a(s) + \sum_{u=s}^{n-1} R(n,u+1)b(u,s) - R(n,s) = 0.$$

Then,

$$\triangle V(s) = |R(n,s+1)| - |R(n,s)|$$

$$+ \sum_{u=s+1}^{n-1} \varphi(u,s+1)|R(n,u+1)| - \sum_{u=s}^{n-1} \varphi(u,s)|R(n,u+1)|$$

$$\geq (-|a(s)|+1)|R(n,s+1)| - \sum_{u=s}^{n-1} |R(n,u+1)||b(u,s)|$$

$$+ \sum_{u=s+1}^{n-1} \varphi(u,s+1)|R(n,u+1)| - \varphi(s,s)|R(n,s+1)|$$

$$- \sum_{u=s+1}^{n-1} \varphi(u,s)|R(n,u+1)|.$$

Or

$$\triangle V(s) = \left[1 - |a(s)| - |b(s,s)| - \varphi(s,s)\right]|R(n,s+1)|$$

$$+ \sum_{u=s+1}^{n-1} \left(\triangle_s \varphi(u,s) - |b(u,s)|\right)|R(n,u+1)|$$

$$\geq \left[1 - |a(s)| - |b(s,s)| - \varphi(s,s)\right]|R(n,s+1)|$$

$$= (-1 + \frac{1}{K})|a(s)||R(n,s+1)|$$

$$+ \frac{1}{K}\left[-|a(s)| + K\left(1 - |b(s,s)| - \varphi(s,s)\right)\right]|R(n,s+1)|$$

$$\geq k\left(1 + |a(s)|\right)|R(n,s+1)| \tag{1.3.20}$$

where $k = \min\left[-1 + \frac{1}{K}, \frac{\beta}{K}\right]$. Summing (1.3.20) from 0 to $n-1$ yields

$$k \sum_{s=0}^{n-1} (1 + |a(s)|)|R(n,s+1)| \leq V(n) - V(0)$$

$$= |R(n,n)| - |R(n,0)|. \tag{1.3.21}$$

Thus,

$$\sup_{n\geq 0} \sum_{s=0}^{n-1} |R(n,s+1)(|1 + |a(s)|) < \frac{1}{k}. \tag{1.3.22}$$

Hence condition (1.3.11) is satisfied and by Theorem 1.3.2 the zero solution of (1.3.16) is (UAS).

Note that the Lyapunov functional $V(s)$ defined by (1.3.19) is of general type. To see this, let $\varphi(u,s) = \sum_{v=0}^{s-1} |b(u,v)|$. Then φ satisfies (1.3.17) and condition (1.3.18) reduces to

$$-|a(n)| + K\left(1 - \sum_{s=0}^{n} |b(n,s)|\right) \geq \beta. \tag{1.3.23}$$

It is easy to see that (1.3.21) implies that $\sum_{s=0}^{n} |b(n,s)| < 1$ and hence (1.3.10) is satisfied. We note that inequality (1.3.22) implies $R(n,s)$ is bounded for $1 \leq s \leq n$. Also, from inequality (1.3.21) $R(n,0)$ is bounded, since $V(s)$ is increasing. We conclude that Lemma 1.3 is not needed for this example.

1.4 Uniform Asymptotic Stability via Resolvent

In this section, we study the stability properties of the zero solution of the nonlinear perturbed Volterra discrete system

$$x(n+1) = A(n)x(n) + \sum_{s=0}^{n} B(n,s)x(s) + g(n,x(n)) \tag{1.4.1}$$

where $g(n,x)$ is continuous in x and satisfies $|g(n,x(n))| \leq \lambda(n)|x|$, where $\lambda(n)$ is such that $0 \leq \lambda(n) \leq N < +\infty$, for some constant N. Moreover, A, B are $k \times k$ matrix functions on $\mathbb{Z}^+$ and $\mathbb{Z}^+ \times \mathbb{Z}^+$, respectively. Recently, several authors have studied the behavior of solutions of variant forms of (1.4.1). Medina [113, 114, 117], Eloe et al. [65], and Raffoul [135] obtained stability and boundedness results of the solutions of the homogenous part of (1.4.1) by means of representing the solution in terms of the resolvent matrix. In addition, Eloe et al. [65] and Elaydi et al. [61] used the notion of total stability and established results on the asymptotic behavior of the zero solution of (1.4.1). Their work heavily depended on showing or assuming the summability of the resolvent matrix. For more results on stability of the zero solution of Volterra discrete system we refer the reader to Crisci, Komanovskii, and Vecchio [36], Elaydi [52], and Agarwal and Pang [5]. This research is a continuation of the research initiated by the authors in [84] and related to the work in [65]. In this section we extend some of the results in [83], and later on we furnish an example as an application to some of our theorems, in which we show the summability of the resolvent matrix.

For $x \in \mathbb{R}^k$, $|x|$ denotes the Euclidean norm of x. For any $k \times k$ matrix A, we define the norm of A by $|A| = \max\{|Ax| : |x| \leq 1\}$. We define the set $C(n) = \{\phi \in \mathbb{R}^k : \phi : [0,n] \to \mathbb{R}^k\}$ with the norm $||\phi|| = \max\{|\phi(s)| : 0 \leq s \leq n\}$ on it.

For each $n_0 \in \mathbb{Z}^+$, and $\phi \in C(n_0)$, there is a unique (vector) function $\phi : \mathbb{Z}^+ \to \mathbb{R}^k$ on $0 \leq s \leq n_0$ on $[n_0, \infty)$ with $x(s) = \phi(s)$ for $0 \leq s \leq n_0$. Such a function $x(n)$ is called a solution of (1.4.1), and is denoted by $x(n,n_0,\phi)$. Throughout this section we write $x(n)$ for $x(n,n_0,\phi)$ unless it is stated otherwise.

Theorem 1.4.1. *Let*

$$Q(n) = |A(n)| + \lambda(n) + \sum_{u=n}^{\infty} |B(u,n)|.$$

Suppose that, for all $n \geq n_0 \geq 0$,

$$\sum_{n=n_0}^{\infty} Q(n)|R(n,n_0)| < \infty, \quad \sum_{u=n}^{\infty} Q(u)|R(u,n+1)| \leq K_1 Q(n) < \infty, \qquad (1.4.2)$$

and for the upper bound N, on $\lambda(n), 0 \leq NK_1 < 1$, hold. Then the zero solution of $(1.4.1)$ is stable.

Proof. Suppose that $x(n)$ is a solution of $(1.4.1)$. If $R(n,s)$ satisfies $(1.3.6)$, then $x(n)$ is given by, see Lemma 1.2,

$$x(n) = R(n,n_0)\phi(n_0) + \sum_{s=n_0}^{n-1} R(n,s+1) \sum_{u=0}^{n_0-1} B(s,u)\phi(u)$$

$$+ \sum_{s=n_0}^{n-1} R(n,s+1) g(s,x(s)). \qquad (1.4.3)$$

First we take the absolute value on both sides of $(1.4.3)$, multiply through by $Q(n)$, and then sum from $n = n_0$ to $n = \infty$ to obtain

$$\sum_{n=n_0}^{\infty} Q(n)|x(n)| \leq \sum_{n=n_0}^{\infty} Q(n)|R(n,n_0)| \,||\phi||$$

$$+ \sum_{n=n_0}^{\infty} Q(n) \sum_{s=n_0}^{n-1} |R(n,s+1)| \sum_{u=0}^{n_0-1} |B(s,u)| \,||\phi||$$

$$+ \sum_{n=n_0}^{\infty} Q(n) \sum_{s=n_0}^{n-1} |R(n,s+1)|\lambda(s)|x(s)|$$

$$\leq \sum_{n=n_0}^{\infty} Q(n)|R(n,n_0)| \,||\phi||$$

$$+ \sum_{n=n_0}^{\infty} \sum_{s=n_0}^{n} Q(n)|R(n,s+1)| \sum_{u=0}^{n_0} |B(s,u)| \,||\phi||$$

$$+ \sum_{n=n_0}^{\infty} \sum_{s=n_0}^{n} Q(n)|R(n,s+1)| \,\lambda(s)|x(s)|.$$

By changing the order of summations, we have

$$\sum_{n=n_0}^{\infty} \sum_{s=n_0}^{n} Q(n)|R(n,s+1)| = \sum_{s=n_0}^{\infty} \sum_{n=s}^{\infty} Q(n)|R(n,s+1)|.$$

Thus, from the above inequality, we obtain

$$\sum_{n=n_0}^{\infty} Q(n)|x(n)| \le \sum_{n=n_0}^{\infty} |R(n,n_0)|Q(n)\,\|\phi\|$$

$$+ \sum_{s=n_0}^{\infty}\sum_{n=s}^{\infty} Q(n)|R(n,s+1)|\sum_{u=0}^{n_0}|B(s,u)|\,\|\phi\|$$

$$+ \sum_{n=n_0}^{\infty}\sum_{u=n}^{\infty} Q(u)|R(u,n+1)|\lambda(n)|x(n)|. \qquad (1.4.4)$$

By (1.4.2) there exists an $M_1 > 0$, and $M_2 > 0$ such that

$$\sum_{n=s}^{\infty} Q(n)|R(n,s)| < M_1, \text{ and } \sum_{s=n}^{\infty} Q(s)|R(s,n+1)| < M_2.$$

Let

$$C^*(n_0) = \sum_{s=n_0}^{\infty}\sum_{u=0}^{n_0} |B(s,u)|.$$

Then from (1.4.4), we have

$$\sum_{n=n_0}^{\infty} Q(n)|x(n)| \le M_1\,\|\phi\| + M_2 \sum_{s=n_0}^{\infty}\sum_{u=0}^{n_0} |B(s,u)|\,\|\phi\|$$

$$+ NK_1 \sum_{n=n_0}^{\infty} Q(n)|x(n)|$$

$$\le M_1\,\|\phi\| + M_2 C^*(n_0)\|\phi\|$$

$$+ NK_1 \sum_{n=n_0}^{\infty} Q(n)|x(n)|.$$

Solving for $\sum_{n=n_0}^{\infty} Q(n)|x(n)|$, we get

$$\sum_{n=n_0}^{\infty} Q(n)|x(n)| \le \frac{\|\phi\|(M_1 + M_2 C^*(n_0))}{1 - NK_1}. \qquad (1.4.5)$$

Next we rewrite (1.4.1) as

$$\triangle x(n) = D(n)x(n) + \sum_{s=0}^{n} B(s,u)x(s) + g(n,x(n))$$

where I is the identity matrix and $D(n) = A(n) - I$. By summing the above equation over n from n_0 to $n-1$ we get

$$\left| \sum_{s=n_0}^{n-1} \triangle x(s) \right| = \left| \sum_{s=n_0}^{n-1} \left[D(s)x(s) + \sum_{u=0}^{s} B(s,u)x(u) + g(s,x(s)) \right] \right|$$

$$\leq \left| \sum_{s=n_0}^{n-1} \left[D(s)x(s) + \sum_{u=n_0+1}^{s} B(s,u)x(u) + g(s,x(s)) \right] \right|$$

$$+ \left| \sum_{s=n_0}^{n-1} \sum_{u=0}^{n_0} B(s,u)x(u) \right|$$

$$\leq \sum_{n=n_0}^{\infty} |D(n)| \, |x(n)| + \sum_{n=n_0}^{\infty} \sum_{u=n_0}^{n} |B(n,u)| \, |x(n)|$$

$$+ \sum_{n=n_0}^{\infty} \lambda(n) |x(n)| + \sum_{n=n_0}^{\infty} \sum_{u=0}^{n_0} |B(n,u)| \, ||\phi||.$$

By interchanging the order of summations in the second term of the right side of the above inequality, we arrive at

$$\left| \sum_{s=n_0}^{n-1} \triangle x(s) \right| \leq \sum_{n=n_0}^{\infty} \left[|D(n)| + \sum_{u=n_0}^{n} |B(n,u)| + \lambda(n) \right] |x(n)| + C^*(n_0) ||\phi||$$

$$\leq \sum_{n=n_0}^{\infty} Q(n) |x(n)| + C^*(n_0) ||\phi||. \tag{1.4.6}$$

By substituting (1.4.5) into (1.4.6), we get

$$\left| \sum_{s=n_0}^{n-1} \triangle x(s) \right| \leq \frac{||\phi|| \, (M_1 + M_2 \, C^*(n_0))}{1 - N K_1} + C^*(n_0) \, ||\phi||.$$

Thus,

$$|x(n) - x(n_0)| \leq ||\phi|| \left[\frac{||\phi|| \, (M_1 + M_2 \, C^*(n_0))}{1 - N K_1} + C^*(n_0) \right].$$

But $|x(n)| - |x(n_0)| \leq |x(n) - x(n_0)|$, and hence we have

$$|x(n)| \leq ||\phi|| \left[\frac{(M_1 + M_2 \, C^*(n_0))}{1 - N K_1} + C^*(n_0) + 1 \right].$$

We remark that if $C^*(n_0)$ is uniformly bounded, then Theorem 1.4.1 implies that the zero solution of (1.4.1) is (US).

Theorem 1.4.2. *Suppose*
i.

$$\sup_{n \geq n_0 \geq 0} \sum_{s=n}^{\infty} |R(s,n_0)| < +\infty, \tag{1.4.7}$$

$$\sup_{n \geq n_0 \geq 0} \left\{ |R(n,n_0)| + \sum_{u=0}^{n_0-1} | \sum_{s=n_0}^{n-1} R(n,s+1) B(s,u) \right\} < +\infty, \tag{1.4.8}$$

$$\sup_{n \geq n_0 \geq 0} \lambda(n) \sum_{s=n}^{\infty} |R(s,n+1)| \leq L < 1, \tag{1.4.9}$$

and there exist a D > 0 such that

$$\sum_{s=n_0}^{\infty} \left(\sum_{n=s}^{\infty} |R(n,s+1)| \sum_{u=0}^{n_0-1} |B(s,u)| \right) \leq D \tag{1.4.10}$$

then the zero solution of (1.4.1) is (US).
ii. If the zero solution of (1.4.1) is (US) and

$$\sup_{n \geq n_0 \geq 0} \sum_{s=n_0}^{n-1} |R(n,s+1)| < \infty \tag{1.4.11}$$

then (1.4.8) holds.

Proof.
i. Suppose that (1.4.8)–(1.4.10) hold. Summing (1.4.3) over n from $n = n_0$ to $n = \infty$ and using $|g(n,x)| \leq \lambda(n)|x|$, we get

$$\sum_{n=n_0}^{\infty} |x(n)| \leq \sum_{n=n_0}^{\infty} |R(n,n_0)| \, ||\phi||$$

$$+ \sum_{n=n_0}^{\infty} \left(\sum_{s=n_0}^{n-1} |R(n,s+1)| \sum_{u=0}^{n_0-1} B(s,u) \, ||\phi|| \right)$$

$$+ \sum_{n=n_0}^{\infty} \left(\sum_{s=n_0}^{n} |R(n,s+1)| \, |g(s,x(s))| \right)$$

$$\leq \sum_{n=n_0}^{\infty} |R(n,n_0)| \, ||\phi||$$

$$+ \sum_{n=n_0}^{\infty} \left(\sum_{s=n_0}^{n} |R(n,s+1)| \sum_{u=0}^{n_0-1} B(s,u) \, ||\phi|| \right)$$

$$+ \sum_{n=n_0}^{\infty} \sum_{s=n}^{\infty} |R(s,n+1)| \lambda(n)|x(n)|.$$

From (1.4.7), there exists a positive constant F such that $\sum_{s=n}^{\infty} |R(n,n_0)| \leq F$. Thus, using (1.4.9) and (1.4.10) we arrive at

$$\sum_{n=n_0}^{\infty} |x(n)| \leq F + \sum_{s=n_0}^{\infty} \left(\sum_{n=s}^{\infty} |R(n,s+1)| \sum_{u=0}^{n_0-1} B(s,u) \right) ||\phi||$$

$$+ L \sum_{n=n_0}^{\infty} |x(n)||. \tag{1.4.12}$$

Hence (1.4.12) yields

$$\sum_{n=n_0}^{\infty} |x(n)| \leq F\|\phi\| + D\|\phi\| + L\sum_{n=n_0}^{\infty} |x(n)|.$$

Thus,

$$\sum_{n=n_0}^{\infty} |x(n)| \leq \frac{(F+D)\|\phi\|}{1-L}. \tag{1.4.13}$$

Using equation (1.4.3), we obtain

$$|x(n)| \leq |R(n,n_0)|\,\|\phi\| + \sum_{u=0}^{n_0-1}\left(|\sum_{s=n_0}^{n-1} R(n,s+1)B(s,u)|\right)\|\phi\|$$

$$+ \sum_{s=n_0}^{n-1} |R(n,s+1)|\lambda(s)|x(s)|. \tag{1.4.14}$$

By (1.4.8) and the fact that $\lambda(n)$ is bounded, there exists a constant $P > 0$ such that $|R(n,s)|\lambda(s) \leq P$ for $0 \leq n_0 \leq s \leq n$. Also by (1.4.8), there exists a constant $E > 0$ such that

$$\sup_{n \geq n_0 \geq 0}\left\{|R(n,n_0)| + \sum_{u=0}^{n_0-1}|\sum_{s=u_0}^{n-1} R(n,s+1)B(s,n)|\right\} < E.$$

Thus (1.4.13) and (1.4.14) yield

$$|x(n)| \leq \left\{|R(n,n_0)| + \sum_{u=0}^{n_0-1}|\sum_{s=n_0}^{n-1} R(n,s+1)B(s,u)|\right\}\|\phi\|$$

$$+ P\sum_{s=n_0}^{n-1} |x(s)|$$

$$\leq E\|\phi\| + P\frac{(F+D)\|\phi\|}{1-L} := J\|\phi\|. \tag{1.4.15}$$

Thus, (1.4.15) implies that the zero solution of (1.4.1) is (US). Suppose that the zero solution of (1.4.1) is (US). Then for $\varepsilon = 1$, there exists a $\delta > 0$ such that $[n_0 \geq 0, \phi \in C(n_0), \|\phi\| \leq \delta, n \geq n_0]$ implies $|x(n,n_0,\phi)| < 1$. Let m be a positive integer and define the sequence of functions ϕ_m by

$$\phi_m(u) = va^{-m(n_0-u)} \text{ on } 0 \leq u \leq n_0. \tag{1.4.16}$$

Let $\psi_m(u) = \frac{\delta}{2}va^{-m(n_0-u)}$ for $0 \leq u \leq n_0$. Then, $|\psi_m(u)| \leq \frac{\delta}{2}$. Hence we have $|x(n,n_0,\psi_m(s))| < \varepsilon$. It is clear from (1.4.16) that $\phi_m(n_0) = v$ and $|\phi_m(s)| \leq 1$ for $0 \leq s \leq n_0$. Thus, from (1.4.3) we have

$$|R(n,n_0)|\frac{\delta}{2} \leq |x(n,n_0,\psi_m)| + \frac{\delta}{2}|\sum_{s=n_0}^{n-1} R(n,s+1)\sum_{u=0}^{n_0-1} B(s,u)\,a^{-m(n_0-u)}|$$

$$+|\sum_{s=n_0}^{n-1} |R(n,s+1)|\,\lambda(s)\,|x(s,n_0,\psi_m(s))|$$

$$\leq 1+\frac{\delta}{2}|\sum_{s=n_0}^{n-1} R(n,s+1)\sum_{u=0}^{n_0-1} B(s,u)\,a^{-m(n_0-u)}|$$

$$+|\sum_{s=n_0}^{n-1} |R(n,s+1)|\,\lambda(s). \tag{1.4.17}$$

Now, for fixed n,

$$|\sum_{s=n_0}^{n-1} R(n,s+1)\sum_{u=0}^{n_0-1} B(s,u)\,a^{-m(n_0-u)}|\to 0 \text{ as } m\to\infty.$$

By (1.4.11), there exists a $G>0$ such that $\sum_{s=n_0}^{n-1}|R(n,s+1)|\lambda(s)\leq G$. Thus from (1.4.17)

$$|R(n,n_0)|\leq\frac{2}{\delta}(1+G). \tag{1.4.18}$$

Next, let $\phi\in C(n_0)$ with $||\phi||<1$. Define $\psi=\delta\phi$. Then $||\psi||<\delta$. Thus, by the definition of δ, we have $|x(n,n_0,\psi)|<1$ for all $n\geq n_0$. It follows from (1.4.3) and (1.4.18) that

$$\left|\sum_{s=n_0}^{n-1} R(n,s+1)\sum_{u=0}^{n_0-1} B(s,u)\,\psi(u)\right|\leq |x(n,n_0,\psi)|+|R(n,n_0)|\,|\psi(n_0)|$$

$$+\sum_{s=n_0}^{n-1}|R(n,s+1)|\lambda(s)|\,|x(s,n_0,\psi)|$$

$$\leq |x(n,n_0,\psi)|+|R(n,n_0)|\,||\psi(n_0)||+G$$

$$\leq 1+2(1+G)+G.$$

Hence,

$$|\sum_{u=0}^{n_0-1}\sum_{s=n_0}^{n-1} R(n,s+1)\,\varphi(u)|\leq\frac{1}{\delta}|\sum_{n=0}^{n_0-1}\sum_{s=n_0}^{n-1} R(n,s+1)\,\psi(n)|\leq\frac{3}{\delta}(1+G)$$

for $n\geq n_0$ and the proof is complete. The next Lemma gives necessary and sufficient conditions for the uniform boundedness of $R(n,s)$.

Lemma 1.5. *There exists a positive constant H such that $|R(n,s)|\leq H$ for $n\geq s\geq 0$ if and only if*

$$\sup_{n\geq s\geq 0}\left|\sum_{u=s}^{n-1} R(n,u+1)\left(D(u)+\sum_{v=s}^{u} R(n,u+1)B(u,v)\right)\right|<\infty. \tag{1.4.19}$$

Proof. By solving equation (1.3.6) for $\triangle_s R(n,s)$, and summing it from s to $n-1$ and then changing the order of summations, we arrive at

$$R(n,s) = I + \sum_{u=s}^{n-1} R(n,u+1)\left[A(u) - I\right] + \sum_{v=s}^{n-1}\sum_{u=v}^{n-1} R(n,u+1)B(u,v)$$

$$= I + \sum_{u=s}^{n-1} R(n,u+1)D(u) + \sum_{u=s}^{n-1}\sum_{v=s}^{u} R(n,u+1)B(u,v)$$

$$= I + \sum_{u=s}^{n-1} R(n,u+1)\left[D(u) + \sum_{v=s}^{u} B(u,v)\right].$$

Hence, the result follows.

For the next theorem, we assume

$$\sup_{n \geq n_0 \geq 0} \sum_{u=0}^{n_0-1} \left| \sum_{s=n_0}^{n} R(n,s+1)\,B(s,u) \right| < +\infty. \tag{1.4.20}$$

Theorem 1.4.3. *Assume that* $|R(n,s+1)| \leq H(s)$ *for* $0 \leq s \leq n < \infty$ *with*

$$\sup_{n \geq 0} \sum_{s=0}^{n-1} H(s)\lambda(s) \leq K \text{ for } K > 0. \tag{1.4.21}$$

Then, the zero solution of (1.4.1) is (US) if and only if (1.4.8) holds .

Proof. If (1.4.8) holds, then for $0 \leq n_0 \leq s \leq n$, we have $|R(n,s+1)| \leq H(s) < \infty$. From (1.4.14), we obtain

$$|x(n)| \leq \left\{ |R(n,n_0)| + \sum_{u=0}^{n_0-1} \left| \sum_{s=n_0}^{n-1} R(n,s+1)\,B(s,u) \right| \right\} \|\phi\|$$

$$+ \sum_{s=n_0}^{n-1} |R(n,s+1)|\,\lambda(s)\,|x(s)|$$

$$\leq E\|\phi\| + \sum_{s=n_0}^{n-1} H(s)\lambda(s)\,|x(s)|.$$

Applying the discrete Gronwall's inequality (Corollary 1.1), we get

$$|x(n)| \leq E\|\phi\| \exp\left(\sum_{s=n_0}^{n-1} H(s)\lambda(s) \right) \leq E\|\phi\| \exp(K) := L\|\phi\|. \tag{1.4.22}$$

This proves that the zero solution of (1.4.1) is (US). The proof of the converse of this theorem is similar to the proof of the converse of Theorem 1.4.2.

Theorem 1.4.4. *(i) If (1.4.8)–(1.4.10) and (1.4.19)–(1.4.20) hold, then the zero solution of (1.4.1) is (US). (ii) If the zero solution of (1.4.1) is (US) and (1.4.11) holds, then (1.4.19)–(1.4.20) hold.*

Proof. Conditions (1.4.19)–(1.4.20) hold if and only if (1.4.8) holds. Therefore the results follow directly from Theorem 1.4.1.

Theorem 1.4.5. *i. Suppose that (1.4.7)–(1.4.10), and*

$$\left\{ |R(n,n_0)| + \sum_{u=0}^{n_0-1} |\sum_{s=n_0}^{n-1} R(n,s+1)B(s,u)| \right\} \to 0 \qquad (1.4.23)$$

as $n - n_0 \to +\infty$ uniformly,

$$\sum_{s=n_0}^{n-1} |R(n,s+1)| \to 0, \qquad (1.4.24)$$

as $n - n_0 \to +\infty$ uniformly hold. Then, the zero solution of (1.4.1) is (UAS).
ii. If the zero solution of (1.4.1) is (UAS) and (1.4.24) hold, then (1.4.23) holds.

Proof. Suppose that (1.4.7)–(1.4.10) hold. Then, by Theorem 1.4.2, the zero solution is obviously (US). Let $B_1 > 0$ be given and $\phi \in C(n_0)$ on $0 \le s \le n_0$ with $\|\phi\| \le B_1$. Then, it follows from (1.4.14) and (1.4.15) that,

$$|x(n)| \le \left\{ |R(n,n_0)| + \sum_{u=0}^{n_0-1} |\sum_{s=n_0}^{n-1} R(n,s+1)B(s,u)| \right\} \|\phi\|$$

$$+ \sum_{s=n_0}^{n-1} |R(n,s+1)| NJ \|\phi\|$$

$$\le \left[|R(n,n_0)| + \sum_{u=0}^{n_0-1} |\sum_{s=n_0}^{n-1} R(n,s+1)B(s,u)| \right] B_1$$

$$+ \left[NJ \sum_{s=n_0}^{n-1} |R(n,s+1)| \right] B_1.$$

From (1.4.23) and (1.4.24), it follows that for any $\varepsilon > 0$, there exists a constant $T > 0$ such that

$$\left[|R(n,n_0)| + \sum_{u=0}^{n_0-1} |\sum_{s=n_0}^{n-1} R(n,s+1)B(s,u)| + NJ \sum_{s=n_0}^{n-1} |R(n,s+1)| \right] < \frac{\varepsilon}{B_1}$$

for all $n \ge T + n_0$. Thus, $|x(n)| < \varepsilon$ for all $n \ge T + n_0$. This implies that the zero solution of (1.4.1) is (UAS). Conversely, suppose that the zero solution of (1.4.1) is (UAS). Then it is (US). Let $\phi \in C(n_0)$ with $\|\phi\| \le 1$. Then, for any $\varepsilon > 0$, there exists $T > 0$ such that $|x(n,n_0,\phi)| < \varepsilon$ for $n \ge T + n_0$. By making use of (1.4.17) and by the argument of Theorem 2.2 (ii), we have $|R(n, n_0)| < \varepsilon$ for all $n \geq T + n_0$. Now using (1.4.14) in (1.4.3), we get

$$|\sum_{u=0}^{n_0-1} \left(\sum_{s=n_0}^{n-1} R(n, s+1) B(s, u) \right) \phi(u)|$$

$$\leq |x(n, n_0, \phi)| + |R(n, n_0)| + NJ \sum_{s=n_0}^{n-1} |R(n, s+1)| < 3\varepsilon$$

for all $n \geq T + n_0$. This implies

$$\sum_{u=0}^{n_0-1} |\sum_{s=n_0}^{n-1} R(n, s+1) B(s, u)| < 3\varepsilon$$

for all $n \geq T + n_0$. This shows that (1.4.23) holds and the proof is complete.

Remark 1.1. The function

$$|R(n, s)| + \sum_{u=s}^{n-1} |\sum_{v=0}^{s-1} R(n, u+1) B(u, v)| \tag{1.4.25}$$

can serve as a Lyapunov functional to directly verify conditions (1.4.8) and (1.4.23)–(1.4.24) as we shall see in the next example.

Lemma 1.6. *The resolvent $R(n, s) \to 0$ as $n - s \to +\infty$ uniformly if and only if*

$$\left\{ I + \sum_{u=s}^{u=n-1} |R(n, u)| \left(D(u) + \sum_{v=s}^{u} B(u, v)| \right) \right\} \to 0 \tag{1.4.26}$$

as $n - s \to +\infty$ uniformly.

Proof. The proof follows directly from Lemma 1.5.

Theorem 1.4.6. *If (1.4.7)–(1.4.10), (1.4.24)–(1.4.26), and*

$$\sum_{u=0}^{n_0-1} |\sum_{s=n_0}^{n-1} R(n, s+1) B(s, u)| \to 0 \tag{1.4.27}$$

as $n - n_0 \to +\infty$ uniformly hold, then the zero solution of (1.4.1) is (UAS).

Proof. The proof of Theorem 1.4.6 follows directly from Lemma 1.6 and Theorem 1.4.5

Using (1.4.22) of Theorem 1.4.3, we obtain the following theorem which is more practical when the sum of $R(n, s+1)$ along with $A(n)$ and $\sum_{s=0}^{n} B(n, s)$ can be estimated.

Theorem 1.4.7. *Suppose that (1.4.8) and (1.4.21) hold. Then, the zero solution of (1.4.1) is (UAS) if and only if (1.4.23) and*

$$\sum_{s=0}^{n-1} |R(n,s+1)| \lambda(s) \to 0 \ \text{for} \ n - n_0 \to +\infty \ \text{uniformly} \tag{1.4.28}$$

hold.

Proof. Suppose that (1.4.8) and (1.4.21) hold. Then, by Theorem 1.4.3, the zero solution is (US).
Let $B_1 > 0$ be a given constant and $\phi \in C(n_0)$ on $0 \le s \le n_0$ with $\|\phi\| < B_1$. Using (1.4.22) in (1.4.3), we obtain

$$|x(n)| \le \left\{ |R(n,n_0)| + \sum_{u=0}^{n_0-1} \left(\sum_{s=n_0}^{n-1} R(n,s+1) B(s,u) \right) \right\} \|\phi\|$$

$$+ L \sum_{s=n_0}^{n-1} |R(n,s+1)| \lambda(s) \|\phi\|.$$

Applying (1.4.23) and (1.4.28) in the above inequality gives the (UAS). The converse of Theorem 1.4.7 follows from the proof of the converse of Theorem 1.4.3.

1.4.1 Application to Scalar Equations

We end the section by furnishing an example in which we show that the zero solution of the scalar nonlinear Volterra discrete equation

$$x(n+1) = a(n)x(n) + \sum_{s=0}^{n} b(n,s)x(s) + g(n,x(n)) \tag{1.4.29}$$

where $|g(n,x)| \le \lambda |x|$, is (UAS).

Example 1.6. Consider equation (1.4.29) and suppose there are positive constants γ, h, B, and K with $K < 1$ satisfying the following conditions for $n \ge 0$:

(i) $-a(n) + K\left(1 - \sum_{s=0}^{n} |b(n,s)|\right) > 0,$

(ii) For each $\gamma > 0$, there exists $h > 0$ such that $\sum_{s=n}^{n+h-1} |a(s)| \ge \gamma,$

(iii) $\dfrac{1}{|a(n)|} \sum_{s=0}^{n_0} |b(n,s)| \to 0$ as $n - n_0 \to +\infty$ uniformly on $\{n \,|\, a(n) \ne 0\},$

(iv) $\sum_{s=0}^{n_0-1} |b(n,s)| \ge \lambda(n)$ for $n_0 \ge 0,$

(v) $\sum_{0}^{n} \lambda(s) < \infty$ for all $n \geq 0$, and

(vi) $\sum_{u=0}^{\infty} |b(u,s)| \leq B.$

Then the zero solution of (1.4.29) is (UAS).

Proof. Define the discrete Lyapunov functional, $V(s)$ on $[0, n-1]$ by

$$V(s) = |R(n,s)| + \sum_{u=s}^{n-1} \sum_{v=0}^{s-1} |R(n, u+1)|\, |b(u,v)|$$

where $R(n,s)$ is the resolvent of (1.4.29) with $g = 0$, satisfying

$$R(n, s+1)\, a(s) + \sum_{u=s}^{n-1} R(n, u+1)\, b(u,s) - R(n,s) = 0, \quad R(n,n) = 1.$$

Then using (i) we have

$$\triangle V(s) = |R(n, s+1)| - |R(n,s)|$$

$$+ \sum_{u=s+1}^{n-1} \sum_{v=0}^{s} |R(n, u+1)|\, |b(u,v)| - \sum_{u=s}^{n-1} \sum_{v=0}^{s-1} |R(n, u+1)|\, |b(u,v)|$$

$$\geq (-|a(s)| + 1)\, |R(n, s+1)|$$

$$- |R(n, s+1)|\, |b(s,s)| - |R(n, s+1)| \sum_{v=0}^{s-1} |b(s,v)|$$

$$= \left(1 - |a(s)| - \sum_{v=0}^{s} |b(s,v)| \right) |R(n, s+1)|$$

$$\geq \left(\frac{1}{K} - 1 \right) |a(s)|\, |R(n, s+1)|. \tag{1.4.30}$$

So we have $\triangle V(s) > 0$.
This yields that for $n \geq n_0 \geq 0$, $V(n_0) \leq V(n) = |R(n,n)| = 1$. That is,

$$|R(n, n_0)| + \sum_{u=n_0}^{n-1} \sum_{v=0}^{n_0-1} |R(n, u+1)|\, |b(u,v)| \leq 1. \tag{1.4.31}$$

Hence (1.4.8) is satisfied.
By summing (1.4.30) from 0 to $n-1$, we obtain

$$\left(\frac{1}{K} - 1 \right) \sum_{s=0}^{n-1} |a(s)|\, |R(n, s+1)| \leq V(n) - V(0) \leq 1 \tag{1.4.32}$$

or

$$\sum_{s=0}^{n-1} |a(s)||R(n,s+1)| \le \frac{K}{1-K} =: D.$$

Note that (1.4.31) implies that there exists a constant $H > 0$ such that

$$\sup_{s \ge n_0 \ge 0} |R(n,s+1)| = H \le 1 \text{ for } 0 \le s \le n < \infty.$$

Hence, by (v) we have

$$\sup_{n \ge 1} \sum_{s=0}^{n-1} H\lambda(s) < \infty.$$

Thus condition (1.4.21) is satisfied and by Theorem 1.4.3, the zero solution of (1.4.29) is (US). By (iii), for any $\varepsilon > 0$ there exists $N_1 > 0$ such that for $u \ge N_1 + s - 1$ implies

$$\sum_{v=0}^{s-1} |b(u,v)| \le \frac{\varepsilon}{(3+B)D} |a(u)|.$$

Thus, for $n \ge N_1 + s - 1$ we have

$$\sum_{u=s}^{s+N_1-1} |R(n,u+1)| \sum_{v=0}^{s-1} |b(u,v)|$$

$$= \sum_{u=s}^{n-1} |R(n,u+1)| \sum_{v=0}^{s-1} |b(u,v)| + \sum_{u=s+N_1}^{n-1} |R(n,u+1)| \sum_{v=0}^{s-1} |b(u,v)|$$

$$\le \frac{\varepsilon}{(3+B)D} \sum_{u=s+N_1}^{n-1} |R(n,u+1)||a(u)| + \sum_{u=s}^{s+N_1-1} |R(n,u+1)| \sum_{v=0}^{s-1} |b(u,v)|$$

$$\le \frac{\varepsilon}{3+B} + \sum_{u=s}^{s+N_1-1} |R(n,u+1)| \sum_{v=0}^{s-1} |b(u,v)|. \qquad (1.4.33)$$

Let $\beta = \frac{K}{1-K}$ and $\alpha = \frac{3+B}{\varepsilon\beta}$. By (ii), there exists an $h > 0$ such that $\sum_{v=s}^{s+h-1} |a(v)| \ge \alpha$, and

$$|R(n,n_s+1)|\beta \sum_{v=s}^{s+h-1} |a(v)| \le \beta \sum_{u=s}^{s+h-1} |R(n,u+1)||a(u)|,$$

for $n_s \in [s, s+h-1]$ and $n \ge s+h$, where

$$|R(n,n_s+1)| = \min_{s \le u \le s+h-1} |R(n,u+1)|.$$

Using (1.4.32) in the above inequality we arrive at

$$|R(n,n_s+1)| \le \frac{1}{\beta \sum_{v=s}^{s+h-1} |a(v)|} < \frac{\varepsilon}{3+B}.$$

Choose $N > 1$ so that $\frac{\beta N \varepsilon}{3+B} > 1$. For each $n_0 \geq 0$ and $n \geq n_0 + (N+1)(N_1 + h - 1)$, define $\{n_j\}$ with

$$n(j-1) + N_1 \leq n_j \leq n(j-1) + N_1 + h - 1, \quad j = 1, 2, 3.....N$$

such that

$$|R(n, n_j + 1)| < \frac{\varepsilon}{3+B}. \tag{1.4.34}$$

It follows that $n_N \leq n_0 + N(N_1 + h - 1)$ and by (1.4.32) we arrive at

$$\sum_{j=1}^{N} \left(\sum_{u=n_j}^{n_j+N_1-1} \beta |R(n, u+1)| \|a(u)\| \right) \leq \sum_{u=n_0}^{n-1} \beta |R(n, u+1)| \|a(u)\| \leq 1.$$

Since $\frac{\beta N \varepsilon}{3+B} > 1$, it follows from the above inequality there exists $n_k, 1 \leq k \leq N$ such that

$$N \sum_{u=n_k}^{n_k+N_1-1} \beta |R(n, u+1)| \|a(u)\| \leq 1.$$

Or,

$$\sum_{u=n_k}^{n_k+N_1-1} |R(n, u+1)| \|a(u)\| < \frac{\varepsilon}{3+B}. \tag{1.4.35}$$

Since $V(s)$ is increasing, we have

$$V(n_k) \leq V(n_k + 1).$$

Hence, using (1.4.33)–(1.4.34) and (vi) we arrive at

$$|R(n, n_k)| + \sum_{u=n_k}^{n-1} \sum_{v=0}^{n_k-1} |R(n, u+1)| \, |b(u, v)|$$

$$\leq |R(n, n_k + 1)| + \sum_{u=n_k}^{n-1} \sum_{v=0}^{n_k-1} |R(n, u+1)| \, |b(u, v)|$$

$$+ \sum_{u=n_k+1}^{n-1} |R(n, u+1)| \, |b(u, n_k)|$$

$$\leq \frac{\varepsilon}{3+B} + \frac{2\varepsilon}{3+B} + \frac{B\varepsilon}{3+B} = \varepsilon.$$

This yields

$$|R(n, n_0)| + \sum_{u=n_0}^{n-1} \sum_{v=0}^{n_0-1} |R(n, u+1)| \, |b(u, v)| = V(n_0) \leq V(n_k) < \varepsilon \tag{1.4.36}$$

for $n \geq n_0 + (N+1)(N_1 + h - 1) \geq n_k, N > \frac{3+B}{\beta \varepsilon}$. Hence condition (1.4.23) is satisfied. Next, for $n \geq n_0 + (N+1)(N_1 + h - 1) \geq n_k$ we have by using condition (iv) in (1.4.36).

$$|R(n,n_0)| + \sum_{u=n_0}^{n-1} |R(n,u+1)| \lambda(u)$$

$$\leq |R(n,n_0)| + \sum_{u=n_0}^{n-1} |R(n,u+1)| \sum_{v=0}^{n_0-1} |b(u,v)|$$

$$\leq |R(n,n_0)| + \sum_{u=n_0}^{n-1} \sum_{v=0}^{n_0-1} |R(n,u+1)||b(u,v)|$$

$$< \varepsilon. \tag{1.4.37}$$

Hence, (1.4.28) follows directly from (1.4.37) and the zero solution of (1.4.29) is (UAS) by Theorem 1.4.7

1.4.2 Homogenous Volterra Equations; $(g(n,x) = 0)$

Theorems of the previous section were complicated due to the presence of the perturbation term $g(n,x)$. In this section, we state necessary and sufficient conditions for the (US) and (UAS) of the unperturbed Volterra difference equation

$$x(n+1) = a(n)x(n) + \sum_{s=0}^{n} b(n,s)x(s). \tag{1.4.38}$$

We will state two parallel theorems concerning (1.4.38). The proof of the next theorem follows along the lines of the proof of Theorem 1.4.2.

Theorem 1.4.8 ([83]). *The zero solution of* (1.4.38) *is (US) if and only if* (1.4.8) *holds.*

Similarly, the proof of the next theorem follows along the lines of the proof of Theorem 1.4.7.

Theorem 1.4.9 ([83]). *The zero solution of* (1.4.38) *is (UAS) if and only if* (1.4.8) *and* (1.4.23) *hold.*

In the next theorems we apply the results of Theorems 1.4.8 and 1.4.9 to (1.4.38).

Theorem 1.4.10 ([83]). *The zero solution of* (1.4.38) *is (US) if and only if*

$$\left| \prod_{i=n_0}^{n-1} a(i) \right| + \sum_{s=n_0}^{n-1} \left| \prod_{i=s+1}^{n-1} a(i) \sum_{u=0}^{n_0-1} b(s,u) \right| \leq L \tag{1.4.39}$$

holds for some positive constant L.

Proof. Let $n_0 \geq 0$, $\phi \in C(n_0)$, and $x(n) = x(n, n_0, \phi)$ be a solution of (1.4.38). By the variation of parameters formula ([52, 135]), we have

$$x(n) = \prod_{i=n_0}^{n-1} a(i)\phi(n_0) + \sum_{s=n_0}^{n-1} \prod_{i=s+1}^{n-1} a(i) \sum_{u=0}^{n_0-1} b(s,u)\phi(u)$$

$$:= R(n,n_0)\phi(n_0) + \sum_{s=n_0}^{n-1} R(n,s+1) \sum_{u=0}^{n_0-1} b(s,u)\phi(u), \qquad (1.4.40)$$

where

$$R(n,n_0) = \prod_{i=n_0}^{n-1} a(i).$$

By taking supremum in (1.4.40), we satisfy (1.4.8) by invoking (1.4.39) and results follow from Theorem 1.4.8.

The next result is very interesting since the growth of $\sum_{s=0}^{n} b(n,s)$ is controlled by the size of $a(n)$.

Theorem 1.4.11 ([83]). *The zero solution of* (1.4.38) *is (US) if and only if there exist positive constant α such that*

$$|a(n)| \leq \alpha < 1, \text{ for all } n \geq 0 \qquad (1.4.41)$$

and

$$\sup_{n \to \infty} \frac{1}{|a(n)|} \Big| \sum_{s=0}^{n_0-1} b(n,s) \Big| < \infty, \text{ provided that } a(n) \neq 0 \text{ for large } n. \qquad (1.4.42)$$

Proof. First we note that condition (1.4.42) implies that there exists a positive constant L such that

$$\Big| \sum_{s=0}^{n_0-1} b(n,s) \Big| \leq L|a(n)|.$$

Let $n_0 \geq 0$, $\phi \in C(n_0)$, and $x(n) = x(n, n_0, \phi)$ be a solution of (1.4.38). Then by (1.4.40) we have that

$$|R(n,n_0)| + \sum_{s=n_0}^{n-1} |R(n,s+1)| \sum_{u=0}^{n_0-1} b(s,u)|$$

$$\leq \Big| \prod_{i=n_0}^{n-1} a(i) \Big| + \sum_{s=n_0}^{n-1} \Big| \prod_{i=s+1}^{n-1} a(i) \Big| \Big| \sum_{u=0}^{n_0-1} b(s,u) \Big|$$

$$\leq \alpha^{n-n_0} + L \sum_{s=n_0}^{n-1} |\alpha^{n-s-1)}| |a(s)|$$

$$\leq \alpha^{n-n_0} + L\alpha \sum_{s=n_0}^{n-1} \alpha^{n-s-1)} |a(s)|$$

$$\leq 1 + \alpha L \frac{1 - \alpha^{n+n_0}}{1 - \alpha}$$

$$\leq 1 + \frac{\alpha L}{1 - \alpha}.$$

Hence the proof follows by Theorem 1.4.8.

The next theorem provides necessary and sufficient conditions for the (UAS) of the zero solution of (1.4.38).

Theorem 1.4.12 ([83]). *The zero solution of* (1.4.38) *is (UAS) if and only if* (1.4.39) *holds together with*

$$\sum_{s=n_0}^{n-1} \Big| \sum_{u=0}^{n_0-1} b(s,u) \Big| \to 0, \ as \ n - n_0 \to \infty \ uniformly. \tag{1.4.43}$$

Proof. Let $n_0 \geq 0$, $\phi \in C(n_0)$, and $x(n) = x(n, n_0, \phi)$ be a solution of (1.4.38). We must show that (1.4.23) holds. For our scalar equation (1.4.38), condition (1.4.23) is equivalent to

$$\Big| \prod_{i=n_0}^{n-1} a(i) \Big| + \sum_{s=n_0}^{n-1} \Big| \prod_{i=s+1}^{n-1} a(i) \Big| \Big| \sum_{u=0}^{n_0-1} b(s,u) \Big| \to 0, \ as \ n - n_0 \to \infty \ \text{uniformly}.$$

Now by (1.4.41), we have that $\Big| \prod_{i=n_0}^{n-1} a(i) \Big| \to 0$. Also, by (1.4.43) we get

$$\sum_{s=n_0}^{n-1} \Big| \prod_{i=s+1}^{n-1} a(i) \Big| \Big| \sum_{u=0}^{n_0-1} b(s,u) \Big| \leq \sum_{s=n_0}^{n-1} \sum_{u=0}^{n_0-1} b(s,u) \Big| \to 0, \ as \ n - n_0 \to \infty \ \text{uniformly}.$$

By Theorem 1.4.12 the proof is complete.

Next, we state some (USA) results concerning the vector Volterra difference equation of convolution type

$$x(n+1) = A(n)x(n) + \sum_{s=0}^{n} B(n-s)x(s) \tag{1.4.44}$$

for all integers $n \geq 0$ and for integers, $0 \leq s \leq n$, where $A(n), B(n-s)$ are $k \times k$ matrix functions, and x is a $k \times 1$ unknown vector. For the next theorem, we let

$$h(n) = \sum_{u=0}^{\infty} \Big| \sum_{s=0}^{n-1} R(n-s-1)B(s+u+1) \Big|.$$

Theorem 1.4.13 ([60]). *Suppose that* $\sum_{j=0}^{\infty} |B(j)| < \infty$ *and let* $R(n)$ *be the resolvent of* (1.4.44) *satisfying* (1.3.5). *Then for* (1.4.44) *the following are equivalent.*

(a) $det(zI - A - B\tilde{(z)}) \neq 0$, for $|z| \geq 1$,
(b) $R(n) \in l^1(\mathbb{Z}^+)$,
(c) The zero solution of (1.4.44) is (UAS),
(d) Both $R(n)$ and $h(n)$ tend to zero as $n \to \infty$.

For more on such equivalent conditions we refer to the interesting survey papers and [59] and [114].

It is crucial to mention that the method of resolvent does not work for nonlinear Volterra difference equations of the form

$$x(n+1) = A(n)x(n) + \sum_{s=0}^{n} B(n,s)f(x(s)) \tag{1.4.45}$$

since the function f enters nonlinearly which makes it impossible to determine the resolvent. In later chapters, we shall employ suitable Lyapunov functionals to qualitatively analyze the solutions of (1.4.45). The following results can be found in [115] and [117]. We consider

$$x(n+1) = A(n)x(n) + \sum_{j=n_0}^{n} B(n,j)x(s), \ x(n_0) = x_0 \tag{1.4.46}$$

and its perturbation

$$y(n+1) = A(n)x(n) + \sum_{j=n_0}^{n} B(n,j)y(s) + F(n) \tag{1.4.47}$$

where A and B are matrices and $R(n,s)$ is given by (1.3.5). Then the solution of (1.4.47) is given by

$$x(n,0,x_0) = R(n,0)x_0 + \sum_{j=0}^{n-1} R(n,j+1)F(j). \tag{1.4.48}$$

Now we state a definition that we need in the sequel.

Definition 1.4.1. The zero solution of (1.4.46) is said to be h-stable if there exists a positive and bounded function h and a constant $c \geq 1$ such that

$$|x(n,0,x_0)| \leq c|x_0|\frac{h(n)}{h(n_0)}, \ n \geq n_0 \geq 0, \text{ and } x_0 \text{ is small enough }.$$

If the boundedness condition on h in Definition 1.4.1 is removed, then (1.4.46) is called an h-system around the null solution. We have the following theorem.

Theorem 1.4.14 ([115]). System (1.4.46) is an h-system, if and only if there exists a positive function h and a constant $c \geq 1$ such that

$$|R(n,n_0)| \leq c\frac{h(n)}{h(n_0)}, \ n \geq n_0 \geq 0. \tag{1.4.49}$$

Theorem 1.4.15 ([115]). *Assume system* (1.4.46) *is an h-system. Let* $\lambda : \mathbb{Z}^+ \to \mathbb{R}^+$. *Then the nonlinear perturbed system*

$$y(n+1) = A(n)x(n) + \sum_{j=n_0}^{n} B(n,j)y(s) + p(n,y(n)) \qquad (1.4.50)$$

with $p(n,0) = 0$, *is an h-system, provided that*

$$|p(n,y(n))| \le \frac{h(n)}{h(n+1)}\lambda(n)|y(n)|,$$

where

$$\sum_{n=n_0}^{\infty} \frac{h(n)}{h(n+1)}\lambda(n) < \infty. \qquad (1.4.51)$$

The proof follows by expressing the solution of (1.4.50) in the form

$$y(n,n_0,y_0) = R(n,n_0)y_0 + \sum_{j=0}^{n-1} R(n,j+1)p(j,y(j)).$$

Remark 1.2. It is clear that condition 1.4.51 translates into

$$\sum_{s=0}^{n-1} R(n,s+1)\lambda(s) < \infty,$$

which is equivalent to our condition (1.4.21). Moreover, a careful reading of [117] and [114] reveals that the resolvent of (1.4.46) has to approach zero as n goes to infinity.

It is obvious that conditions (1.4.49) and (1.4.51) require the boundedness and summability of the resolvent . Unlike others, we were successfully able to verify such conditions in Sections 1.3.1 and 1.4.1 with the aid of Lyapunov functionals.

1.5 Open Problems

We just saw that the resolvent method could not be directly applied to nonlinear Volterra difference equations. However, if we impose Almost-Linear conditions on the nonlinear functions, then it might be possible to carry out some of the work of the chapter to nonlinear Volterra difference equations. To be precise, we consider the Volterra difference equation

$$y(n+1) = A(n)y(n) + \sum_{s=0}^{n} C(n,s)y(s) + p(n), \ y(0) = y_0. \qquad (1.5.1)$$

If the resolvent $R(n,s)$ satisfies (1.3.5), then the solution of (1.5.1) is given by

$$y(n) = R(n,0)y_0 + \sum_{s=0}^{n-1} R(n,s+1)p(s). \tag{1.5.2}$$

By imposing conditions on the resolvent (see [3]), one may show that all solutions of (1.5.1) are bounded for every bounded $p(n)$. Next, we consider the nonlinear Volterra difference equation

$$x(n+1) = A(n)h(x(n)) + \sum_{s=0}^{n} C(n,s)g(x(s)) \tag{1.5.3}$$

in which h and g are continuous and real valued functions. Equation (1.5.3) is Almost-Linear in the sense that there is a positive constant K such that

$$|h(x) - x| \le K, \ \ |g(x) - x| \le K.$$

We write (1.5.3) as

$$x(n+1) = A(n)h(x(n)) + \sum_{s=0}^{n} C(n,s)x(s)$$

$$- A(n)[x(n) - h(x(n))] - \sum_{s=0}^{n} C(n,s)[x(s) - g(x(s))] \tag{1.5.4}$$

and express its solution as

$$x(n) = R(n,0)x_0 + \sum_{s=0}^{n-1} R(n,s+1)[A(s)(x(s) - h(x(s)))]$$

$$+ \sum_{s=0}^{n-1} R(n,s+1) \sum_{u=0}^{s} C(s,u)[x(u) - g(x(u)). \tag{1.5.5}$$

The results say: in (1.5.3) replace $g(x)$ and $h(x)$ by x to obtain

$$y(n+1) = A(n)y(n) + \sum_{s=0}^{n} C(n,s)y(s). \tag{1.5.6}$$

Now an interested researcher can pursue different paths. First he or she can follow the study of the last section of this chapter and prove parallel results concerning (1.5.4), by making use of (1.5.5). Another path is using fixed point theory (see Chapter 3), namely contraction mapping principle and obtain results regarding stability and boundedness of solutions, by working directly with (1.5.5). For more on this we ask that you consult with [28].

Next, we explore the possibility of existence of a periodic solution. Again, consider (1.5.1) and suppose we have proved that every solution of (1.5.1) is bounded. Suppose $y_1(n)$ is a fixed solution of (1.5.1) and that $y_2(n)$ is any other solution, then

$$y(n) := y_1(n) - y_2(n)$$

solves (1.5.6). Say we can show that every solution of (1.5.6) tends to zero. Hence, if $y_1(n)$ is a fixed solution of (1.5.1), then every other solution converges to it. Thus, if A and p are periodic, might $y_1(n)$ be periodic under more assumption on C?

Chapter 2
Functional Difference Equations

In this chapter we consider functional difference equations that we apply to all types of Volterra difference equations. Our general theorems will require the construction of suitable Lyapunov functionals, a task that is difficult but possible. As we have seen in Chapter 1, the concept of resolvent can only apply to linear Volterra difference systems. The theorems on functional difference equations will enable us to qualitatively analyze the theory of boundedness, uniform ultimate boundedness, and stability of solutions of vectors and scalars Volterra difference equations. We extend and prove parallel theorems regarding functional difference equations with finite or infinite delay, and provide many applications. In addition, we will point out the need of more research in delay difference equations. In the second part of the chapter, we state and prove theorems that guide us on how to systematically construct suitable Lyapunov functionals for a specific nonlinear Volterra difference equation. We end the chapter with open problems. Most of the results of this chapter can be found in [37, 38, 128, 133, 135, 141, 147, 181], and [182].

2.1 Uniform Boundedness and Uniform Ultimate Boundedness

We begin by considering Lyapunov functionals to prove general theorems about boundedness, uniform ultimate boundedness of solutions, and stability of the zero solution of the nonlinear functional discrete system

$$x(n+1) = G(n, x(s); \ 0 \le s \le n) \overset{def}{=} G(n, x(\cdot)) \qquad (2.1.1)$$

where $G : \mathbb{Z}^+ \times \mathbb{R}^k \to \mathbb{R}^k$ is continuous in x. When Lyapunov functionals are used to study the behavior of solutions of functional difference equations of the form of (2.1.1), we often end up with a pair of inequalities of the form

© Springer Nature Switzerland AG 2018
Y. N. Raffoul, *Qualitative Theory of Volterra Difference Equations*,
https://doi.org/10.1007/978-3-319-97190-2_2

$$V(n, x(\cdot)) = W_1(x(n)) + \sum_{s=0}^{n-1} K(n, s) W_2(x(s)), \tag{2.1.2}$$

$$\triangle V(n, x(\cdot)) \leq -W_3(x(n)) + F(n) \tag{2.1.3}$$

where V is a Lyapunov functional bounded below, x is the known solution of the functional difference equation, and K, F, and $W_i, i = 1, 2, 3$, are scalar positive functions. Inequalities (2.1.2) and (2.1.3) are rich in information regarding the qualitative behavior of the solutions of (2.1.1).

The goal is to use inequalities (2.1.2) and (2.1.3) to conclude boundedness of $x(n)$ when F is bounded. Also, we obtain stability results about the zero solution of (2.1.1) when $F = 0$ and $G(n, 0) = 0$. In the celebrated paper of Kolmanovskii et al. [36], the authors investigated the boundedness of solutions of Volterra difference equations by means of Lyapunov functionals. Also, in [37] the same authors constructed general theorems for the stability of the zero solution of Volterra type difference equations.

As we have seen in Chapter 1, several authors like Medina [113, 115, 116], Islam and Raffoul [83], and Raffoul [135] obtained stability and boundedness results of the solutions of discrete Volterra equations by means of representing the solution in terms of the resolvent matrix of the corresponding system of difference Volterra equations. Eloe et al. [65] and Elaydi et al. [61] used the notion of total stability and established results on the asymptotic behavior of the solutions of discrete Volterra system with nonlinear perturbation. Their work heavily depended on the summability of the resolvent matrix. For more results on stability of the zero solution of Volterra discrete system we refer the reader to Elaydi [52] and Agarwal and Pang [5] and [117].

Boundedness of solutions of linear and nonlinear discrete Volterra equations was also studied by Diblik and Schmeidel [47], Gronek and Schmeidel [72], and the references therein. A survey of the fundamental results on the stability of linear Volterra difference equations, of both convolution and non-convolution type, can be found in Elaydi [59].

We say that $x(n) = x(n, n_0, \phi)$ is a solution of (2.1.1) with a bounded initial function $\phi : [0, n_0] \to \mathbb{R}^k$ if it satisfies (2.1.1) for $n > n_0$ and $x(j) = \phi(j)$ for $j \leq n_0$.

If D is a matrix or a vector, $|D|$ means the sum of the absolute values of the elements. Since we are now dealing with functional difference equations, we restate the following stability definitions.

Definition 2.1.1. Solutions of (2.1.1) are uniformly bounded (UB) if for each $B_1 > 0$ there is $B_2 > 0$ such that $[n_0 \geq 0, \phi : [0, n_0] \to \mathbb{R}^k$ with $|\phi(n)| < B_1$ on $[0, n_0], n > n_0]$ implies $|x(n, n_0, \phi)| < B_2$.

Definition 2.1.2. Solutions of (2.1.1) are uniformly ultimately bounded (UUB) for bound B if there is a $B > 0$ and if for each $M > 0$ there exists $N > 0$ such that $[n_0 \geq 0, \phi : [0, n_0] \to \mathbb{R}^k$ with $|\phi(n)| < M$ on $[0, n_0], n > n_0, n > n_0 + N]$ implies $|x(n, n_0, \phi)| < B$.

If $G(n, 0) = 0$, then $x(n) = 0$ is a solution of (2.1.1). In this case we state the following definitions.

Definition 2.1.3. The zero solution of (2.1.1) is stable (S) if for each $\varepsilon > 0$, there is a $\delta = \delta(\varepsilon) > 0$ such that $\left[\phi : [0,n_0] \to \mathbb{R}^k \text{ with } |\phi(n)| < \delta \text{ on } [0,n_0], n \geq n_0\right]$ implies $|x(n,n_0,\phi)| < \varepsilon$. It is uniformly stable (US) if δ is independent of n_0.

Definition 2.1.4. The zero solution of (2.1.1) is uniformly asymptotically stable (UAS) if it is (US) and there exists a $\gamma > 0$ with the property that for each $\mu > 0$ there exists $N > 0$ such that $\left[n_0 \geq 0, \phi : [0,n_0] \to \mathbb{R}^k \text{ with } |\phi(n)| < \gamma \text{ on } [0,n_0], n \geq n_0 + N\right]$ implies $|x(n,n_0,\phi)| < \mu$.

We begin by proving general theorems regarding boundedness and stability of solutions of (2.1.1).

Theorem 2.1.1 ([133]). *Let $\varphi(n,s)$ be a scalar sequence for $0 \leq s \leq n < \infty$ and suppose that $\varphi(n,s) \geq 0, \triangle_n \varphi(n,s) \leq 0, \triangle_s \varphi(n,s) \geq 0$ and there are constants B and J such that $\sum_{s=0}^{n} \varphi(n,s) \leq B$ and $\varphi(0,s) \leq J$. Also, suppose that for each $n_0 \geq 0$ and each bounded initial function $\phi : [0,n_0] \to \mathbb{R}^k$, every solution $x(n) = x(n,n_0,\phi)$ of (2.1.1) satisfies*

$$W_1(|x(n)|) \leq V(n,x(\cdot)) \leq W_2(|x(n)|) + \sum_{s=0}^{n-1} \varphi(n,s)W_3(|x(s)|) \tag{2.1.4}$$

and

$$\triangle V_{(2.1.1)}(n,x(\cdot)) \leq -\rho W_3(|x(n)|) + K \tag{2.1.5}$$

for some constants ρ and $K \geq 0$. Then solutions of (2.1.1) are (UB).

Proof. $H > 0$ and $|\phi(n)| < H$ on $[0,n_0]$, and set $V(n) = V(n,x(\cdot))$. Let $V(n^*) = \max_{0 \leq n \leq n_0} V(n)$. If $V(n) \leq V(n^*)$ for all $n \geq n_0$, then by (2.1.4) we have

$$W_1(|x(n)|) \leq V(n) \leq V(n^*)$$

$$\leq W_2(|x(n^*)|) + \sum_{s=0}^{n^*-1} \varphi(n^*,s)W_3(|\phi(s)|)$$

$$\leq W_2(|\phi(n^*)|) + \sum_{s=0}^{n^*-1} \varphi(n^*,s)W_3(|\phi(s)|)$$

$$\leq W_2(H) + BW_3(H).$$

From which it follows that

$$|x(n)| \leq W_1^{-1}\left[W_2(H) + BW_3(H)\right].$$

On the other hand, if $V(n) > V(n^*)$ for some $n \geq n_0$, so that $V(n) = \max_{0 \leq s \leq n} V(s)$. We multiply both sides of (2.1.5) by $\varphi(n,s)$ and then sum from $s = n_0$ to $s = n-1$, we obtain

$$\sum_{s=n_0}^{n-1} (\triangle V(s))\varphi(n,s) \leq -\rho \sum_{s=n_0}^{n-1} \varphi(n,s)W_3(|x(s)|) + KB.$$

Summing by parts the left side we arrive at

$$V(n)\varphi(n,n) - V(n_0)\varphi(n,n_0) - \sum_{s=n_0}^{n-1} V(s+1)\triangle_s\varphi(n,s)$$

$$\leq -\rho \sum_{s=n_0}^{n-1} \varphi(n,s)W_3(|x(s)|) + KB.$$

Hence

$$\rho \sum_{s=n_0}^{n-1} \varphi(n,s)W_3(|x(s)|) \leq V(n)\varphi(n,n) + V(n_0)\varphi(n,n_0)$$

$$+ \sum_{s=n_0}^{n-1} V(s+1)\triangle_s\varphi(n,s) + KB. \qquad (2.1.6)$$

Since $\triangle_s\varphi(n,s) \geq 0$, we have for $V(n) = \max_{0 \leq s \leq n-1} V(s+1)$,

$$\sum_{s=n_0}^{n-1} V(s+1)\triangle_s\varphi(n,s) \leq V(n) \sum_{s=n_0}^{n-1} \triangle_s\varphi(n,s)$$

$$= V(n)[\varphi(n,n) - \varphi(n,n_0)].$$

Thus, from inequality (2.1.6) we have

$$\rho \sum_{s=n_0}^{n-1} \varphi(n,s)W_3(|x(s)|) \leq V(n)[\varphi(n,n) - \varphi(n,n_0)]$$

$$- V(n)\varphi(n,n) + V(n_0)\varphi(n,n_0) + KB$$

$$\leq V(n_0)\varphi(n,n_0) + KB$$

$$\leq V(n_0)\varphi(0,n_0) + KB$$

$$\leq V(n_0)J + KB. \qquad (2.1.7)$$

In view of (2.1.4), we have

$$V(n_0) \leq W_2(|\phi(n_0)|) + \sum_{s=0}^{n_0-1} \varphi(n_0,s)W_3(|\phi(s)|)$$

$$\leq W_2(H) + BW_3(H).$$

As a result, inequality (2.1.7) yields

$$\sum_{s=n_0}^{n-1} \varphi(n,s)W_3(|x(s)|) \leq \frac{W_2(H) + BW_3(H)}{\rho} + \frac{KB}{\rho}.$$

Now, inequality (2.1.4) implies that

$$V(n) \leq W_2(|x(n)|) + \sum_{s=0}^{n_0-1} \varphi(n,s)W_3(|x(s)|) + \sum_{s=n_0}^{n-1} \varphi(n,s)W_3(|x(s)|)$$

$$\leq W_2(|x(n)|) + BW_3(H) + \frac{W_2(H) + BW_3(H)}{\rho} + \frac{KB}{\rho}$$

$$\leq W_2(|x(n)|) + D(H),$$

where $D(H) = BW_3(H) + \frac{W_2(H)+BW_3(H)}{\rho} + \frac{KB}{\rho}$.

As $W_3(r) \to \infty$ as $r \to \infty$, there exists an $L > 0$ such that $W_3(L) = \frac{K}{\rho}$. Now, by (2.1.5), if $|x| > L$, then $\triangle V < 0$. Thus, $V(n)$ attains its maximum when $|x| \leq L$. Hence we have

$$W_1(|x(n)|) \leq V(n) \leq W_2(|x(n)|) + D(H)$$
$$\leq W_2(L) + DH.$$

Finally, from the above inequality we arrive at

$$|x(n)| \leq W_1^{-1}[W_2(L) + D(H)].$$

This completes the proof.

The next theorem extends Theorem 2.1.1.

Theorem 2.1.2 ([133]). *Let $\varphi_i(n,s)$ be a scalar sequence for $0 \leq s \leq n < \infty$ and suppose that $\varphi_i(n,s) \geq 0, \triangle_n \varphi_i(n,s) \leq 0, \triangle_s \varphi_i(n,s) \geq 0$ and there are constants B_i and J_i such that $\sum_{s=0}^{n} \varphi_i(n,s) \leq B_i$ and $\varphi_i(0,s) \leq J_i$. Also, suppose that for each $n_0 \geq 0$ and each bounded initial function $\phi : [0,n_0] \to \mathbb{R}^k$, every solution $x(n) = x(n,n_0,\phi)$ of (2.1.1) satisfies*

$$W_1(|x(n)|) \leq V(n,x(\cdot))$$

$$\leq W_2(|x(n)|) + \sum_{s=0}^{n-1} \varphi_1(n,s)W_3(|x(s)|)$$

$$+ \sum_{s=0}^{n-1} \varphi_2(n,s)W_4(|x(s)|) \qquad (2.1.8)$$

and

$$\triangle V_{(2.1.1)}(n,x(\cdot)) \leq -\rho_1 W_3(|x(n)|) - \rho_2 W_4(|x(n)|) + K \qquad (2.1.9)$$

for some constants $\rho_i \geq 0, i = 1,2$ and $K \geq 0$. Then solutions of (2.1.1) are (UB).

Proof. We follow the proof of the previous theorem. Let $V(n) = \max_{0 \leq s \leq n} V(s)$, $n \geq n_0$. If the *max* of $V(n)$ occurs on $[0,n_0]$, then it is trivial. Multiply both sides of (2.1.9) by $\varphi_i(n,s)$ and then sum from $s = n_0$ to $s = n-1$ to obtain

$$\rho_i \sum_{s=n_0}^{n-1} \varphi_i(n,s)W_3(|x(s)|) \leq V(n_0)J + KB, \quad i = 1,2. \qquad (2.1.10)$$

For $H > 0$ and $|\phi(n)| < H$, we have

$$V(n_0) \leq W_2(H) + BW_3(H) + B_2 W_4(H) \overset{def}{=} R(H),$$

and

$$\sum_{s=0}^{n_0-1} \varphi_i(n,s) W_{i+2}(|x(s)|) \leq W_{i+2} B_i.$$

Thus, inequality (2.1.10) yields,

$$\sum_{s=0}^{n-1} \varphi_i(n,s) W_{i+2}(|x(s)|) \leq \frac{R(H)J_i + KB_i}{\rho_i} + W_{i+2} B_i \overset{def}{=} S_i(H).$$

Now, by (2.1.9), if $|x| > L$, then $\triangle V < 0$. Thus, $V(n)$ attains its maximum when $|x| \leq L$ and hence

$$W_1(|x(n)|) \leq V(n) \leq W_2\big[|x(H)| + S_1(H)\big] + S_2(H)$$
$$\leq W_2\big[L + S_1(H)\big] + S_2(H).$$

From the above inequality we obtain

$$|x(n)| \leq W_1^{-1}\Big[W_2\big[L + S_1(H)\big] + S_2(H)\Big].$$

This completes the proof.

In the next theorem we obtain boundedness and stability results about solutions and the zero solution of (2.1.1).

Theorem 2.1.3 ([133]). *Let $\varphi(n) \geq 0$ be a scalar sequence for $n \geq 0$ and V and $W_i, i = 1, 2$, be defined as before. Also, suppose that for each $n_0 \geq 0$ and each bounded initial function $\phi : [0, n_0] \to \mathbb{R}^k$, every solution $x(n) = x(n, n_0, \phi)$ of (2.1.1) satisfies*

$$W_1(|x(n)|) \leq V(n, x(\cdot)) \leq \alpha W_2(|x(n)|) + \sum_{s=0}^{n-1} \varphi(n-s-1) W_2(|x(s)|) \quad (2.1.11)$$

and

$$\triangle V_{(2.1.1)}(n, x(\cdot)) \leq -\rho W_2(|x(n)|) \quad (2.1.12)$$

for some constants ρ and $\alpha > 0$.

    a) *If $\sum_{s=0}^{\infty} \varphi(s) = B$, then solutions of (2.1.1) are (UB) and the zero solution of (2.1.1) is (US).*

b) If $\sum_{n=0}^{\infty} \sum_{s=n}^{\infty} \varphi(s) = J$, then solutions of (2.1.1) are (UAB) and the zero solution of (2.1.1) is (UAS).

Proof. Let $H > 0$ and $|\phi(n)| < H$ on $[0, n_0]$, and set $V(n) = V(n, x(\cdot))$. By (2.1.12), $V(n)$ is monotonically decreasing and hence, by (2.1.11), we have

$$W_1(|x(n)|) \leq V(n) \leq V(n_0)$$

$$\leq \alpha W_2(H) + W_2(H) \sum_{u=0}^{n_0-1} \varphi(u)$$

$$\leq W_2(H)(\alpha + B). \tag{2.1.13}$$

Let $\varepsilon > 0$ be given. Choose H such that $H < \varepsilon$ and

$$W_2(H)(\alpha + B) < W_1(\varepsilon).$$

Hence from (2.1.13), we have $|x(n)| < \varepsilon$, for $n \geq n_0$. Consequently, the zero solution of (2.1.1) is US). Also, it follows from (2.1.13) that

$$|x(n)| < W_1^{-1}\left[W_2(H)(\alpha + B)\right],$$

which implies solutions of (2.1.1) are (UB).

Sum (2.1.12) from $s = n_0$ to $s = n - 1$ to obtain

$$-V(n_0) \leq V(n) - V(n_0) \leq -\rho \sum_{s=n_0}^{n-1} W_2(|x(s)|)$$

and hence

$$\sum_{s=n_0}^{n-1} W_2(|x(s)|) \leq \frac{V(n_0)}{\rho} \leq \frac{(\alpha + B)W_2(H)}{\rho}.$$

On the other hand, if we sum (2.1.11) from $s = n_0$ to $s = n - 1$ we arrive at

$$\sum_{s=n_0}^{n-1} V(s) \leq \alpha \frac{(\alpha + B)W_2(H)}{\rho} + \sum_{u=n_0}^{n-1} \sum_{s=0}^{u-1} \varphi(u - s - 1)W_2(|x(s)|)$$

$$\leq \frac{(\alpha + B)W_2(H)}{\rho} + \sum_{s=0}^{n_0-1} \sum_{u=n_0}^{n-1} \varphi(u - s - 1)W_2(|x(s)|)$$

$$+ \sum_{s=n_0}^{n-1} \sum_{u=s}^{n-1} \varphi(u - s - 1)W_2(|x(s)|)$$

$$\leq \frac{(\alpha + B)W_2(H)}{\rho} + \sum_{s=0}^{n_0-1} W_2(|x(s)|) \sum_{u=n_0}^{n-1} \varphi(u - s - 1)$$

$$+ \sum_{s=n_0}^{n-1} W_2(|x(s)|) \sum_{u=s}^{n-1} \varphi(u - s - 1)$$

$$\leq \frac{(\alpha + B)W_2(H)}{\rho} + W_2(H) \sum_{s=0}^{n_0-1} \sum_{u=n_0}^{n-1} \varphi(u-s-1)$$

$$+ B \sum_{s=n_0}^{n-1} W_2(|x(s)|)$$

$$\leq \frac{(\alpha + B)^2 W_2(H)}{\rho} + W_2(H) \sum_{s=0}^{n_0-1} \sum_{r=n_0-s-1}^{u-n} \varphi(r)$$

$$\leq \frac{(\alpha + B)^2 W_2(H)}{\rho} + W_2(H) \sum_{\xi=0}^{\infty} \sum_{r=\xi}^{\infty} \varphi(r)$$

$$\leq \frac{(\alpha + B)^2 W_2(H)}{\rho} + W_2(H)J$$

$$\leq \left[J + \frac{(\alpha + B)^2}{\rho} \right] W_2(H) \overset{def}{=} aW_2(H). \qquad (2.1.14)$$

Since $V(n)$ is positive and decreasing for all $n \geq n_0 \geq 0$, we have

$$\sum_{s=n_0}^{n-1} V(s) \geq V(n)(n-n_0).$$

Let $\varepsilon > 0$ be given. Then, for $n \geq n_0 + \frac{aW_2(H)}{W_1(\varepsilon)}$ we have form (2.1.11) and (2.1.14) that

$$W_1(|x(n)|) \leq V(n) \leq \frac{aW_2(H)}{n-n_0} < W_1(\varepsilon). \qquad (2.1.15)$$

Hence, inequality (2.1.15) implies that

$$|x(n)| \leq W_1^{-1}\left(\frac{aW_2(H)}{n-n_0} \right) < \varepsilon.$$

From this we have the (UAB) and the (UAS).

2.2 Functional Delay Difference Equations

Next, we discuss the papers by Zhang [181], the paper [182] by Zhang, and MinG-Po, and the papers by Raffoul [133, 145], in which the authors prove general theorems regarding boundedness and stability of functional difference equations with infinite or finite delay. In [145], the author proves general theorem in which he offers necessary and sufficient conditions for the uniform boundedness of all solutions. We begin by noting that Definition 2.1.4 can be easily extended to accommodate infinite or finite delays systems by considering the initial sequence $\phi : [-\alpha, n_0] \to \mathbb{R}^k$ where α can either be taken finite or infinite. We consider the functional delay difference equation

$$x(t+1) = F(t, x_t). \tag{2.2.1}$$

We assume that F is continuous in x and that $F : \mathbb{Z} \times C \to \mathbb{R}^n$ where C is the set of sequences $\phi : [-\alpha, 0] \to \mathbb{R}^n$, $\alpha > 0$. Let

$$C(t) = \{\phi : [t - \alpha, t] \to \mathbb{R}^n\}.$$

It is to be understood that $C(t)$ is C when $t = 0$. Also ϕ_t denotes $\phi \in C(t)$ and $||\phi_t|| = \max_{t-\alpha \leq s \leq t} |\phi(t)|$, where $|\cdot|$ is a convenient norm on $\mathbb{R}^n$. For $t = 0$,

$$C(0) = \{\phi : [-\alpha, 0] \to \mathbb{R}^n\}.$$

Theorem 2.2.1 ([181]). *Let $\varphi(n) \geq 0$ be a scalar sequence for $n \geq 0$ and V and $W_i, i = 1, 2$, be defined as before. Also, suppose that for each $n_0 \geq 0$ and each bounded initial function $\phi : [0, n_0] \to [0, \infty)$, every solution $x(n) = x(n, n_0, \phi)$ of (2.1.1) satisfies*

$$W_1(|x(n)|) \leq V(n, x(\cdot)) \leq W_2(|x(n)|) + W_3\left(\sum_{s=l}^{n} \varphi(n-s)W_4(|x(s)|)\right) \tag{2.2.2}$$

and

$$\Delta V_{(2.2.1)}(n, x(\cdot)) \leq -W_5(|x(n)|). \tag{2.2.3}$$

In addition, if $\sum \varphi(s) = J$, then the zero solution of (2.2.1) is (UAS).

It is widely known that there are two methods in studying the qualitative theory of delay differential or difference equations. The basic and more natural one is what we call the Razumikhin Lyapunov method and the most popular one is the direct method of Lyapunov function or functional. In some cases one has an advantage over the other and that all depends on the system being studied. It is the opinion of the author that Razumikhin Lyapunov method is easier to use since the Lyapunov function is readily available. Moreover, the imposed conditions are less restrictives. We consider (2.2.1) for $n \in \mathbb{Z}^+$. We assume $F : \mathbb{Z}^+ \times C_H \to \mathbb{R}^n$, where

$$C_H = \{\phi \in C(0) : ||\phi|| < H\},$$

for some positive constant H. Also, $x_t(s) = x(t + s)$, $s \in C(0)$. We assume that $F(t, 0) = 0$, so that $x = 0$ is a solution. It is assumed that for any $t_0 \in \mathbb{Z}^+$ and a given function $\phi \in C_H$, there is a unique solution of (2.2.1), denoted by $x(t, t_0, \phi)$, such that it satisfies (2.2.1) for all integers $t \geq t_0$, and $x(t_0, t_0, \phi) = \phi$. Lastly, we assume there is a constant $L > 0$ such that

$$|F(t, \phi)| \leq L||\phi||, \text{ for } t \in \mathbb{Z}^+ \text{ and } \phi \in C_H.$$

In the next theorem, we use Lyapunov-Razumikhin method type functions to prove the (UAS) of the zero solution of (2.2.1). Its proof is too long for our purpose and it can be found in [182].

Theorem 2.2.2 ([182]). *In addition to the above assumptions, suppose there exists a continuous Lyapunov function* $V : \mathbb{Z}^+ \times B_H \rightarrow \mathbb{R}^+$ *with* $B_H = \{x \in \mathbb{R}^k : |x| < H\}$, *such that*

$$W_1(|x|) \leq V(t,x) \leq W_2(|x|), \tag{2.2.4}$$

and

$$\triangle V(t,x(.)) \leq -W_3(|x(t+1)|). \tag{2.2.5}$$

If $P\big(V(t+1,x(t+1))\big) \geq V(t+s,x(t+s))$, *for* $s \in C(0)$ *and* $P : \mathbb{R}^+ \rightarrow \mathbb{R}^+$ *is continuous function with* $P(s) > s$, *for* $s > 0$, *then the zero solution of* (2.2.1) *is (UAS).*

Next, we display an example in the form of a theorem to show the application of Theorem 2.2.2. Consider the Volterra difference equation with multiple delays

$$x(t+1) = a(t)x(t) + \sum_{i=1}^{k} b_i(t)x(t-h_i), \tag{2.2.6}$$

where the delays h_i are positive integers for $i = 1,2,3,....k$.

Theorem 2.2.3. *Let*

$$a^* = \max_{t \in \mathbb{Z}^+}|a(t)|, \quad b_i^* = \max_{t \in \mathbb{Z}^+}|b_i(t)|, \quad i = 1,2,3,...k.$$

If

$$a^* + \sum_{i=1}^{k} b_i^* < 1, \tag{2.2.7}$$

then the zero solution of (2.2.6) *is (UAS).*

Proof. Consider the Razumikhin type Lyapunov function

$$V(t,x) = |x(t)|.$$

Then along the solutions of (2.2.6) we have that

$$\triangle V(t,x) = |x(t+1)| - |x(t)|. \tag{2.2.8}$$

Due to condition (2.2.7), there exists a constant $\mu \in (0,1)$ such that

$$\sum_{i=1}^{k} b_i^* < \mu(1-a^*).$$

Set $P(t) = \dfrac{1}{\mu}t$, for $t \geq 0$. Let $t_0 \in \mathbb{Z}^+$, whenever

$$P\big(V(t+1,x(t+1))\big) \geq V(t+s,x(t+s)), \text{ for } s \in C(0),$$

or

$$\frac{1}{\mu}|x(t+1)| > |x(t+s)|, \text{ for } s \in C(0)$$

then, by (2.2.6), we have

$$|x(t+1)| \leq |a(t)||x(t)| + \sum_{i=1}^{k} |b_i(t)||x(t-h_i)|$$

$$\leq a^*|x(t)| + 1/\mu \left(\sum_{i=1}^{k} |b_i^*|\right)|x(t+1)|,$$

which implies that

$$|x(t)| \geq \frac{\mu - \sum_{i=1}^{k}|b_i^*|}{a^*\mu}|x(t+1)|.$$

Thus, by (2.2.8) we have

$$\triangle V(t,x) \leq \left(1 - \frac{\mu - \sum_{i=1}^{k}|b_i^*|}{a^*\mu}\right)|x(t+1)|$$

$$\leq -\left(-\frac{(\mu(1-a^*) - \sum_{i=1}^{k}|b_i^*|)}{a^*\mu}\right)|x(t+1)|,$$

if $P(V(t+1,x(t+1))) \geq V(t+s,x(t+s))$, for $s \in C(0)$. Thus the conditions of Theorem 2.2.2 are satisfied with

$$W_1(|x|) = W_2(|x|) = |x|$$

and

$$W_3(|x(t+1)|) = \left(-\frac{(\mu(1-a^*) - \sum_{i=1}^{k}|b_i^*|)}{a^*\mu}\right)|x(t+1)|,$$

and the zero solution is (UAS). This completes the proof.

Equations of the form (2.2.6) play a leading role in modeling additive neural networks. It is worth mentioning that Theorem 2.2.2 cannot be applied to Volterra equations of the form

$$x(n+1) = A(n)x(n) + \sum_{s=0}^{n} C(n,s)x(s).$$

We will revisit such equation later in the chapter using Razumihkin-Lyapunov type functions. The next theorem offers easily verifiable conditions. Its proof can be found in [145]

Theorem 2.2.4 ([145]). *Let $D > 0$ and there is a scalar functional $V(t, \psi_t)$ that is continuous in ψ and locally Lipschitz in ψ_t when $t \geq t_0$ and $\psi_t \in C(t)$ with $||\psi_t|| < D$. In addition we assume that if $x : [t_0 - \alpha, \infty) \to \mathbb{R}^n$ is a bounded sequence, then $F(t,x_t)$ is bounded on $[t_0, \infty)$. Suppose there is a function V such that $V(t,0) = 0$,*

$$W_1(|\psi(t)|) \leq V(t, \psi_t) \leq W_2(||\psi_t||),$$

and

$$\triangle V(t, \psi_t) \leq -W_3(|\psi(t)|),$$

then the zero solution of (2.2.1) is (UAS).

It is noted that Theorem 2.2.4 requires Lyapunov functional, unlike Theorem 2.2.2. In the next theorem we use a Lyapunov functional, and with the aid of Theorem 2.2.4, we show the zero solution of (2.2.6) is (UAS).

Theorem 2.2.5 ([145]). *Assume there exists a $\delta > 0$ such that*

$$|a(t)| - 1 + k\delta < 0 \text{ and } \delta \geq \sum_{i=1}^{k} |b_i(t)|.$$

Then the zero solution of (2.2.6) is (UAS).

Proof. Consider the Lyapunov functional

$$V(t, x_t) = |x(t)| + \delta \sum_{i=1}^{k} \sum_{s=t-h_i}^{t-1} |x(s)|. \tag{2.2.9}$$

Then along the solutions of (2.2.6) we have

$$\triangle V(t, x_t) = |x(t+1)| - |x(t)| + \delta \sum_{i=1}^{k} \Big[\sum_{s=t-h_i+1}^{t} |x(s)| - \sum_{s=t-h_i}^{t-1} |x(s)| \Big]$$

$$= |x(t+1)| - |x(t)| + \delta \sum_{i=1}^{k} |x(t)| - \delta \sum_{i=1}^{k} |x(t-h_i)|$$

$$\leq (|a(t)| - 1 + k\delta)|x(t)| + \sum_{i=1}^{k} (|b_i(t)| - \delta)|x(t-h_i)|$$

$$\leq -\alpha |x(t)|, \text{ for some positive constant } \alpha.$$

To make sure the conditions of Theorem 2.2.4 are satisfied, we note that

$$|x(t)| \leq V(t, x_t) = |x(t)| + \delta \sum_{i=1}^{k} \sum_{s=t-h_i}^{t-1} |x(s)|$$

$$= |x(t)| + \delta \sum_{i=1}^{k} \sum_{u=-h_i}^{-1} |x(u+t)|.$$

Hence, if we take $W_1(|\psi(t)|) = |\psi(t)|$, $W_3(|\psi(t)|) = \alpha|\psi(t)|$, and $W_2(|\psi_t|) = \delta \sum_{i=1}^{k} \sum_{u=-h_i}^{-1} |\psi(u+t)|$, then we satisfy all the requirements of Theorem 2.2.4, and hence we have the zero solution of (2.2.6) is (UAS). This completes the proof.

It is very obvious that Theorem 2.2.1 would not work for our Lyapunov functional in Theorem 2.2.5. Theorem 2.2.1 is suitable for Volterra difference equations of convolution types. For example, if we consider

$$x(n+1) = a(n)x(n) + \sum_{s=0}^{n-1} b(n-s)g(n,x(n)),$$

then a typical Lyapunov functional would be

$$V(n,x) = |x(n)| + \sum_{s=0}^{n-1} \sum_{u=n-s}^{\infty} |b(u)||x(s)|,$$

where the function g satisfies $|g(n,x)| \leq \lambda|x|$ for all $n \in \mathbb{Z}^+$ for positive constant $\lambda < 1$. By assuming that

$$|a(n)| + \lambda|b(0)| \sum_{u=1}^{\infty} |b(u)| - 1 \leq -\alpha, \ \alpha > 0,$$

we have along the solutions that

$$\triangle V(n,x) \leq \left(|a(n)| + \lambda|b(0)| \sum_{u=1}^{\infty} |b(u)| - 1\right)|x(n)| + (\lambda - 1) \sum_{s=0}^{n-1} |b(n-s)||x(s)|$$
$$\leq -\alpha|x(n)|.$$

Thus all the conditions of Theorem 2.2.1 are satisfied for $W_1 = W_2 = W_3 = W_4 = |x|$, $W_5 = \alpha|x|$, $l = 0$, and $\phi(n-s) = \sum_{u=n-s}^{\infty} |b(u)|$.

We end this section with an application to the second order difference equation with constant delay, $r > 0$

$$x(t+2) + ax(t+1) + bx(t-r) = 0, \ t \in \mathbb{Z}, \tag{2.2.10}$$

where a and b are constants.

Theorem 2.2.6 ([145]). *Suppose there are positive constants* $\eta_1, \eta_2, \alpha_1, \alpha_2$ *and* γ *such that*

$$\alpha_1|b| - \alpha_2 + \gamma r \leq -\eta_1, \ \alpha_1|a| - \alpha_1 + \alpha_2 + \gamma r \leq -\eta_2,$$

and

$$|b| - \gamma \leq 0.$$

Then the zero solution of (2.2.10) is (UAS).

Proof. First we write (2.2.10) into a system by letting $y(t) = x(t+1)$. Then by noting that

$$\triangle x(t) = y(t) - x(t),$$

we have

$$b \sum_{s=-r}^{-1} (y(t+s) - x(t+s)) = b \sum_{s=-r}^{-1} \triangle x(t+s) = bx(t) - bx(t-r).$$

This implies that Equation (2.2.10) is equivalent to the system

$$\begin{cases} x(t+1) = y(t) \\ y(t+1) = -bx(t) - ay(t) + b\sum_{s=-r}^{-1}\left(y(t+s) - x(t+s)\right) \end{cases} \tag{2.2.11}$$

Let

$$\beta = \max\{\eta_1, \eta_2\}$$

and define the Lyapunov functional

$$V(x_t, y_t) = \alpha_1|y(t)| + \alpha_2|x(t)| + \gamma \sum_{s=-r}^{-1}\sum_{u=t+s}^{t-1}\left(|y(u)| + |x(u)|\right).$$

Then along the solutions of (2.2.11) we have

$$\begin{aligned} \triangle V(x_t, y_t) &\leq (\alpha_1|b| - \alpha_2 + \gamma r)|x(t)| + (\alpha_1|a| - \alpha_1 + \alpha_2 + \gamma r)|y(t)| \\ &\quad + (|b| - \gamma)\sum_{s=-r}^{-1}\left(|y(t+s)| + |x(t+s)|\right) \\ &\leq -\beta\left(|x| + |y|\right). \end{aligned} \tag{2.2.12}$$

The results follow from Theorem 2.2.4.

Remark 2.1. In Theorem 2.2.6 we saw that the stability depended on the size of the delay, which was not the case in Theorem 2.2.5.

In [112] the authors considered the linear difference system with diagonal delay

$$\begin{aligned} x(n+1) &= ax(n-h) + by(n) \\ y(n+1) &= cx(n) + ay(n-h) \end{aligned} \tag{2.2.13}$$

where a, b, and c are real numbers and h is a positive integer. They used the method of characteristics and proved two theorems on the asymptotic stability of the zero solution of (2.2.13) by imposing conditions on the size of the delay; that is $\sqrt{|ab|} < (h+1)/h$. Also, they required b and c be of the same sign and the delay h is odd. The above system has some limitations by considering all constant coefficients and diagonal entries have the same coefficient. Next we shall display a Lyapunov functional to obtain (UAS) of the zero solution of (2.2.13) by appealing to Theorem 2.2.4.

Theorem 2.2.7 ([145]). *Let δ be a positive constant such that*

$$|a| - \delta \leq 0, \ |c| + \delta - 1 < 0, \ and \ |b| + \delta - 1 < 0.$$

Then the zero solution of (2.2.13) is (UAS).

Proof. Let

$$\beta = \min\{|c| + \delta - 1, |b| + \delta - 1\}$$

and define the Lyapunov functional

$$V(x,y) = |x(n)| + |y(n)| + \delta \sum_{s=n-h}^{n-1} \left(|x(s)| + |y(s)| \right).$$

Then along the solutions of (2.2.13)

$$\begin{aligned}
\triangle V(x,y) &\le (|a| - \delta)|x(n-h))| + (|c| + \delta - 1)|x(n)| \\
&\quad + (|a| - \delta)|y(n-h)| + (|c| + \delta - 1)|y(n)| \\
&\le \beta(|x| + |y|).
\end{aligned}$$

The (UAS) follows from Theorem 2.2.4.

Next we extend Theorem 2.2.7 to the Volterra delay system

$$x(n+1) = ax(n-h) + by(n) + d_1 \sum_{s=-h}^{-1} y(n+s)$$

$$y(n+1) = cx(n) + ay(n-h) + d_2 \sum_{s=-h}^{-1} x(n+s) \qquad (2.2.14)$$

Theorem 2.2.8 ([145]). *Let δ be a positive constant such that*

$$|a| + h\delta_1 + \delta_2 - 1 < 0, \quad |b| + h\delta_1 + \delta_2 - 1 < 0,$$

and

$$|a| - \delta_2 \le 0, \quad \text{and} \quad |d_1| + |d_2| - \delta_1 \le 0.$$

Then the zero solution of (2.2.14) is (UAS).

Proof. Let

$$\beta = \min\{|a| + h\delta_1 + \delta_2 - 1, \ |b| + h\delta_1 + \delta_2 - 1\}$$

and define the Lyapunov functional

$$V(x,y) = |x(n)| + |y(n)| + \delta_2 \sum_{s=n-h}^{n-1} \left(|x(s)| + |y(s)| \right)$$

$$+ \delta_1 \sum_{s=-h}^{-1} \sum_{u=n+s}^{n-1} \left(|x(u)| + |y(u)| \right).$$

Then along the solutions of (2.2.14)

$$\begin{aligned}
\triangle V(x,y) &\le (|a| - \delta_2)|x(n-h))| + (|a| + h\delta_1 + \delta_2 - 1)|x(n)| \\
&\quad + (|a| - \delta_2)|y(n-h)| + (|b| + h\delta_1 + \delta_2 - 1)|y(n)| \\
&\le \beta(|x| + |y|) \\
&\quad + (|d_1| + |d_2| - \delta_1) \sum_{s=-h}^{-1} \left(|x(n+s)| + |y(n+s)| \right) \\
&\le \beta(|x| + |y|).
\end{aligned}$$

Hence, the result of (UAS) follows from Theorem 2.2.4.

The next theorem shows that the zero solution of the nonlinear delay difference equation

$$x(n+1) = a(n)f(x(n)) + b(n)g(x(n-\tau)), \ h \in \mathbb{Z}^+ \qquad (2.2.15)$$

is (UAS).

Theorem 2.2.9. *Suppose f and g are continuous and there are positive constants α, β, and γ with $\gamma > 1 + \alpha$ such that*

$$\gamma |a(n)||f(x)| \geq |x|, \ |f(x)| \geq |g(x)| \ for \ 0 < |x| < \beta, \qquad (2.2.16)$$

and

$$(1-\gamma)|a(n)| + |b(n+\tau)| \leq -\alpha|a(n)|. \qquad (2.2.17)$$

Then the zero solution of (2.2.15) is (UAS).

Proof. Define the Lyapunov functional

$$V(n,x_n) = |x(n)| + \sum_{s=n-\tau}^{n-1} \big(|b(s+\tau)||g(x(s))|\big). \qquad (2.2.18)$$

Then along the solutions of (2.2.15) we have

$$\triangle V(n,x_n) \leq |a(n)||f(x(n))| + |b(n)||g(x(n-\tau))| - |x(n)|$$
$$+ |b(n+\tau)||g(x(n))| - |b(n)||g(x(n-\tau))|.$$

Using (2.2.16) and (2.2.17) we arrive at

$$\triangle V(n,x_n) \leq \big(|a(n)| + |b(n+\tau)| - \gamma|a(n)|\big)|f(x(n))|$$
$$\leq -\alpha|a(n)||f(x(n))|. \qquad (2.2.19)$$

The (UAS) follows from Theorem 2.2.4. This completes the proof.

2.2.1 Application to Volterra Difference Equations

In this section we apply Theorems 2.1.1, 2.1.2, and 2.1.3 to establish stability and boundedness results regarding the nonlinear Volterra discrete system

$$x(n+1) = A(n)x(n) + \sum_{s=0}^{n} C(n,s)f(x(s)) + g(n,x(n)) \qquad (2.2.20)$$

where A, C, are $k \times k$ matrices, g, f are $k \times 1$ vector functions with $|g(n,x(n))| \leq N$ and $|f(x)| \leq \lambda|x|$, for some positive constants N and λ.
In the case of $f(x) = x$, Medina [117], showed that if the zero solution of the homogenous equation associated with (2.2.20) is uniformly asymptotically stable, then

all solutions of (2.2.20) are bounded when $C(n,s) = C(n-s)$ and $g(n,x) = g(n)$ is bounded. In proving his results, Medina used the notion of the resolvent matrix coupled with the variation of parameters formula. Also, the author in [128] used Lyapunov functionals of convolution type coupled with the z-transform and obtained results about boundedness of solutions of (2.2.20) when $g(n,x(n)) = g(n)$. Moreover, we saw in Chapter 1 that when f is linear in x, unlike the case here, we used total stability and under suitable conditions, we showed that the zero solution of (2.2.20) is uniformly asymptotically stable, when $|g(n,x)| \leq \lambda(n)|x|$. We remark that the notion of the resolvent cannot be used to obtain boundedness of solutions of (2.2.20), since the summation term in (2.2.20) is nonlinear.

Theorem 2.2.10 ([133]). *Suppose $A(n) = A$ is a $k \times k$ constant matrix, and $C^T(n,s) = C(n,s)$. Let I be the $k \times k$ identity matrix. Also, suppose there exist positive constants ρ, μ, and a constant $k \times k$ symmetric matrix B such that*

$$A^T BA - B = -\mu I, \qquad (2.2.21)$$

$$\lambda |A^T B| \sum_{s=0}^{n} |C(n,s)| + |B| \sum_{s=n}^{\infty} |C(n,s)| + N^2 - \mu \leq -\rho, \qquad (2.2.22)$$

and

$$\lambda |A^T B| \sum_{s=0}^{n} |C(n,s)| + \lambda^2 |B| \sum_{s=0}^{n} |C(n,s)| + \lambda - |B| \leq 0. \qquad (2.2.23)$$

Then solutions of (2.2.20) are (UB).

Proof. Define the Lyapunov functional $V(n) = V(n,x(n,\cdot))$ by

$$V(n,x(\cdot)) = x^T(n)Bx(n) + |B| \sum_{j=0}^{n-1} \sum_{s=n}^{\infty} |C(s,j)|x^2(j), \qquad (2.2.24)$$

where $x^2(j) = x^T(j)x(j)$. Then along solutions of (2.2.20) we have

$$\triangle V_{(2.2.20)}(n) = x^T(n)\left[A^T BA - B\right]x(n) + 2x^T(n)A^T B \sum_{s=0}^{n} C(n,s)f(x(s))$$

$$+ 2x^T(n)A^T Bg(n,x(n)) + 2g^T(n,x(n))B \sum_{s=0}^{n} C(n,s)f(x(s))$$

$$+ \sum_{s=0}^{n} f^T(x(s))C(n,s)^T B \sum_{s=0}^{n} C(n,s)f(x(s))$$

$$+ |B| \sum_{s=n+1}^{\infty} |C(n,s)|x^2(n) - |B| \sum_{s=0}^{n-1} |C(n,s)|x^2(s)$$

$$+ g^T(n,x(n))Bg(n,x(n)). \qquad (2.2.25)$$

Using (2.2.21)–(2.2.23) and the fact that for any two real numbers a and b, $2ab \leq a^2 + b^2$, equation (2.2.25) reduces to

$$\triangle V_{(2.2.20)}(n) \leq \left[\lambda |A^T B| \sum_{s=0}^{n} |C(n,s)| + |B| \sum_{s=n}^{\infty} |C(n,s)| + N^2 - \mu\right] x^2(n)$$

$$+ \left[\lambda |A^T B| \sum_{s=0}^{n} |C(n,s)| + \lambda^2 |B| \sum_{s=0}^{n} |C(n,s)| + \lambda - |B|\right] \sum_{s=0}^{n} |C(n,s)| x^2(s)$$

$$+ |A^T B|^2 + \lambda N^2 |B|^2 \sum_{s=0}^{n} |C(n,s)| + |g^T Bg|$$

$$\leq -\rho x^2(n) + K,$$

where $K = |A^T B|^2 + \lambda N^2 |B|^2 \sum_{s=0}^{n} |C(n,s)| + |g^T Bg|$. Thus, by Theorem 2.1.1 all solutions of (2.2.20) are (UB).

In the next theorem, we use Theorem 2.1.3 to establish (UB) and (UAS) for (2.2.20), when $g(n,x(n))$ is identically zero.

Theorem 2.2.11 ([133]). *Assume $g(n,x(n)) = 0$ and suppose there is a function $\varphi(n) \geq 0$, with $\triangle \varphi(n) \leq 0$ for $n \geq 0$, $\triangle_n \varphi(n-s-1) + |C(n,s)| \leq 0$ for $0 \leq s < n < \infty$. Also, suppose that for $n \geq 0$, $|A(n)| + |C(n,n)| + \varphi(0) \leq 1 - \rho$ for some $\rho \in (0,1)$.*

a) If $\sum_{s=0}^{\infty} \varphi(s) = B$, then solutions of (2.2.20) are (UB) and the zero solution of (2.2.20) is (US).

b) If $\sum_{n=0}^{\infty} \sum_{s=n}^{\infty} \varphi(s) = J$, then solutions of (2.2.20) are (UUB) and the zero solution of (2.2.20) is (UAS).

Proof. Define the Lyapunov functional $V(n) = V(n,x(n,\cdot))$, by

$$V(n) = |x(n)| + \sum_{s=0}^{n-1} \varphi(n-s-1)|x(s)|, \quad n \geq 0. \tag{2.2.26}$$

Then along solutions of (2.2.20) we have

$$\triangle V_{(2.2.20)}(n) \leq \left(|A(n)| + |C(n,n)| + \varphi(0) - 1\right)|x(n)|$$

$$+ \sum_{s=0}^{n-1} \left(\triangle_n \varphi(n-s-1) + |C(n,s)|\right)|x(s)|$$

$$\leq -\rho |x(n)|,$$

and the results follow from Theorem 2.1.3.

2.3 Necessary and Sufficient Conditions

We prove a general theorem in which necessary and sufficient conditions for obtaining uniform boundedness for functional difference system are present. We apply our results to finite and infinite delay Volterra difference equations. In the analysis we state and prove a discrete Jensen's type inequality. Thus we consider the system of functional difference equations of the form

$$x(n+1) = G(n, x_n), \; x \in \mathbb{R}^k \tag{2.3.1}$$

where $G : \mathbb{Z}^+ \times \mathbb{R}^k \to \mathbb{R}^k$ is continuous in x. Let C be set of bounded sequences $\phi : (-\infty, 0] \to \mathbb{R}^k$ with the maximum norm. For $n \in \mathbb{Z}^+, C(n)$ denotes the set of sequences $\psi : [0, n] \to \mathbb{R}^k$ with $||\psi|| = \max\{|\psi(s)| : 0 \le s \le n\}$, where $|\cdot|$ is the Euclidean norm on $\mathbb{R}^k$.

We assume that for each $n_0 \ge 0$, and each $\phi \in C(n_0)$ there is at least one solution $x(n, n_0, \phi)$ of (2.3.1) defined on an interval $[n_0, \alpha]$ with $x_{n_0} = \phi$. If the solution remains bounded for all n, then $\alpha = \infty$. Notation wise, $x_n(s) = x(n+s)$ for $s \le 0$. If D is a matrix or a vector, $|D|$ means the sum of the absolute values of the elements.

Definition 2.3.1. Solutions of (2.3.1) are (UB) if for each $B_1 > 0$ there is $B_2 > 0$ such that $[n_0 \ge 0, \phi \in C, ||\phi|| < B_1, n \ge n_0]$ implies $|x(n, n_0, \phi)| < B_2$.

Definition 2.3.2. Solutions of (2.3.1) are (UUB) if there is a $B > 0$ and for each $B_3 > 0$ there is $N > 0$ such that $[n_0 \ge 0, \phi \in C, ||\phi|| < B_3, n \ge n_0 + N]$ implies $|x(n, n_0, \phi)| < B$.

Theorem 2.3.1 ([133]). *Let $\mathbb{R}^+ = [0, \infty)$ and assume there is a scalar sequence $\Phi : \mathbb{Z}^+ \to \mathbb{R}^+$ that satisfy $\Phi \in l^\infty(\mathbb{R}^+)$. Assume the existence of wedges $W_j, j = 1, 2, \cdot, \cdot, , 5$. with $W_1(r) \to \infty$, and positive constants K, M with $W_5(K) > M$. Suppose there is a functional $V : \mathbb{R}^+ \times C \to \mathbb{R}^+$ such that for each $x \in C(n)$, we have:*

$$W_1(|x(n)|) \le V(n, x_n) \le W_2(|x(n)|) + W_3 \left(\sum_{s=0}^{n-1} \Phi(n-s) W_4(|x(s)|) \right) \tag{2.3.2}$$

and

$$\triangle V(n, x_n) \le -W_5(|x(n)|) + M. \tag{2.3.3}$$

Then solutions of (2.3.1) are (UB) if and only if for each $K_1 > 0$, there exists $K_2 > 0$ such that if $x(n) = x(n, n_0, \phi)$ is a solution of (2.3.1) with $||\phi|| \le K_1$, then

$$\sum_{s=n_0}^{n^*-1} \Phi(n^* - s) W_4(|x(s)|) \le K_2 \tag{2.3.4}$$

whenever $v(s) < v(n^*)$ for $n_0 \le s < n^*$, where $v(s) = V(s, x_s)$.

Proof. Let $x(n) = x(n, n_0, \phi)$ be a solution of (2.3.1) that is (UB). Then, for ever $B_1 > 0$ there exists a $B_2 > 0$, say $B_2 > B_1$ so that for $n_0 \ge 0, ||\phi|| < B_1, n \ge n_0$, we have $|x(n, n_0, \phi)| < B_2$. Let $J := \sum_{u=0}^{\infty} \Phi(u)$. Then, for $n \ge n_0$ we have that

$$\sum_{s=0}^{n-1} \Phi(n-s) W_4(|x(s)|) \le \sum_{s=0}^{n-1} \Phi(n-s) W_4(B_2) \le J W_4(B_2).$$

This completes the proof of (2.3.4).

Conversely, suppose that (2.3.4) holds. Then for $x(n) = x(n, n_0, \phi)$ and $v(n) = V(n, x_n)$ with $\|\phi\| < B_1$, we have the two cases:

(i) $v(n) \leq v(n_0)$ for all $n \geq n_0$ or

(ii) $v(s) \leq v(n^*)$ for some $n^* > n_0$ and all $n_0 \leq s < n^*$.

If (i) holds, then

$$W_1(|x(n)|) \leq V(n) \leq V(n_0)$$

$$\leq W_2(|x(n_0)|) + \sum_{s=0}^{n_0-1} \varphi(n_0, s) W_3(|\phi(s)|)$$

$$\leq W_2(|\phi(n_0)|) + \sum_{s=0}^{n_0-1} \varphi(n_0, s) W_3(|\phi(s)|)$$

$$\leq W_2(H) + BW_3(H).$$

From which it follows that

$$|x(n)| \leq W_1^{-1}\left[W_2(H) + BW_3(H)\right].$$

On the other hand, if (ii) holds, then $V(n, \cdot)$ is increasing and hence we have $0 \leq -W_5(|x(n^*)|) + M$. Or, $W_5(|x(n^*)|) \leq M$. Now since $W_5(K) > M$, we get $|x(n^*)| \leq W_5^{-1}(M)$. It follows from (i) and (2.3.2) that

$$v(n^*) \leq W_2(|x(n^*)|) + W_3\left(\sum_{s=0}^{n_0-1} \Phi(n^* - s) W_4(|x(s)|) + \sum_{s=n_0}^{n^*-1} \Phi(n^* - s) W_4(|x(s)|)\right)$$

$$\leq W_2\left[W_5^{-1}(M)\right] + W_3\left[JW_4(K_1) + K_2\right].$$

Since n^* is arbitrary, we have for all $n \geq n_0$ that

$$v(n) \leq W_2\left[W_5^{-1}(M)\right] + W_3\left[JW_4(K_1) + K_2\right] + v(n_0).$$

$$\leq W_2\left[W_5^{-1}(M)\right] + W_3\left[JW_4(K_1) + K_2\right] + W_2(K_1) + W_3(JW_4(K_1)).$$

On the other hand, from (2.3.2) we have $W_1(|x(n)|) \leq v(n)$, which implies that

$$|x(n)| \leq W_1^{-1}\left[W_2\left[W_5^{-1}(M)\right] + W_3\left[JW_4(K_1) + K_2\right] + W_2(K_1) + W_3(JW_4(K_1))\right].$$

$$(2.3.5)$$

Finally, for all $n \geq n_0$, we have $|x(n)| \leq B_2$, where B_2 is given by the right side of (2.3.5).

This completes the proof.

The next Lemma is needed for our next results. One could say it is a Jensen's type inequality.

Lemma 2.1 (Raffoul Jensen's Type Discrete Inequality [133]). *Assume $\Phi : \mathbb{Z}^+ \to \mathbb{R}^+$ such that $\Phi(u+1), \triangle\Phi(u)u \in l^1(\mathbb{Z}^+)$. Also, assume that $q : [n_0, n] \to \mathbb{R}^+$ is such that for constants α and β, we have*

$$\frac{1}{n-s} \sum_{u=s}^{n-1} q(u) \le \alpha + \frac{\beta}{n-s} \tag{2.3.6}$$

for all $n_0 \le s < n$. Then,

$$\sum_{s=n_0}^{n-1} \Phi(n-s)q(s) \le \alpha \max_{n>0}\left\{ \Phi(n)n + \sum_{u=0}^{n-1} |\triangle\Phi(u)|u \right\} + \beta \sum_{u=0}^{\infty} |\triangle\Phi(u)|. \tag{2.3.7}$$

Proof. Let b be any positive integer. Then since $\Phi(u+1), \triangle\Phi(u)u \in l^1(\mathbb{Z}^+)$, we have $\Phi(\infty) = 0$. Moreover,

$$\sum_{u=b}^{\infty} |\triangle\Phi(u)| \ge |\sum_{u=b}^{\infty} \triangle\Phi(u)| = |\Phi(\infty) - \Phi(b)|,$$

from which we have

$$\Phi(b) \le \sum_{u=b}^{\infty} |\triangle\Phi(u)|. \tag{2.3.8}$$

We claim that since $\Phi(u+1), \triangle\Phi(u)u \in l^1(\mathbb{Z}^+)$, we have $\Phi(\infty) = 0$. To see this for any two sequences y and z, we use the summation by parts formula

$$\sum(\triangle z)y = yz - \sum Ez\triangle y,$$

where $Ez(n) = z(n+1)$. With this in mind, we have

$$\sum_{u=0}^{\infty} \triangle\Phi(u)u = \Phi(u)u \mid_{u=0}^{\infty} - \sum_{u=0}^{\infty} \Phi(u+1).$$

From which we get

$$\Phi(u)u \mid_{u=0}^{\infty} = \sum_{u=0}^{\infty} \triangle\Phi(u)u + \sum_{u=0}^{\infty} \Phi(u+1) < \infty.$$

Suppose the contrary; that is $\Phi(\infty) \nrightarrow 0$. Then, there exists a T large enough so that $\Phi(p) > \varepsilon$, for $p > T$. Thus,

$$\lim_{p\to\infty} p\Phi(p) \ge \lim_{p\to\infty} p\varepsilon = \infty,$$

which contradict the fact that $\triangle\Phi(u)u, \Phi(u+1) \in l^1(\mathbb{Z}^+)$. This completes the proof of the claim.

Again, using summation by parts, we get

$$\sum_{u=0}^{b-1} |\triangle\Phi(u)u| \geq |\sum_{u=0}^{b-1} \triangle\Phi(u)u| = |\Phi(b)b - \sum_{u=0}^{b-1} \Phi(u+1)|.$$

From this inequality we arrive at,

$$|\Phi(b)b| - |\sum_{u=0}^{b-1} \Phi(u+1)| \leq |\Phi(b)b - \sum_{u=0}^{b-1} \Phi(u+1)|$$

$$\leq \sum_{u=0}^{b-1} |\triangle\Phi(u)u|,$$

from which we get

$$\Phi(b)b \leq \sum_{u=0}^{\infty} [\Phi(u+1) + |\triangle\Phi(u)u|] < \infty.$$

Let $y = \Phi(n-s)$ and $\triangle z = q(s)$. Then we have $z = -\sum_{u=s}^{n-1} q(u)$. Hence,

$$\sum_{s=n_0}^{n-1} \Phi(n-s)q(s) \leq \Phi(n-s)\left(-\sum_{u=s}^{n-1} q(u)\right)|_{s=n_0}^{n} - \sum_{s=n_0}^{n-1} \triangle\Phi(n-s) \sum_{u=s+1}^{n-1} q(u)$$

$$= \Phi(n-n_0) \sum_{u=n_0}^{n-1} q(u) - \sum_{s=n_0}^{n-1} \triangle\Phi(n-s) \sum_{u=s+1}^{n-1} q(u)$$

$$\leq \Phi(n-n_0) \sum_{u=n_0}^{n-1} q(u) + \sum_{s=n_0}^{n-1} |\triangle\Phi(n-s)| \sum_{u=s}^{n-1} q(u)$$

$$\leq \Phi(n-n_0)\Big((n-n_0)\alpha+\beta\Big) + \sum_{s=n_0}^{n-1} |\triangle\Phi(n-s)|\Big((n-s)\alpha+\beta\Big)$$

$$= \left[\Phi(n-n_0)(n-n_0) + \sum_{s=n_0}^{n-1} |\triangle\Phi(n-s)|(n-s)\right]\alpha$$

$$+ \left[\Phi(n-n_0) + \sum_{s=n_0}^{n-1} |\triangle\Phi(n-s)|\right]\beta$$

$$= \left[\Phi(n-n_0)(n-n_0) + \sum_{u=1}^{n-n_0} |\triangle\Phi(u)|u\right]\alpha \text{ (letting } u = n-s)$$

$$+ \left[\Phi(n-n_0)(n-n_0) + \sum_{u=1}^{n-n_0} |\triangle\Phi(u)|\right]\beta$$

$$\leq \left[\Phi(n-n_0)(n-n_0) + \sum_{u=0}^{n-n_0} |\triangle\Phi(u)|u\right] \alpha$$

$$+ \left[\Phi(n-n_0)(n-n_0) + \sum_{u=0}^{n-n_0} |\triangle\Phi(u)|\right] \beta$$

$$\leq \alpha \max_{n\geq 0}\{\Phi(n)n + \sum_{u=0}^{n-1} |\triangle\Phi(u)|u\}$$

$$+ \Phi(n-n_0)(n-n_0) + \left[\sum_{u=0}^{\infty} |\triangle\Phi(u)| - \sum_{u=n-n_0}^{\infty} |\triangle\Phi(u)|\right] \beta$$

$$\leq \alpha \max_{n\geq 0}\{\Phi(n)n + \sum_{u=0}^{n-1} |\triangle\Phi(u)|u\} + \beta \sum_{u=0}^{\infty} |\triangle\Phi(u)| < \infty,$$

where we have used (2.3.8). This completes the proof.

Theorem 2.3.2 ([133]). *Assume there is a scalar sequence* $\Phi : \mathbb{Z}^+ \to \mathbb{R}^+$ *that satisfies* $\Phi \in l^\infty(\mathbb{R}^+)$. *Assume the existence of wedges* $W_j, j = 1, 2, \cdot, \cdot, , 5.$ *with* $W_1(r) \to \infty$, *and positive constants* K, M *with* $W_5(K) > M$. *Suppose there is a functional* $V : \mathbb{R}^+ \times C \to \mathbb{R}^+$ *such that for each* $x \in C(n)$, (2.3.2) *and* (2.3.3) *hold. Suppose that for every* $\alpha > 0$, *there exists* $\alpha^* > 0$ *such that for* $0 \leq s < n$,

$$\frac{1}{n-s} \sum_{u=s}^{n-1} W_5(|x(u)|) \leq \alpha \Rightarrow \frac{1}{n-s} \sum_{u=s}^{n-1} W_4(|x(u)|) \leq \alpha^*. \tag{2.3.9}$$

Then solutions of (2.3.1) *are (UB.)*

Proof. Let $x(n) = x(n, n_0, \phi)$ be a solution of (2.3.1) and $B_1 > 0$ with $||\phi|| < B_1$. Set $v(n) = V(n, x_n)$. In the case of (ii) we sum (2.3.3) from s to $n^* - 1$ to get

$$v(n^*) - v(s) \leq - \sum_{u=s}^{n^*-1} W_5(|x(u)|) + M(n^* - s).$$

This yields that

$$\frac{1}{n-s} \sum_{u=s}^{n^*-1} W_5(|x(u)|) \leq M. \tag{2.3.10}$$

Then by (2.3.9), there exists $M^* > 0$ such that

$$\frac{1}{n-s} \sum_{u=s}^{n^*-1} W_4(|x(u)|) \leq M^*.$$

If we let $q(u) = W_4(|x(u)|), \alpha = M^*$ then an application of Lemma 1 with $\beta = 0$, we obtain (2.3.4) and hence by Theorem 2.3.1, solutions are (UB).

It is worth noting that in general (2.3.9) does not hold for arbitrary wedges. To see this, we let $W_5(r) = \frac{1}{r}, r > 0$ and $W_4(r) = r$. Let $x(u) = \frac{1}{a^{u-1}}, |a| > 1$. Then

$$\frac{1}{n-1} \sum_{u=1}^{n} W_4(|x(u)|) = \frac{1}{n-1} \sum_{u=1}^{n} (\frac{1}{a})^{u-1} \to \infty, \text{ as } n \to \infty.$$

On the other hand

$$\frac{1}{n-1} \sum_{u=1}^{n} W_5(|x(u)|) = \frac{1}{n-1} \sum_{u=1}^{n} a^{u-1} = \frac{1}{n-1} \frac{1 - (\frac{1}{a})^n}{1 - \frac{1}{a}} \leq 1,$$

for $n > 1$.

Remark 2.2. If $\Phi : \mathbb{Z}^+ \to \mathbb{R}^+$ with $\triangle \Phi(n) \leq 0$, for all $n \in \mathbb{Z}^+$ and $\Phi(u+1) \in l^1(\mathbb{Z}^+)$, then $\triangle \Phi(u)u \in l^1(\mathbb{Z}^+)$. As a matter of fact,

$$\Phi(n)n + \sum_{u=0}^{n-1} |\triangle \Phi(u)|u = \Phi(n)n - \sum_{u=0}^{n-1} \triangle \Phi(u)u = \sum_{u=0}^{n-1} \Phi(u+1).$$

For example, if we take $\Phi(n) = a^n sin(n\pi/2)$ for $0 < a < 1$, then

$$\triangle \Phi(n) = a^{n+1} cos(n\pi/2) - a^n sin(n\pi/2).$$

It is easy to see that $\triangle \Phi(3) = a^3 > 0$. On the other hand

$$\sum_{n=0}^{\infty} \triangle \Phi(n)n < \infty.$$

In Theorem 2.1.1 we asked that $\triangle_n \Phi(n)(n) \leq 0$.

We end the section with the following theorem.

Theorem 2.3.3 ([133]). *Solutions of the scalar difference equation*

$$x(n+1) = a(n)x(n) + \sum_{s=0}^{n} D(n-s)x(s) + p(n), \qquad (2.3.11)$$

are (UB) if and only if

$$-1 + |a(n)| + \sum_{u=0}^{\infty} |D(u)| \leq -\beta, \ \beta > 0. \qquad (2.3.12)$$

Proof. First it is obvious that (2.3.12) implies $\sum_{u=0}^{\infty} |D(u)|$ is bounded. Consider the Lyapunov functional

$$V(n, x_n) = |x(n)| + \sum_{s=0}^{n-1} \sum_{u=n-s}^{\infty} |D(u)||x(s)|. \qquad (2.3.13)$$

Let $M = |p(n)|$. Then along the solutions of (2.3.11) we have

$$\triangle V(n, x_n) = (|x(n+1)| - |x(n)|) + \sum_{s=0}^{n} \sum_{u=n-s+1}^{\infty} |D(u)||x(s)|$$

$$- \sum_{s=0}^{n-1} \sum_{u=n-s}^{\infty} |D(u)||x(s)| + |p(n)|$$

$$\leq \left(-1 + |a(n)| + \sum_{u=0}^{\infty} |D(u)| \right) |x(n)| + |p(n)|$$

$$\leq -\beta |x(n)| + M \qquad (2.3.14)$$

if and only if (2.3.12) holds (by noting that (2.3.12) is the condition given by (2.3.4)). We have from equation (2.3.13) and inequality (2.3.14) that

$$|x(n)| \leq V(n, x_n) \leq |x(n)| + \sum_{s=0}^{n-1} \Theta(n-s)|x(s)| \qquad (2.3.15)$$

and $\Theta(n) = \sum_{u=n}^{\infty} |D(u)|$. The results follow from Theorem 2.3.1.

2.4 More on Boundedness

In this section, we state and prove general theorems that guarantee boundedness of all solutions of (2.1.1). Then we utilize the theorems and use nonnegative definite Lyapunov functionals to obtain sufficient conditions that guarantee boundedness of solutions of (2.1.1). The theory is illustrated with several examples. A stereotype of equation (2.1.1) is the Volterra discrete system

$$x(n+1) = A(n)x(n) + \sum_{s=0}^{n} B(n,s)f(s, x(s)). \qquad (2.4.1)$$

Also, in [85], the author studied the exponential stability and boundedness of solutions of the nonlinear discrete system

$$x(n+1) = F(n, x(n)); \ n \geq 0$$
$$x(n_0) = x_0; \ n_0 \geq 0.$$

We emphasize that the results of [85] do not apply to equations similar to (2.4.1). We are mainly interested in applying our results to Volterra discrete systems of the form of (2.4.1)) with $f(x) = x^n$ where n is positive and rational. This section offers a new perspective at looking at the notion of constructing Lyapunov functionals that can be effectively used to obtain existence results. For this section, we use a slightly different boundedness definition.

Definition 2.4.1. We say that solutions of system (2.1.1) are bounded, if any solution $x(n,n_0,\phi)$ of (2.1.1) satisfies

$$||x(n,n_0,\phi)|| \leq C\Big(||\phi||,n_0\Big), \quad \text{for all } n \geq n_0,$$

where $C : \mathbb{R}^+ \times \mathbb{R}^+ \to \mathbb{R}^+$ is a constant that depends on n_0 and ϕ is a given bounded initial function. We say that solutions of system (2.1.1) are uniformly bounded if C is independent of n_0.

Theorem 2.4.1 ([147]). *Let D be a set in $\mathbb{R}^k$. Suppose there exists a Lyapunov functional $V : \mathbb{Z}^+ \times D \to \mathbb{R}^+$ that satisfies*

$$\lambda_1 W_1(|x|) \leq V(n,x(.)) \leq \lambda_2 W_2(|x|) + \lambda_2 \sum_{s=0}^{n-1} \varphi_1(n,s) W_3(|x(s)|) \qquad (2.4.2)$$

and

$$\triangle V(n,x(.)) \leq -\lambda_3 W_4(|x|) - \lambda_3 \sum_{s=0}^{n-1} \varphi_2(n,s) W_5(|x(s)|) + L \qquad (2.4.3)$$

for some positive constants $\lambda_1, \lambda_2, \lambda_3$ and L, and $\lambda_2 > \lambda_3$, where $\varphi_i(n,s) \geq 0$ is a scalar sequence for $0 \leq s \leq t < \infty, i = 1, 2$, such that for some constant $\gamma \geq 0$ the inequality

$$W_2(|x|) - W_4(|x|) + \sum_{s=0}^{n-1} \Big(\varphi_1(n,s) W_3(|x(s)|) - \varphi_2(n,s) W_5(|x(s)|) \Big) \leq \gamma \qquad (2.4.4)$$

holds. Moreover, if $\sum_{s=0}^{n-1} \phi_1(n,s) \leq B$ for some positive constant B, then all solutions of (2.1.1) that stay in D are uniformly bounded.

Proof. Let $M = -\ln(1 - \lambda_3/\lambda_2) > 0$. For any initial value $n_0 \in \mathbb{Z}^+$, let $x(n)$ be any solution of (2.1.1) with $x(n) = \phi(n)$, for $0 \leq n \leq n_0$. Taking the difference of the function $V(n,x)e^{M(n-n_0)}$, we have

$$\triangle\Big(V(n,x)e^{M(n-n_0)}\Big) = \Big[V(n+1,x)e^M - V(n,x)\Big]e^{M(n-n_0)}.$$

For $x \in D$, using (2.4.2) and (2.4.3) we get

$$\triangle\Big(V(n,x)e^{M(n-n_0)}\Big)$$

$$\leq \Big[V(n,x)e^M - \lambda_3 W_4(|x|)e^M - \lambda_3 \sum_{s=0}^{n-1} \varphi_2(n,s) W_5(|x(s)|)e^M + Le^M - V(n,x)\Big]e^{M(n-n_0)}$$

$$= \Big[V(n,x)(e^M - 1) - \lambda_3 e^M\Big(W_4(|x|) + \sum_{s=0}^{n-1} \varphi_2(n,s) W_5(|x(s)|)\Big) + Le^M\Big]e^{M(n-n_0)}$$

$$\leq \left[(e^M-1)\lambda_2\left(W_2(|x|)+\sum_{s=0}^{n-1}\varphi_1(n,s)W_3(|x(s)|)\right)-\lambda_3 e^M\left((W_4(|x|)\right.\right.$$

$$\left.\left.+\sum_{s=0}^{n-1}\varphi_2(n,s)W_5(|x(s)|)\right)+Le^M\right]e^{M(n-n_0)}. \tag{2.4.5}$$

Since $M=-\ln(1-\lambda_3/\lambda_2)>0$, we have $\lambda_2(e^M-1)=\lambda_3 e^M$. Thus, the above inequality reduces to

$$\triangle\left(V(n,x)e^{M(n-n_0)}\right)\leq\left[(e^M-1)\lambda_2\left(W_2(|x|)-(W_4(|x|)+\sum_{s=0}^{n-1}\left(\varphi_1(n,s)W_3(|x(s)|)\right.\right.\right.$$

$$\left.\left.\left.-\varphi_2(n,s)W_5(|x(s)|)\right)+Le^M\right]e^{M(n-n_0)}. \tag{2.4.6}$$

By invoking condition (2.4.4), the inequality (2.4.6) takes the form

$$\triangle\left(V(n,x)e^{M(n-n_0)}\right)\leq\left((e^M-1)\lambda_2\gamma+Le^M\right)e^{M(n-n_0)}\leq\alpha e^{M(n-n_0)},$$

where $\alpha=(e^M-1)\lambda_2\gamma+Le^M$. Summing the above inequality from n_0 to $n-1$, we obtain,

$$V(n,x)e^{M(n-n_0)}-V(n_0,\phi)\leq\alpha\sum_{s=0}^{n-1}e^{M(s-n_0)}\leq\alpha e^{-Mn_0}\sum_{s=0}^{n-1}(e^M)^s\leq\frac{\alpha}{e^M-1}e^{M(n-n_0)},$$

that is,

$$V(n,x)\leq V(n_0,\phi)e^{-M(n-n_0)}+\frac{\alpha}{e^M-1}\leq V(n_0,\phi)+\frac{\alpha}{e^M-1}.$$

From condition (2.4.2), we have

$$\|x\|\leq W_1^{-1}\left[\frac{1}{\lambda_1}\left(\lambda_2 W_2(|\phi|)+\lambda_2 W_3(|\phi|)\sum_{s=0}^{n_0-1}\varphi_1(n_0,s)\right)+\frac{\alpha}{e^M-1}\right]; \text{for all } n\geq n_0.$$

This completes the proof.

In the next theorems, we consider variables $\lambda_i(n)$, $i=1,2,3,4,5$.

Theorem 2.4.2 ([147]). *Let D be a set in $\mathbb{R}^k$. Suppose there exist a Lyapunov functional $V:\mathbb{Z}^+\times D\to\mathbb{R}^+$ that satisfies*

$$\lambda_1(n)W_1(|x|)\leq V(n,x(.))\leq\lambda_2(n)W_2(|x|)+\lambda_2(n)\sum_{s=0}^{n-1}\varphi_1(n,s)W_3(|x(s)|) \tag{2.4.7}$$

and

$$\triangle V(n,x(.)) \leq -\lambda_3(n)W_4(|x|) - \lambda_3(n)\sum_{s=0}^{n-1}\varphi_2(n,s)W_5(|x(s)|) + L \quad (2.4.8)$$

for some positive constant L, and positive sequence $\lambda_1(n),\lambda_2(n),\lambda_3(n)$, where $\lambda_1(n)$ is nondecreasing sequence, and $\varphi_i(n,s) \geq 0$ is a scalar sequence for $0 \leq s \leq n < \infty, i = 1,2$. Assume that for some positive constants θ, and γ the inequality with

$$0 < \frac{\lambda_3(n)}{\lambda_2(n)} \leq \theta < 1, \quad (2.4.9)$$

and

$$W_2(|x|) - W_4(|x|) + \sum_{s=0}^{n-1}\left(\varphi_1(n,s)W_3(|x(s)|) - \varphi_2(n,s)W_5(|x(s)|)\right) \leq \gamma \quad (2.4.10)$$

holds. If $\sum_{s=0}^{n-1}\phi_1(n,s) \leq B$, $\lambda_2(n) \leq N$ for some positive constants B and N, then all solutions of (2.1.1) that stay in D are uniformly bounded.

Proof. First we note that due to condition (2.4.9), $\lambda_3(n)$ is bounded for all $n \geq n_0 \geq 0$. For any initial value $n_0 > 0$, let $x(n)$ be any solution of (2.1.1) with $x(n) = \phi(n)$ for $0 \leq n \leq n_0$. Taking the difference of the function $V(n,x)e^{M(n-n_0)}$ with

$$M = \inf_{n \in \mathbb{Z}^+}\left(-\ln(1 - \frac{\lambda_3(n)}{\lambda_2(n)})\right) > 0,$$

we have

$$\triangle\left(V(n,x)e^{M(n-n_0)}\right) = \left[V(n+1,x)e^M - V(n,x)\right]e^{M(n-n_0)}.$$

By a similar argument as in Theorem 2.4.1 we obtain,

$$\triangle\left(V(n,x)e^{M(n-n_0)}\right) \leq \left((e^M - 1)\lambda_2(n)\gamma + Le^M\right)e^{M(n-n_0)}$$

$$\leq \left((e^M - 1)N\gamma + Le^M\right)e^{M(n-n_0)}.$$

We let $\beta = (e^M - 1)N\gamma + Le^M$, and summing the above inequality from n_0 to $n-1$, we obtain

$$V(n,x)e^{M(n-n_0)} \leq V(n_0,\phi(n_0)) + \beta\sum_{s=0}^{n-1}e^{M(s-n_0)}$$

$$\leq V(n_0,\phi(n_0)) + \frac{\beta}{e^M - 1}e^{M(n-n_0)}.$$

By condition (2.4.7), we have

$$\|x\| \leq W_1^{-1} \left[\frac{1}{\lambda_1(n_0)} \left(V(n_0, \phi(n_0)) e^{M(n-n_0)} + \frac{\beta}{e^M - 1} \right) \right]$$

$$\leq W_1^{-1} \left[\frac{1}{\lambda_1(n_0)} \left(\lambda_2(n_0) W_2(|\phi|) + \lambda_2(n_0) W_3(|\phi|) \sum_{s=0}^{n_0-1} \phi_1(n_0, s) + \frac{\beta}{e^M - 1} \right) \right],$$

for all $n \geq n_0$. Hence, the solutions of (2.1.1) that start in D are uniformly bounded. This completes the proof.

Theorems 2.4.1 and 2.4.2 are of special importance since, by the aid of constructing a suitable Lyapunov functionals, the theorems can be applied to nonlinear Volterra systems of the form

$$x(n+1) = \sigma(n)x(n) + \sum_{s=0}^{n} B(n,s)x^{2/3}(s), \ n \geq 0, \ x(n) = \phi(n) \ \text{ for } 0 \leq n \leq n_0,$$

$$(2.4.11)$$

where $\phi(n)$ is a given bounded initial sequence.

2.5 Applications to Nonlinear Volterra Difference Equations

In this section, we apply the results of the previous section to nonlinear Volterra difference equations. As we shall see that Theorems 2.4.1 and 2.4.2 will guide us step by step on how such Lyapunov functionals are constructed.

Theorem 2.5.1 ([147]). *Consider the scalar nonlinear Volterra difference equation given by* (2.4.11). *Suppose there are constants* $\beta_1, \ \beta_2 \in (0,1)$ *such that*

$$\sigma^2(n) + \frac{2}{3} \left(|\sigma(n)| + \sum_{s=0}^{n} |B(n,s)| \right) |B(n,n)| + \sum_{u=n+1}^{\infty} |B(u,n)| - 1 \leq -\beta_1,$$

and

$$\frac{2}{3} \left(|\sigma(n)| + \sum_{s=0}^{n} |B(n,s)| \right) - 1 \leq -\beta_2. \qquad (2.5.1)$$

If

$$\sum_{s=0}^{n-1} \sum_{u=n}^{\infty} |B(u,s)|, \ \sum_{s=0}^{n} |B(n,s)| < \infty,$$

and

$$|B(n,s)| \geq \sum_{u=n}^{\infty} |B(u,s)|,$$

then all solutions of (2.4.11) *are uniformly bounded.*

Proof. To see this, we consider the Lyapunov functional

$$V(n,x) = x^2(n) + \sum_{s=0}^{n-1} \sum_{u=n}^{\infty} |B(u,s)| x^2(s).$$

Then along solutions of (2.4.11) we have

$$\triangle V(n,x) = x^2(n+1) - x^2(n) + \sum_{u=n+1}^{\infty} |B(u,n)| x^2(n) - \sum_{s=0}^{n-1} |B(u,s)| x^2(s)$$

$$= \left(\sigma(n)x(n) + \sum_{s=0}^{n} B(n,s) x^{2/3}(s) \right)^2 - x^2(n)$$

$$+ 2 \sum_{u=n+1}^{\infty} |B(u,n)| x^2(n) - \sum_{s=0}^{n-1} |B(u,s)| x^2(s)$$

$$= \left(\sigma^2(n) + \sum_{u=n+1}^{\infty} |B(u,n)| - 1 \right) x^2(t)$$

$$+ 2\sigma(n)x(n) \sum_{s=0}^{n} |B(n,s)| x^{2/3}(s) + \left(\sum_{s=0}^{n} |B(n,s)| x^{2/3}(s) \right)^2$$

$$- \sum_{s=0}^{n-1} |B(n,s)| x^2(s). \tag{2.5.2}$$

To further simplify the above inequality we perform the following calculations. Using the fact that $ab \le a^2/2 + b^2/2$,

$$2\sigma(n)x(n) \sum_{s=0}^{n} |B(n,s)| x^{2/3}(s) \le 2|\sigma(n)||x(n)| \sum_{s=0}^{n} |B(n,s)| x^{2/3}(s)$$

$$\le |\sigma(n)| \sum_{s=0}^{n} |B(n,s)| \left(x^2(n) + x^{4/3}(s) \right).$$

Using the Cauchy-Schwartz inequalities for series, one obtains

$$\left(\sum_{s=0}^{n} |B(n,s)| x^{2/3}(s) \right)^2 \le \left(\sum_{s=0}^{n} |B(n,s)|^{1/2} |B(n,s)|^{1/2} x^{2/3}(s) \right)^2$$

$$\le \sum_{s=0}^{n} |B(n,s)| \sum_{s=0}^{n} |B(n,s)| x^{4/3}(s).$$

Adding the above two inequalities yields

$$2\sigma(n)x(n) \sum_{s=0}^{n} |B(n,s)| x^{2/3}(s) + \left(\sum_{s=0}^{n} |B(n,s)| x^{2/3}(s) \right)^2$$

$$\le |\sigma(n)| \sum_{s=0}^{n} |B(n,s)| \left(x^2(n) + x^{4/3}(s) \right) + \sum_{s=0}^{n} |B(n,s)| \sum_{s=0}^{n} |B(n,s)| x^{4/3}(s)$$

$$= |\sigma(n)| \sum_{s=0}^{n} |B(n,s)| x^2(n)$$

$$+ \left(|\sigma(n)| + \sum_{s=0}^{n} |B(n,s)| \right) \sum_{s=0}^{n} |B(n,s)| x^{4/3}(s). \tag{2.5.3}$$

Finally, we make use of Young's inequality, which says for any two nonnegative real numbers ω and ϖ, we have

$$\omega\varpi \leq \frac{\omega^e}{e} + \frac{\varpi^f}{f}, \quad \text{with } 1/e + 1/f = 1.$$

Thus, for $e = 3/2$ and $f = 3$, we get

$$\sum_{s=0}^{n} |B(n,s)| x^{4/3}(s) = \sum_{s=0}^{n} |B(n,s)|^{1/3} |B(n,s)|^{2/3} x^{4/3}(s)$$

$$\leq \sum_{s=0}^{n} \left(\frac{|B(n,s)|}{3} + \frac{2}{3} |B(n,s)| x^2(s) \right)$$

$$= \sum_{s=0}^{n} \frac{|B(n,s)|}{3} + \frac{2}{3} |B(n,s)| x^2(s) + \frac{2}{3} \sum_{s=0}^{n-1} |B(n,s)| x^2(s).$$

With this in mind, inequality (2.5.3) reduces to

$$2\sigma(n)x(n) \sum_{s=0}^{n} |B(n,s)| x^{2/3}(s) + \left(\sum_{s=0}^{n} |B(n,s)| x^{2/3}(s) \right)^2$$

$$\leq \frac{2}{3} \left(|\sigma(n)| + \sum_{s=0}^{n} |B(n,s)| \right) |B(n,n)|$$

$$+ \frac{1}{3} \left(|\sigma(n)| + \sum_{s=0}^{n} |B(n,s)| \right) \sum_{s=0}^{n} |B(n,s)|$$

$$+ \frac{2}{3} \left(|\sigma(n)| + \sum_{s=0}^{n} |B(n,s)| \right) \sum_{s=0}^{n-1} |B(n,s)| x^2(s). \tag{2.5.4}$$

By substituting (2.5.4) into (2.5.2), we arrive at

$$\Delta V(n,x) \leq \left[\sigma^2(n) + \frac{2}{3} \left(|\sigma(n)| + \sum_{s=0}^{n} |B(n,s)| \right) |B(n,n)| + \sum_{u=n+1}^{\infty} |B(u,n)| - 1 \right] x^2(n)$$

$$+ \frac{2}{3} \left(|\sigma(n)| + \sum_{s=0}^{n} |B(n,s)| - 1 \right) \sum_{s=0}^{n-1} |B(n,s)| x^2(s)$$

$$+ \frac{1}{3} \left(|\sigma(n)| + \sum_{s=0}^{n} |B(n,s)| \right) \sum_{s=0}^{n} |B(n,s)|. \tag{2.5.5}$$

Let $L = \frac{1}{3}\Big(|\sigma(n)| + \sum_{s=0}^{n} |B(n,s)| \Big) \sum_{s=0}^{n} |B(n,s)|$. Take $W_1 = W_2 = W_4 = x^2(n)$, $W_3 = W_5 = x^2(s)$, $\lambda_1 = \lambda_2 = 1$ and $\lambda_3 = \min\{\beta_1, \beta_2\}$. Also, we choose $\varphi_1(n,s) = \sum_{u=n}^{\infty} |B(u,s)|$, and $\varphi_2(n,s) = |B(n,s)|$, we see that conditions (2.4.7) and (2.4.8) of Theorem 2.4.2 are satisfied. Next we make sure condition (2.4.10) is satisfied. To see this,

$$
W_2(|x|) - W_4(|x|) + \sum_{s=0}^{n-1} \Big(\varphi_1(n,s) W_3(|x(s)|) - \varphi_2(n,s) W_5(|x(s)|) \Big)
$$

$$
= x^2(n) - x^2(n) + \sum_{s=0}^{n-1} \Big(\sum_{u=n}^{\infty} |B(u,s)| - |B(u,s)| \Big) x^2(s)
$$

$$
= \sum_{s=0}^{n-1} \Big(\sum_{u=n}^{\infty} |B(u,s)| - |B(u,s)| \Big) x^2(s) \leq 0.
$$

Thus, condition (2.4.10) is satisfied for $\gamma = 0$. An application of Theorem 2.4.2 yields the results.

In the next theorem we establish sufficient conditions that guarantee the boundedness of all solutions of the vector Volterra difference equation

$$
\triangle x(t) = Ax(t) + \sum_{s=0}^{t-1} C(t,s)x(s) + g(t), \tag{2.5.6}
$$

where $t \geq 0$, $x(t) = \phi(t)$ for $0 \leq t \leq t_0$, $\phi(t)$ is a given bounded continuous initial $k \times 1$ vector function. Also, A and C are $k \times k$ matrices and g is $k \times 1$ vector functions that is continuous in x If D is a matrix, $|D|$ means the sum of the absolute values of the elements. For what to follow we write g and x for $g(t)$ and $x(t)$, respectively.

Theorem 2.5.2. *Suppose $C^T(t,s) = C(t,s)$. Let I be the $k \times k$ identity matrix. Assume there exist positive constants $L, \nu, \xi, \beta_1, \beta_2, \lambda_3$, and $k \times k$ positive definite constant symmetric matrix B such that*

$$
\Big[A^T B + BA + A^T BA \Big] \leq -\xi I, \tag{2.5.7}
$$

$$
\Big[-\xi + |Bg| + \sum_{s=0}^{t-1} |B||C(t,s)| + \sum_{s=0}^{t-1} |A^T B||C(t,s)|
$$
$$
+ \nu \sum_{u=t+1}^{\infty} |B(u,t)| \Big] (1 + \lambda_3) \leq -\beta_1, \tag{2.5.8}
$$

$$
\Big[|B| - \nu + \Big((g^T B)^2 + 1 + |A^T B|
$$
$$
+ \sum_{s=0}^{t-1} |C(t,s)| \Big) \Big] (1 + \lambda_3) \leq -\beta_2, \tag{2.5.9}
$$

$$(|g^T g| + |Bg|)(1 + \lambda_3) = L, \tag{2.5.10}$$

$$|C(t,s)| \geq v \sum_{u=t+1}^{\infty} |C(u,s)|, \tag{2.5.11}$$

and

$$\sum_{s=0}^{t-1} \sum_{u=t}^{\infty} |C(u,s)|, \ \sum_{s=0}^{t-1} |C(t,s)| < \infty. \tag{2.5.12}$$

Then solutions of (2.5.6) are uniformly bounded.

Proof. Since B is $k \times k$ positive definite constant symmetric matrix, then there exists an $r_1 \in (0,1]$ and $r_2 > 0$ such that

$$r_1 x^T x \leq x^T B x \leq r_2 x^T x. \tag{2.5.13}$$

Define

$$V(t,x) = x^T B x + v \sum_{s=0}^{t-1} \sum_{u=t}^{\infty} |C(u,s)| x^2(s).$$

Here $x^T x = x^2 = (x_1^2 + x_2^2 + \cdots + x_k^2)$. We have along the solutions that

$$\Delta V(t,x) = \left[Ax + \sum_{0}^{t-1} C(t,s)x(s) + g \right]^T Bx \tag{2.5.14}$$

$$+ x^T B \left[Ax + \sum_{s=0}^{t-1} C(t,s)x(s) + g \right]$$

$$+ \left[Ax + \sum_{s=0}^{t-1} C(t,s)x(s) + g \right]^T B \left[Ax + \sum_{0}^{t-1} C(t,s)x(s) + g \right]$$

$$- v \sum_{s=0}^{t-1} |B(t,s)| x^2(s) + v \sum_{u=t+1}^{\infty} |B(u,t)| x^2.$$

By noting that the right side of (2.5.14) is scalar and by recalling that B is a symmetric matrix, expression (2.5.14) simplifies to

$$\Delta V(t,x) = x^T \left(A^T B + BA + A^T BA \right) x + 2x^T Bg \tag{2.5.15}$$

$$+ 2 \sum_{s=0}^{t-1} x^T BC(t,s)x(s)$$

$$+ \left[x^T A^T Bg + 2g^T B \sum_{s=0}^{t-1} C(t,s)x(s) + 2x^T A^T B \sum_{s=0}^{t-1} C(t,s)x(s) \right.$$

$$+ \left. \sum_{s=0}^{t-1} x^T(s)C(t,s)\Delta_s B \sum_{s=0}^{t-1} C(t,s)x(s) + g^T B g \right]$$

$$- v \sum_{s=0}^{t-1} |C(t,s)| x^2(s) + v \sum_{u=t+1}^{\infty} |C(u,t)| x^2$$

$$\leq -\xi x^2 + 2|x^T| |Bg| + 2 \sum_{s=0}^{t-1} |x^T| |B| |C(t,s)| |x(s)|$$

$$+ \Big[\sum_{s=0}^{t-1} |C(t,s)| 2 |g^T B| |x(s)| + 2 \sum_{s=0}^{t-1} |x^T| |A^T B| |C(t,s)| |x(s)|$$

$$+ \sum_{s=0}^{t-1} x^T(s) C(t,s) \, B \sum_{s=0}^{t-1} C(t,s) x(s) + |g^T g| \Big]$$

$$- v \sum_{s=0}^{t-1} |C(t,s)| x^2(s) + v \sum_{u=t+1}^{\infty} |C(u,t)| x^2.$$

Next, we perform some calculations to simplify inequality (2.5.15).

$$2|x^T| |Bg| = 2|x^T| |Bg|^{1/2} |Bg|^{1/2} \leq x^2 |Bg| + |Bg|,$$

$$2 \sum_{s=0}^{t-1} |x^T| |B| |C(t,s)| |x(s)| \leq \sum_{s=0}^{t-1} |B| |C(t,s)| (x^2 + x^2(s)),$$

$$\sum_{s=0}^{t-1} |C(t,s)| 2 |g^T B| |x(s)| \leq \sum_{s=0}^{t-1} |C(t,s)| (|g^T B|^2 + x^2(s)),$$

and

$$2 \sum_{s=0}^{t-1} |x^T| |A^T B| |C(t,s)| |x(s)| \leq \sum_{s=0}^{t-1} |A^T B| |C(t,s)| (x^2 + x^2(s)).$$

Finally,

$$\sum_{s=0}^{t-1} x^T(s) C(t,s) \, B \sum_{s=0}^{t-1} C(t,s) x(s)$$

$$|B| \, | \sum_{s=0}^{t-1} x^T(s) C(t,s)| | \sum_{s=0}^{t-1} C(t,s) x(s)|$$

$$\leq |B| \Big(\sum_{s=0}^{t-1} x^T(s) C(t,s) \Big)^2 \Big/ 2 + |B| \Big(\sum_{s=0}^{t-1} C(t,s) x(s) \Big)^2 \Big/ 2$$

$$= |B| \Big(\sum_{s=0}^{t-1} C(t,s) x(s) \Big)^2$$

$$= |B| \Big(\sum_{s=0}^{t-1} |C(t,s)|^{\frac{1}{2}} |C(t,s)|^{\frac{1}{2}} |x(s)| \Big)^2$$

$$\leq |B| \sum_{s=0}^{t-1} |C(t,s)| \sum_{s=0}^{t-1} |C(t,s)| x^2(s).$$

Substitution of the above inequalities into (2.5.15) yields

$$\triangle V(t,x) \leq \Big[-\xi + |Bg| + \sum_{s=0}^{t-1}|B||C(t,s)|$$

$$+ \sum_{s=0}^{t-1}|A^T B||C(t,s)| + v\sum_{u=t+1}^{\infty}|C(u,t)|\Big]x^2$$

$$+ \Big[|B| - v + \big((g^T B)^2 + 1 + |A^T B| + |B|\sum_{s=0}^{t-1}|C(t,s)|\big)\Big]\sum_{s=0}^{t-1}|C(t,s)|x^2(s)$$

$$+ (\mu(t)|g^T Bg| + |Bg|)(1 + \lambda_3).$$

Applying conditions (2.5.8), (2.5.9), and (2.5.10), $\triangle V(t,x)$ reduces to

$$\triangle V(t,x) \leq -\beta_1 x^2 - \beta_2 \sum_{s=0}^{t-1}|C(t,s)|x^2(s) + L,$$

where $L = (\mu(t)|g^T Bg| + |Bg|)(1 + \lambda_3)$. By taking $W_1 = r_1 x^T x, W_2 = x^T Bx, W_4 = r_2 x^T x, W_3 = W_5 = x^2(s), \lambda_1 = \lambda_2 = 1$ and $\lambda_3 = \min\{\beta_1,\beta_2\}, \phi_1(t,s) = v\sum_{u=t}^{\infty}|C(u,s)|$, and $\phi_2(t,s) = |C(t,s)|$, we see that conditions (2.4.7) and (2.4.8) of Theorem 2.4.2 are satisfied. Next we make sure condition (2.4.10) is satisfied. Using (2.5.11) and (2.5.13) we obtain

$$W_2(|x|) - W_4(|x|) + \sum_{s=0}^{t-1}(\phi_1(t,s)W_3(|x|) - \phi_2(t,s)W_5(|x(s)|))$$

$$= x^T Bx - r_2 x^T x + \sum_{s=0}^{t-1}\Big(v\sum_{u=t}^{\infty}|C(u,s)| - |C(t,s)|\Big)x^2(s) \leq 0.$$

Thus condition (2.4.10) is satisfied with $\gamma = 0$. An application of Theorem 2.4.2 yields the results.

2.6 Open Problems

Open Problem 1.
Reformulate Theorems 2.4.1 and 2.4.2 to obtain results concerning the exponential stability of the zero solution of (2.1.1).
For Open Problem 2, we consider the functional delay difference equation

$$x(t+1) = F(t,x_t). \tag{2.6.1}$$

We assume that F is continuous in x and that $F : \mathbb{Z} \times C \to \mathbb{R}^n$ where C is the set of sequences $\phi : [-\alpha,0] \to \mathbb{R}^n, \alpha > 0$. Let

$$C(t) = \{\phi : [t - \alpha, t] \to \mathbb{R}^n\}.$$

It is to be understood that $C(t)$ is C when $t = 0$. For $\phi \in C(t)$ we denote

$$|||\phi_t||| = \left[\sum_{i=1}^{n} \sum_{s=t-\alpha}^{t-1} \phi_i^2(s)\right]^{1/2}$$

where $\phi(t) = (\phi_1(t), \cdots, \phi_n(t))$.

Open Problem 2.

Theorem 2.6.1. *Let $D > 0$ and there is a scalar functional $V(t, \psi_t)$ that is continuous in ψ and locally Lipschitz in ψ_t when $t \geq t_0$ and $\psi_t \in C(t)$ with $\|\psi_t\| < D$. In addition we assume that if $x : [t_0 - \alpha, \infty) \to \mathbb{R}^n$ is a bounded sequence, then $F(t, x_t)$ is bounded on $[t_0, \infty)$. If V such that $V(t, 0) = 0$,*

$$W_1(|\phi(t)|) \leq V(t, \phi_t) \leq W_2(|\phi_t|) + W_3(|||\phi_t|||),$$

and

$$\Delta V(t, \phi_t) \leq -W_4(|\phi(t)|),$$

then the zero solution of (2.6.1) is uniformly asymptotically stable .

Open Problem 3.

General theorem in the spirit of Theorems 2.1.1, 2.1.2, and 2.1.3 regarding functional delay difference equations is nowhere to be found and hence there is a desperate need of such theorems. In particular, for $h > 0$ and constant, we ask that parallel theorems to Theorems 2.1.1, 2.1.2, and 2.1.3 should be developed regarding the functional discrete system

$$x(n+1) = \big(G(n, x(s); \; -h \leq s \leq n\big) \overset{def}{=} G(n, x(\cdot)) \tag{2.6.2}$$

where $G : \mathbb{Z}^+ \times \mathbb{R}^k \to \mathbb{R}^k$ is continuous in x. Then such theorems can be applied to Volterra difference systems of the form

$$x(n+1) = b(n)x(n) + \sum_{s=-h}^{n-1} C(n, s)g(x(s)), \tag{2.6.3}$$

and

$$x(n+1) = b(n)x(n) + \sum_{s=n-h}^{n-1} C(n, s)g(x(s)). \tag{2.6.4}$$

In the next theorem we establish sufficient conditions that guarantee the boundedness of all solutions of the vector Volterra difference equation by using Lyapunov-Razumikhini method. It should serve as a guidance to formulate and prove boundedness results concerning functional difference equations. Thus, we consider the Volterra difference equation

$$x(t+1) = Ax(t) + \sum_{s=0}^{t-1} C(t,s)x(s) + g(t), \qquad (2.6.5)$$

where $t \geq 0$, $x(t) = \phi(t)$ for $0 \leq t \leq t_0$, $\phi(t)$ is a given bounded initial $k \times 1$ vector functions. Also, A and C are $k \times k$ matrices and g is $k \times 1$ vector functions. If D is a matrix, $|D|$ means the sum of the absolute values of the elements. Let $||g||^{[0,\infty)}$ denote the norm of g.

Theorem 2.6.2 ([144]). *Let I be the $k \times k$ identity matrix. Assume there exists a $k \times k$ positive definite constant symmetric matrix B such that*

$$A^T B + BA = -I, \qquad (2.6.6)$$

Suppose that there is a positive constant M such that

$$\sum_{s=0}^{t-1} |BC(t,s)| \leq M,$$

so that

$$\frac{2\beta hM}{\alpha} < 1,$$

where α, β, and h are all positive constants to specify in the proof. If in addition, g is bounded, then all solutions of of (2.6.5) are uniformly bounded.

Proof. Since B is $k \times k$ positive definite constant symmetric matrix, then there exists an $\alpha, \beta \in (0,1]$ such that

$$\alpha^2 |x|^2 \leq x^T Bx \leq \beta^2 |x|^2.$$

Define the Lyapunov-Razumikhini function

$$V(t,x) = x^T Bx.$$

Then clearly

$$\alpha^2 |x|^2 \leq V(t,x) \leq \beta^2 |x|^2.$$

Then along the solutions of (2.6.5) we have

$$\Delta V(t,x) = \left[Ax + \sum_{0}^{t-1} C(t,s)x(s) + g \right]^T Bx$$

$$+ x^T B \left[Ax + \sum_{s=0}^{t-1} C(t,s)x(s) + g \right]$$

$$= x^T \left(A^T B + BA \right) x + 2x^T Bg + 2 \sum_{s=0}^{t-1} x^T BC(t,s)x(s)$$

$$\leq -|x|^2 + 2|x||B|||g||^{[0,\infty)} + 2 \sum_{s=0}^{t-1} |BC(t,s)||x(s)|.$$

Now, if $h^2 V(t,x(t)) > V(s,x(s))$ for $0 \leq s \leq t-1$, where $h > 1$ is a constant to be determined, then

$$\alpha^2 |x(s)|^2 \leq V(s,x(s)) \leq h^2 V(t,x(t)) \leq h^2 \beta^2 |x|^2,$$

and

$$\frac{h\beta}{\alpha} |x(t)| \geq |x(s)|, \quad s \leq t-1.$$

Thus,

$$\triangle V(t,x) \leq -|x|^2 + 2\frac{h\beta}{\alpha}|x|^2 \sum_{s=0}^{t-1} |BC(t,s)| + 2|x||B|||g||^{[0,\infty)}$$

$$\leq -|x|^2 + 2\frac{h\beta M}{\alpha}|x|^2 + 2|x||B|||g||^{[0,\infty)}.$$

Since $\dfrac{2\beta h M}{\alpha} < 1$, h maybe chosen so that $h > 1$ and $\dfrac{2h\beta h M}{\alpha} < 1$, yielding

$$\triangle V(t,x) \leq (2\frac{h\beta}{\alpha} - 1)|x|^2 + 2|x||B|||g||^{[0,\infty)} \leq 0$$

provided that

$$|x| \geq \frac{2|B|||g||^{[0,\infty)}}{1 - \frac{2\beta h M}{\alpha}} := K.$$

Now we summarize what we have

(a) $W_1(|x|) \leq V(t,x) \leq W_2(|x|)$,
(b) there exists $K > 0$ so that if $x(t)$ is a solution of (2.6.5) with $|x(t)| \geq K$ for some $t \geq 0$ and $V(s,x(s)) < p(V(t,x))$ for $0 \leq s \leq t-1$ and $p(u) > u$, then $\triangle V(t,x) \leq 0$, where $p(u) = h^2 u$.

Now choose any solution $x(t)$ such that $|\phi(t)| < H$ for $0 \leq t \leq n_0$ for some $H > 0$. Let $L > \max\{H,K\}$ and choose $D > 0$ with $W_2(L) < W_1(D)$. If this solution is unbounded, then there is $t_1 > 0$ such that $|x(t_1)| > D$, $|x(t)| \leq D$ for $0 < t \leq t_1 - 1$. If $V(t_1,x(t_1)) \leq V(t_0,x(t_0))$, then we would have

$$W_1(|x(t_1)|) \leq V(t_1,x(t_1)) \leq V(t_0,\phi(t_0) \leq W_2(|\phi(t_0)|)$$
$$\leq W_2(L) < W_1(D),$$

from which we get $|x(t_1)| < D$, a contradiction. Thus, $V(t_1,x(t_1)) > V(t_0,x(t_0))$. We leave it for the reader to complete the proof using (a) and (b) as guidance.

Open Problem 4.
We propose that the reader develops general theorems for the boundedness and stability of functional difference equations using Lyapunov-Razumikhini method. Conditions (a) and (b) should serve as guidance for stating and proving such theorems. For more on the subject we refer to [181] and [182].

Chapter 3
Fixed Point Theory in Stability and Boundedness

In the past hundred and fifty years, Lyapunov functions/functionals have been exclusively and successfully used in the study of stability and existence of periodic and bounded solutions. The author has extensively used Lyapunov functions/functionals for the purpose of analyzing solutions of functional equations, and each time the suitable Lyapunov functional presented us with unique difficulties, that could only overcome by the imposition of severe conditions on the given coefficients. In practice, Lyapunov direct method requires pointwise conditions, while as so many real-life problems call for averages. Moreover, it is rare that we encounter a problem for which a suitable Lyapunov functional can be easily constructed. It is a common knowledge among researchers that results on stability and boundedness go hand in hand with the constructed Lyapunov functional.

In this chapter, we begin a systematic study of stability theory for ordinary and functional difference equations by means of fixed point theory. The study of fixed point theory is motivated by a number of difficulties encountered in the study of stability by means of Lyapunov's direct method. We notice that these difficulties frequently vanish when we apply fixed point theory. We provide a brief introduction on topics in Cauchy sequences, metric spaces, compactness, contraction mapping principle, and Banach spaces. In some cases, contraction mapping principle fails to produce any results. This forces us to look for other alternatives, namely the concept of Large Contraction. We will restate the contraction mapping principle and Krasnoselskii's fixed point theorems in which the regular contraction is replaced with Large Contraction. Most of the work in this chapter can be found in [4, 140, 142, 150, 166], and [167].

© Springer Nature Switzerland AG 2018
Y. N. Raffoul, *Qualitative Theory of Volterra Difference Equations*,
https://doi.org/10.1007/978-3-319-97190-2_3

3.1 Motivation

We begin by offering an example that exposes the difficulties encountered by the use
of Lyapunov functionals. Fixed point theory was first used in difference equations
by Raffoul in [136] to study the stability and the existence of periodic solutions of
the linear delay difference equation

$$\triangle x(t) = -a(t)x(t-\tau).$$

It was followed by a series of papers in which different authors considered the same
idea and analyzed various types of difference and Volterra difference equations. For
example, in [134] the author initiated the use of fixed point theory to alleviate some
of the difficulties that arise from the deployment of Lyapunov functionals to study
boundedness and stability of the neutral nonlinear delay differential equation

$$x'(t) = -a(t)x(t) + c(t)x'(t-g(t)) + q(t,x(t),x(t-g(t))),$$

where $a(t), b(t), g(t)$, and q are continuous in their respective arguments. Later on,
Islam and Yankson [87] extended the work of [134] to the neutral nonlinear delay
difference equation

$$x(t+1) = a(t)x(t) + c(t)\triangle x(t-g(t)) + q(x(t),x(t-g(t))),$$

where $a, c : \mathbb{Z} \to \mathbb{R}, q : \mathbb{R} \times \mathbb{R} \to \mathbb{R}$, and $g : \mathbb{Z} \to \mathbb{Z}$.
To illustrate some of the difficulties that arise from the deployment of Lyapunov
functionals, we consider the delay difference equation

$$x(t+1) = a(t)x(t) + b(t)x(t-\tau) + p(t), \; t \in \mathbb{Z}^+, \tag{3.1.1}$$

where $a, b, p : \mathbb{Z}^+ \to \mathbb{R}, \tau$ is a positive integer. Assume

$$|a(t)| < 1, \text{ for all } t \in \mathbb{Z}^+ \tag{3.1.2}$$

and there is a $\delta > 0$ such that

$$|b(t)| + \delta < 1, \; t \in \mathbb{Z}^+ \tag{3.1.3}$$

and
$$|a(t)| \le \delta, \text{ and } |p(t)| \le K, \text{ for some positive constant } K. \tag{3.1.4}$$

Then all solutions of (3.1.1) are bounded. If $p(t) = 0$ for all t, then the zero solution
of (3.1.1) is (UAS). To see this we consider the Lyapunov functional

$$V(t,x(\cdot)) = |x(t)| + \delta \sum_{s=t-\tau}^{t-1} |x(s)|.$$

Then along solutions of (3.1.1) we have

$$\triangle V = |x(t+1)| - |x(t)| + \delta \sum_{s=t+1-\tau}^{t} |x(s)| - \delta \sum_{s=t-\tau}^{t-1} |x(s)|$$

$$\leq |a(t)||x(t)| - |x(t)| + |b(t)||x(t-\tau)| + \delta \sum_{s=t+1-\tau}^{t} |x(s)| - \delta \sum_{s=t-\tau}^{t-1} |x(s)| + |p(t)|$$

$$= (|a(t)| + \delta - 1)|x(t)| + (|b(t)| - \delta)|x(t-\tau)| + |p(t)|$$

$$\leq (|a(t)| + \delta - 1)|x(t)| + |p(t)|$$

$$\leq -\gamma |x(t)|, \text{ for some positive constant } \gamma.$$

The results follow from Chapter 2. It is severe to ask that a, b be bounded and that $|b(t)|$ is bounded by a all of the time. For another illustration, we consider the nonlinear delay difference equation

$$x(t+1) = a(t)g(x(t)) + b(t)h(x(t-r)), \tag{3.1.5}$$

where the functions g and h are continuous. Define the Lyapunov functional V by

$$V(t) = |x(t)| + \sum_{s=t-r}^{t-1} |b(s+r)||h(x(s))|.$$

We assume that there are positive constants γ_1 and γ_2 such that $|g(x)| \leq \gamma_1 |x|$ and $|h(x)| \leq \gamma_2 |x|$, so that

$$\gamma_1 |a(t)| + \gamma_2 |b(t+r)| - 1 \leq -\beta, \ \beta > 0.$$

Then along solutions of (3.1.5) we have

$$\triangle V = |x(t+1)| - |x(t)| + |b(t+r)||h(x(t))| - |b(t)||h(x(t-r))|$$

$$\leq |a(t)||g(x(t))| + |b(t)||h(x(t-r))| - |x(t)|$$

$$+ |b(t+r)||h(x(t))| - |b(t)||h(x(t-r))|$$

$$\leq (\gamma_1 |a(t)| + \gamma_2 |b(t+r)| - 1)|x(t)|$$

$$\leq -\beta |x(t)|.$$

Now one may refer to Chapter 2 and argue that the zero solution of (3.1.5) is asymptotically stable.

3.2 Metrics and Banach Spaces

This section is devoted to introductory materials related to Cauchy sequences, metric spaces, contraction, compactness, contraction mapping principle, and Banach spaces. Materials in this section are taken from class notes that the author have used in graduate course on real analysis. For an excellent reference, we refer the reader to [23].

Definition 3.2.1. A pair (E,ρ) is a metric space if E is a set and $\rho : E \times E \to [0,\infty)$ such that when y, z, and u are in E, then

(a) $\rho(y,z) \geq 0$, $\rho(y,y) = 0$, and $\rho(y,z) = 0$ implies $y = z$.
(b) $\rho(y,z) = \rho(z,y)$, and
(c) $\rho(y,z) \leq \rho(y,u) + \rho(u,z)$.

Definition 3.2.2 (Cauchy Sequence). A sequence $\{x_n\} \subseteq E$ is a Cauchy sequence if for each $\varepsilon > 0$ there exists an $N \in \mathbb{N}$ such that $n, m > N \implies \rho(x_n, x_m) < \varepsilon$.

Definition 3.2.3 (Completeness of Metric Space). A metric space (E,ρ) is said to be complete if every Cauchy sequence in E converges to a point in E.

Definition 3.2.4. A set L in a metric space (E,ρ) is compact if each sequence in L has a subsequence with a limit in L.

Definition 3.2.5. Let $\{f_n\}$ be a sequence of real functions with $f_n : [a,b] \to \mathbb{R}$.

1. $\{f_n\}$ is uniformly bounded on $[a,b]$ if there exists $M > 0$ such that $|f_n(t)| \leq M$ for all $n \in \mathbb{N}$ and for all $t \in [a,b]$.
2. $\{f_n\}$ is equicontinuous at t_0 if for each $\varepsilon > 0$ $\delta > 0$ such that for all $n \in \mathbb{N}$, if $t \in [a,b]$ and $|t_0 - t| < \delta$, then $|f_n(t_0) - f_n(t)| < \varepsilon$. Also, $\{f_n\}$ is equicontinuous if $\{f_n\}$ is equicontinuous at each $t_0 \in [a,b]$.
3. $\{f_n\}$ is uniformly equicontinuous if for each $\varepsilon > 0$ there exists $d > 0$ such that for all $n \in \mathbb{N}$, if $t_1, t_2 \in [a,b]$ and $|t_1 - t_2| < \delta$, then $|f_n(t_1) - f_n(t_2)| < \varepsilon$.

Easy to see that $\{f_n\} = \{x^n\}$ is not an equicontinuous sequence of functions on $[0,1]$ but each f_n is uniformly continuous.

Proposition 3.1 (Cauchy Criterion for Uniform Convergence). *If $\{F_n\}$ is a sequence of bounded functions that is Cauchy in the uniform norm, then $\{F_n\}$ converges uniformly.*

Definition 3.2.6. A real-valued function f defined on $E \subseteq \mathbb{R}$ is said to be Lipschitz continuous with Lipschitz constant M if $|f(x) - f(y)| \leq M|x - y|$ for all $x, y \in E$.

Remark 3.1. It is an easy exercise that a Lipschitz continuous function is uniformly continuous. Also, if each f_n in a sequence of functions $\{f_n\}$ has the same Lipschitz constant, then the sequence is uniformly equicontinuous.

Lemma 3.1. *If $\{f_n\}$ is an equicontinuous sequence of functions on a closed bounded interval, then $\{f_n\}$ is uniformly equicontinuous.*

Proof. Suppose $\{f_n\}$ is equicontinuous defined on $[a,b]$ (which is contraction). Let $\varepsilon > 0$. For each $x \in K$, let $\delta_x > 0$ be such that $|y - x| < \delta_x \implies |f_n(x) - f_n(y)| < \varepsilon/2$ for all $n \in \mathbb{N}$. The collection $\{B(x, \delta_x/2) : x \in [a,b]\}$ is an open cover of $[a,b]$ so has a finite subcover $\{B(x_i, \delta_{x_i}/2) : i = 1, \ldots, k\}$. Let $\delta = \min\{\delta_{x_i}/2 : i = 1, \ldots, k\}$. Then, if $x, y \in [a,b]$ with $|x - y| < \delta$, there is some i with $x \in B(x_i, \delta_{x_i}/2)$. Since $|x - y| < \delta \leq \delta_{x_i}/2$, we have $|x_i - y| \leq |x_i - x| + |x - y| < \delta_{x_i}/2 + \delta_{x_i}/2 = \delta_{x_i}$. Hence $|x_i - y| < \delta_{x_i}$ and $|x_i - x| < \delta_{x_i}$. So, for any $n \in \mathbb{N}$ we have $|f_n(x) - f_n(y)| \leq |f_n(x) - f_n(x_i)| + |f_n(x_i) - f_n(y)| < \varepsilon/2 + \varepsilon/2 = \varepsilon$. So, $\{f_n\}$ is uniformly equicontinuous.

The next theorem gives us the main method of proving compactness in the spaces in which we are interested.

Theorem 3.2.1 (Ascoli-Arzelà). *If $\{f_n(t)\}$ is a uniformly bounded and equicontinuous sequence of real valued functions on an interval $[a,b]$, then there is a subsequence which converges uniformly on $[a,b]$ to a continuous function.*

Proof. Since $\{f_n(t)\}$ is equicontinuous on $[a,b]$, by Lemma 3.1, $\{f_n(t)\}$ is uniformly equicontinuous. Let $t_1, t_2, \ldots$ be a listing of the rational numbers in $[a,b]$ (note, the set rational numbers is countable, so this enumeration is possible). The sequence $\{f_n(t_1)\}_{n=1}^{\infty}$ is a bounded sequence of real numbers (since $\{f_n\}$ is uniformly bounded) so, it has a subsequence $\{f_{n_k}(t_1)\}$ converging to a number which we call $\phi(t_1)$. It will be more convenient to represent this subsequence without subsubscripts, so we write f_k^1 for f_{n_k} and switch the index from k to n. So, the subsequence is written as $\{f_n^1(t_1)\}_{n=1}^{\infty}$. Now, the sequence $\{f_n^1(t_2)\}$ is bounded, so it has a convergent subsequence, say $\{f_n^2(t_2)\}$, with limit $\phi(t_2)$. We continue in this way obtaining a sequence of sequences $\{f_n^m(t)\}_{n=1}^{\infty}$ (one sequence for each m) each of which is a subsequence of the previous. Furthermore, we have $f_n^m(t_m) \to \phi(t_m)$ as $n \to \infty$ for each $m \in \mathbb{N}$. Now, consider the "diagonal" functions defined $F_k(t) = f_k^k(t)$. Since $f_n^m(t_m) \to \phi(t_m)$, it follows that $F_r(t_m) \to \phi(t_m)$ as $r \to \infty$ for each $m \in \mathbb{N}$ (in other words, the sequence $\{F_r(t)\}$ converges pointwise at each t_m). We now show that $\{F_k(t)\}$ converges uniformly on $[a,b]$, by showing it is Cauchy in the uniform norm. Let $\varepsilon > 0$. Let $\delta > 0$ be as in the definition of uniformly equicontinuous for $\{f_n(t)\}$ applied with $\varepsilon/3$. Divide $[a,b]$ into p intervals where $p > \frac{b-a}{\delta}$. Let ξ_j be a rational number in the jth interval, for $j = 1, \ldots, p$. Remember, $\{F_r(t)\}$ converges at each of the points ξ_j, since they are rational numbers. So, for each j, there is $M_j \in \mathbb{N}$ such that $|F_r(\xi_j) - F_s(\xi_j)| < \varepsilon/3$ whenever $r, s > M_j$. Let $M = \max\{M_j : j = 1, \ldots, p\}$. If $t \in [a,b]$, then it is in one of the p intervals, say the jth. So, $|t - \xi_j| < \delta$ and so $|f_r^r(t) - f_r^r(\xi_j)| = |F_r(t) - F_r(\xi_j)| < \varepsilon/3$ for every r. Also, if $r, s > M$, then $|F_r(\xi_j) - F_s(\xi_j)| < \varepsilon/3$ (since M is the max of the M_i's). So, we have for $r, s > M$,

$$|F_r(t) - F_s(t)| = |F_r(t) - F_r(\xi_j) + F_r(\xi_j) - F_s(\xi_j) + F_s(\xi_j) - F_s(t)|$$

$$\leq |F_r(t) - F_r(\xi_j)| + |F_r(\xi_j) - F_s(\xi_j)| + |F_s(\xi_j) - F_s(t)|$$

$$\leq \frac{\varepsilon}{3} + \frac{\varepsilon}{3} + \frac{\varepsilon}{3} = \varepsilon.$$

By the Cauchy Criterion for convergence, the sequence $\{F_r(t)\}$converges uniformly on $[a,b]$. Since each $F_r(t)$ is continuous, the limit function $\phi(t)$ is also continuous.

Remark 3.2. The Ascoli-Arzelà Theorem can be generalized to a sequence of functions from $[a,b]$ to $\mathbb{R}^n$. You apply the Ascoli-Arzelà to the first coordinate function to get a uniformly convergent subsequence. Then, apply the theorem again, this time to the corresponding subsequence of functions restricted to the second coordinate, getting a sub-subsequence, and so on.

Banach spaces form an important class of metric spaces. We now define Banach spaces in several steps.

Definition 3.2.7. A triple $(V,+,\cdot)$ is said to be a linear (or vector) space over a field F if V is a set and the following are true.

1. Properties of $+$

    a. $+$ is a function from $V \times V$ to V. Outputs are denoted $x+y$.
    b. for all $x,y \in V$, $x+y = y+x$. ($+$ is commutative)
    c. for all $x,y,w \in V$, $x+(y+w) = (x+y)+w$. ($+$ is associative)
    d. there is a unique element of V which we denote 0 such that for all $x \in V$, $0+x = x+0 = x$. (additive identity)
    e. for each $x \in V$ there is a unique element of V which we denote $-x$ such that $x+(-x) = -x+x = 0$. (additive inverse)

2. Scalar multiplication

    a. $\cdot$ is a function from $F \times V$ to V. Outputs are denoted $\alpha \cdot x$, or αx.
    b. for all $\alpha, \beta \in F$ and $x \in V$, $\alpha(\beta x) = (\alpha\beta)x$.
    c. for all $x \in V$, $1 \cdot x = x$.
    d. for all $\alpha, \beta \in F$ and $x \in V$, $(\alpha+\beta)x = \alpha x + \beta x$.
    e. for all $\alpha \in F$ and $x,y \in V$, $\alpha(x+y) = \alpha x + \alpha y$.

Commonly, the real numbers or complex numbers are the field in the above definition. For our purposes, we only consider the field of real numbers $F = \mathbb{R}$.

Definition 3.2.8 (Normed Spaces). A vector space $(V,+,\cdot)$ is a **normed space** if for each $x \in V$ there is a nonnegative real number $\|x\|$, called the **norm** of x, such that for each $x,y \in V$ and $\alpha \in \mathbb{R}$

1. $\|x\| = 0$ if and only if $x = 0$
2. $\|\alpha x\| = |\alpha|\|x\|$
3. $\|x+y\| \le \|x\| + \|y\|$

Remark 3.3. A norm on a vector space always defines a metric $\rho(x,y) = \|x-y\|$ on the vector space. Given a metric ρ defined on a vector space, it is tempting to define $\|v\| = \rho(v,0)$. But this is not always a norm.

Definition 3.2.9. A Banach space is a complete normed vector space. That is, a vector space $(X,+,\cdot)$ with norm $\|\cdot\|$ for which the metric $\rho(x,y) = \|x-y\|$ is complete.

Example 3.1. The space $(\mathbb{R}^n, +, \cdot)$ over the field $\mathbb{R}$ is a vector space (with the usual vector addition, $+$ and scalar multiplication, $\cdot$) and there are many suitable norms for it. For example, if $x = (x_1, x_2, \ldots, x_n)$, then

1. $\|x\| = \max\limits_{1 \le i \le n} |x_i|$,

2. $\|x\| = \sqrt{\sum\limits_{i=1}^{n} x_i^2}$, or

3. $\|x\| = \sum\limits_{i=1}^{n} |x_i|$

are all suitable norms. Norm 2. is the Euclidean norm: the norm of a vector is its Euclidean distance to the zero vector and the metric defined from this norm is the usual Euclidean metric. Norm 3. generates the "taxi-cab" metric on $\mathbb{R}^2$.

Remark 3.4. Consider the vector space $(\mathbb{R}^n, +, \cdot)$ as a metric space with its metric defined $\rho(x, y) = \|x - y\|$ where $\|\cdot\|$ is any of the norms as in Example 3.1. The completeness of this metric space comes directly from the completeness of $\mathbb{R}$, hence $(\mathbb{R}^n, \|\cdot\|)$ is a Banach space.

Remark 3.5. In the Euclidean space $\mathbb{R}^n$, compactness is equivalent to closed and bounded (Heine-Borel Theorem). In fact, the metrics generated from any of the norms in Example 3.1 are equivalent in the sense that they generate the same topologies. Moreover, compactness is equivalent to closed and bounded in each of those metrics.

Example 3.2. Let $C([a, b], \mathbb{R}^n)$ denote the space of all continuous functions $f : [a, b] \to \mathbb{R}^n$.

1. $C([a, b], \mathbb{R}^n)$ is a vector space over $\mathbb{R}$.
2. If $\|f\| = \max\limits_{a \le t \le b} |f(t)|$ where $|\cdot|$ is a norm on $\mathbb{R}^n$, then $(C([a, b], \mathbb{R}^n), \|\cdot\|)$ is a Banach space.
3. Let M and K be two positive constants and define

$$L = \{f \in C([a, b], \mathbb{R}^n) : \|f\| \le M; |f(u) - f(v)| \le K|u - v|\}$$

then L is compact.

Proof. (of Part 3.) Let $\{f_n\}$ be any sequence in L. The functions are uniformly bounded by M and have the same Lipschitz constant, K. So, the sequence is uniformly equicontinuous. By the Ascoli-Arzelà Theorem, there is a subsequence, $\{f_{n_k}\}$, that converges uniformly to a continuous function $f : [a, b] \to \mathbb{R}^n$. We now show that $f \in L$. Well, $|f_n(t)| \le M$ for each $t \in [a, b]$, so $|f(t)| \le M$ for each $t \in [a, b]$ and hence $\|f\| \le M$. Now, fix $u, v \in [a, b]$ and fix $\varepsilon > 0$. Since $\{f_{n_k}\}$ converges uniformly to f, there is $N \in \mathbb{N}$ such that $|f_{n_k}(t) - f(t)| < \varepsilon/2$ for all $t \in [a, b]$ and all $k \ge N$. So, fix any $k \ge N$ and we have

$$|f(u) - f(v)| = |f(u) - f_{n_k}(u) + f_{n_k}(u) - f_{n_k}(v) + f_{n_k}(v) - f(v)|$$

$$\leq |f(u) - f_{n_k}(u)| + |f_{n_k}(u) - f_{n_k}(v)| + |f_{n_k}(v) - f(v)|$$
$$< \varepsilon/2 + K|u-v| + \varepsilon/2 = K|u-v| + \varepsilon.$$

Since $\varepsilon > 0$ was arbitrary, $|f(u) - f(v)| \leq K|u-v|$. Hence $f \in L$. We have demonstrated that $\{f_n\}$ has a subsequence converging to an element of L. Hence, L is compact.

Example 3.3. Consider $\mathbb{R}$ as a vector space over $\mathbb{R}$ and define the metric $d(x,y) = \dfrac{|x-y|}{1+|x-y|}$. For each $x \in \mathbb{R}$, we can define $\|x\| = d(x,0)$. Explain why $\|\cdot\|$ is not a norm on $\mathbb{R}$.

Example 3.4. Let $\phi : [a,b] \to \mathbb{R}^n$ be continuous and let S be the set of continuous functions $f : [a,c] \to \mathbb{R}^n$ with $c > b$ and with $f(t) = \phi(t)$ for $a \leq t \leq b$. Define $\rho(f,g) = \|f - g\| = \sup_{a \leq t \leq c <} |f(t) - g(t)|$ for $f, g \in S$. Then (S, ρ) is a complete metric space but not a Banach space since $f + g$ is not in S.

Example 3.5. Let (S, ρ) be the space of continuous bounded functions $f : (-\infty, 0] \to \mathbb{R}$ with $\rho(f,g) = \|f - g\| = \sup_{-\infty < t \leq 0} |f(t) - g(t)|$.

1. Show that (S, ρ) is a Banach space.
2. The set $L = \{f \in S : \|f\| \leq 1, |f(u) - f(v)| \leq |u-v|\}$ is not compact in (S, ρ).

Proof. (of 2.) Consider the sequence of functions defined

$$f_n(t) = \begin{cases} 0 & \text{if } t \leq -n \\ \frac{t}{n} + 1 & \text{if } -n < t \leq 0 \end{cases}$$

Then, the sequence converges pointwise to $f = 1$, but $\rho(f_n, f) = 1$ for all $n \in \mathbb{N}$. So, there is no subsequence of $\{f_n\}$ converging in the norm $\|\cdot\|$ (i.e., converging uniformly) to f.

Example 3.6. Let (S, ρ) be the space of continuous functions $f : (-\infty, 0] \to \mathbb{R}^n$ with

$$\rho(f,g) = \sum_{n=1}^{\infty} 2^{-n} \rho_n(f,g) / \{1 + \rho_n(f,g)\}$$

where

$$\rho_n(f,g) = \max_{-n \leq s \leq 0} |f(s) - g(s)|$$

and $|\cdot|$ is the Euclidean norm on $\mathbb{R}^n$

1. Then (S, ρ) is a complete metric space. The distance between all function is bounded by 1.
2. $(S, +, \cdot)$ is a vector space over $\mathbb{R}$.
3. (S, ρ) is not a Banach space because ρ does not define a norm, since $\rho(x,0) = \|x\|$ does not satisfy $\|\alpha x\| = |\alpha| \|x\|$.

4. Let M and K be given positive constants. Then the set

$$L = \{f \in S : \|f\| \le M \text{ on } (-\infty, 0], |f(u) - f(v)| \le K|u - v|\}$$

is compact in (S, ρ).

Proof. (of 4.) Let $\{f_n\}$ be a sequence in L. It is clear that if $f_n \to f$ uniformly on compact subsets of $(-\infty, 0]$, then we have $\rho(f_n, f) \to 0$ as $n \to \infty$. Let's begin by considering $\{f_n\}$ on $[-1, 0]$. Then the sequence is uniformly bounded and equicontinuous and so there is a subsequence, say $\{f_n^1\}$ converging uniformly to some continuous f on $[-1, 0]$. Moreover the argument of Example 3.2 shows that $|f(t)| \le M$, and $|f(u) - f(v)| \le K|u - v|$. Next we consider $\{f_n^1\}$ on $[-2, 0]$. Then the sequence is uniformly bounded and equicontinuous and so there is a subsequence, say $\{f_n^2\}$ converging uniformly, say, to some continuous f on $[-2, 0]$. Continuing this way we arrive at $F_n = f_n^n$ which has a subsequence of $\{f_n\}$ and it converges uniformly on compact subsets of $(-\infty, 0]$ to a function $f \in L$. This proves L is compact.

The next result is stated in the form of a theorem that we leave its proof to the reader.

Theorem 3.2.2. *Let $g : (-\infty, 0] \to [1, \infty)$ be a continuous strictly decreasing function with $g(0) = 1$ and $g(r) \to \infty$ as $r \to -\infty$. Let $(S, |\cdot|_g)$ be the space of continuous functions $f : (-\infty, 0] \to \mathbb{R}^n$ for which*

$$|f|_g := \sup_{-\infty < t \le 0} \frac{|f(t)|}{|g(t)|}$$

exists. Then

1. $(S, |\cdot|_g)$ *is a Banach space.*
2. *Let M and K be given positive constants. Then the set*

$$L = \{f \in S : \|f\| \le M \text{ on } (-\infty, 0], |f(u) - f(v)| \le K|u - v|\}$$

is compact in (S, ρ).

Definition 3.2.10. Let (E, ρ) be a metric space and $D : E \to E$. The operator or mapping D is a contraction if there exists an $\alpha \in (0, 1)$ such that

$$\rho\left(D(x), D(y)\right) \le \alpha \rho(x, y).$$

Theorem 3.2.3 (Contraction Mapping Principle). *Let (E, ρ) be a complete metric space and $D : E \to E$ a contraction operator. Then there exists a unique $\phi \in E$ with $D(\phi) = \phi$. Moreover, if $\psi \in E$ and if $\{\psi_n\}$ is defined inductively by $\psi_1 = D(\psi)$ and $\psi_{n+1}D(\psi_n)$, then $\psi_n \to \phi$, the unique fixed point.*

Proof. Let $y_0 \in E$ and define a sequence $\{y_n\}$ in E by $y_1 = Dy_0, y_2 = Dy_1 = D(Dy_0) = D^2 y_0, \ldots, y_n = Dy_{n-1} = D^n y_0$. Next we show that $\{y_n\}$ is a Cauchy sequence. To see this, if $m > n$, then

$$\begin{aligned}
\rho(y_n, y_m) &= \rho(D^n y_0, D^m y_0) \\
&\le \alpha \rho(D^{n-1} y_0, D^{m-1} y_0) \\
&\ \ \vdots \\
&\le \alpha^n \rho(y_0, y_{m-1}) \\
&\le \alpha^n \{\rho(y_0, y_1) + \rho(y_1, y_2) + \ldots + \rho(y_{m-n-1}, y_{m-n})\} \\
&\le \alpha^n \{\rho(y_0, y_1) + \alpha \rho(y_0, y_1) + \ldots + \alpha^{m-n-1} \rho(y_0, y_1)\} \\
&\le \alpha^n \rho(y_0, y_1) \{1 + \alpha + \ldots + \alpha^{m-n-1}\} \\
&\le \alpha^n \rho(y_0, y_1) \frac{1}{1 - \alpha}.
\end{aligned}$$

Thus, since $\alpha \in (0, 1)$, we have that

$$\rho(y_n, y_m) \to 0, \text{ as, } n \to \infty.$$

This shows the sequence $\{y_n\}$ is Cauchy. Since (E, ρ) is a complete metric space, $\{y_n\}$ has a limit, say y in E. Since the mapping D is continuous we have that

$$D(x) = D(\lim_{n \to \infty} y_n) = \lim_{n \to \infty} D(y_n) = \lim_{n \to \infty} y_{n+1} = y,$$

and y is a fixed point. Left to show y is unique. Let $x, y \in E$ such that $D(x) = x$ and $D(y) = y$. Then

$$0 \le \rho(x, y) = \rho(D(x), D(y)) \le \alpha \rho(x, y),$$

which implies that

$$0 \le (1 - \alpha) \rho(x, y) \le 0.$$

Since $1 - \alpha \ne 0$, we must have $\rho(x, y) = 0$ and hence $x = y$. This completes the proof.

Another form of the contraction mapping principle.

Theorem 3.2.4 (Contraction Mapping Principle, Banach Fixed Point Theorem). *Let (E, ρ) be a complete metric space and $P : E \to E$ such that P^m is a contraction for some fixed positive integer m. Then there is a unique $x \in E$ with $P(x) = x$.*

3.3 Highly Nonlinear Delay Equations

We limit our study to the highly nonlinear delay difference equation, typified by

$$x(t + 1) = a(t)g(x(t - r)) \tag{3.3.1}$$

where $a : \mathbb{Z}^+ \to \mathbb{R}$ and r is a positive integer. More conditions on g are forthcoming. Results of this section can be found in [140]. In the paper of Raffoul [136], the author considered the linear difference equation

$$\triangle x(t) = -a(t)x(t-r)$$

and used fixed point theory and obtained asymptotic and periodicity results using fixed point theory. It is worth mentioning here that (3.3.1) has fundamental differ-ence from the above-mentioned equation due to the nonlinearity that the function g presents. Moreover, when inverting (3.3.1) in order to construct a mapping that is suitable for fixed point theory, one will have to introduce a linear term which results in the addition term of $x - g(x)$. Also, the results of this section offer the use of nonconventional metric in order to avoid that the contraction constant not to depend on the Lipschitz constant K that g will be required to satisfy.

First we rewrite (3.3.1) and have it ready for inversion so that fixed point theory can be used. Rewrite (3.3.1) as

$$x(t+1) = a(t+r)g(x(t)) - \triangle_t \sum_{s=t-r}^{t-1} a(s+r)g(x(s)),$$

where $\triangle_t$ represents the difference with respect to t. We must create a linear tern in x in order to be able to invert. Thus, we add and subtract $a(t+r)x(t)$ and get,

$$x(t+1) = a(t+r)x(t) - a(t+r)[x(t) - g(x(t))] - \triangle_t \sum_{s=t-r}^{t-1} a(s+r)g(x(s)). \quad (3.3.2)$$

For each $t_0 \geq 0$, equation (3.3.2) requires initial function $\psi : [t_0 - r, t_0] \to \mathbb{R}$ in order to specify a solution $x(t, t_0, \psi)$. The computation is the same for any $t_0 \geq 0$ and so we take $t_0 = 0$. Thus, we say $x(t) := x(t, 0, \psi)$ is a solution of (3.3.2) if $x(t) = \psi(t)$ on $[-r, 0]$ and $x(t)$ satisfies (3.3.2) for $t \geq 0$. We begin with the following lemma which we omit its proof.

Lemma 3.2. *Suppose that $a(t+r) \neq 0$ for all $t \in \mathbb{Z}^+$. Then $x(t)$ is a solution of equation (3.3.2) if and only if*

$$x(t) = \psi(0) \prod_{s=0}^{t-1} a(s+r) - \sum_{s=t-r}^{t-1} a(s+r)g(x(s)) + \prod_{u=0}^{t-1} a(u+r) \sum_{s=-r}^{-1} a(s+r)g(\psi(s)$$

$$+ \sum_{s=0}^{t-1} \left(a(s+r) \prod_{k=s+1}^{t-1} a(k+r) \sum_{u=s-r}^{s-1} a(u+r)g(x(u)) \right)$$

$$- \sum_{s=0}^{t-1} \left(\prod_{u=s+1}^{t-1} a(u+r) \right) a(s+r)[x(s) - g(x(s))], \ t \geq 0. \quad (3.3.3)$$

The proof of lemma 3.2 is easily obtained from the variation of parameters formula followed with summation by parts. It is assumed that the function g is continuous, locally Lipschitz with Lipschitz constant K and odd. On the other hand, we assume that $x - g(x)$ is nondecreasing and $g(x)$ is increasing on an interval $[0, L]$ for some $L > 0$. Due to these assumptions, it is obvious that the functions $g(x)$ and $x - g(x)$ are locally Lipschitz with the same Lipschitz constant $K > 0$.

Note that if $0 < L_1 < L$, then the conditions on g hold on $[-L_1, L_1]$. Also note that if $\phi : [-r, \infty) \to \mathbb{R}$ with $\phi_0 = \psi$, and if $|\phi(t)| \leq L$, then for $t \geq 0$ we have

$$|\phi(t) - g(\phi(t))| \leq L - g(L),$$

since $x - g(x)$ is odd and nondecreasing on $[0, L]$. Here $\phi_0 = \psi(s)$ for $-r \leq s \leq 0$. Let

$$S = \left\{ \phi : [-r, \infty) \to \mathbb{R} : \phi_0 = \psi, |\phi(t)| \leq L \right\}.$$

For $\phi \in S$, we define $P : S \to S$ by

$$(P\phi)(t) = \psi(t) \text{ if } -r \leq t \leq 0$$

and

$$(P\phi)(t) = \psi(0) \prod_{s=0}^{t-1} a(s+r) - \sum_{s=t-r}^{t-1} a(s+r)g(\phi(s)) + \prod_{u=0}^{t-1} a(u+r) \sum_{s=-r}^{-1} a(s+r)g(\psi(s)$$

$$+ \sum_{s=0}^{t-1} \left(a(s+r) \prod_{k=s+1}^{t-1} a(k+r) \sum_{u=s-r}^{s-1} a(u+r)g(\phi(u)) \right)$$

$$- \sum_{s=0}^{t-1} \left(\prod_{u=s+1}^{t-1} a(u+r) \right) a(s+r)[\phi(s) - g(\phi(s))], \ t \geq 0. \tag{3.3.4}$$

Let g be odd, increasing on $[0, L]$, satisfy a Lipschitz condition, and let $x - g(x)$ be nondecreasing on $[0, L]$. Suppose that if $L_1 \in (0, L]$, then

$$|L_1 - g(L_1)| \max_{t \geq 0} \sum_{s=0}^{t-1} |(\prod_{u=s+1}^{t-1} a(u+r))a(s+r)| + g(L_1) \sum_{s=t-r}^{t-1} |a(s+r)|$$

$$+ g(L_1) \max_{t \geq 0} \sum_{s=0}^{t-1} |(a(s+r) \prod_{k=s+1}^{t-1} a(k+r)| \sum_{u=s-r}^{s-1} |a(u+r)| < L_1. \tag{3.3.5}$$

We note that since $g(x)$ is Lipschitz with Lipschitz constant K and $g(0) = 0$, then $|g(x)| \leq K|x|$.

Theorem 3.3.1 ([140]). *Let g be odd, increasing on $[0, L]$, satisfy a Lipschitz condition, and let $x - g(x)$ be nondecreasing on $[0, L]$. Suppose that $a(t + r) \neq 0$ for all $t \in \mathbb{Z}^+$. If (3.3.5) hold, then every solution $x(t, 0, \psi)$ of (3.3.2) with small initial function $\psi(t)$, is bounded provided P is a contraction.*

Proof. Let $\phi \in S$. Then, by (3.3.5), there exists an $\alpha \in (0, 1)$ such that for $t \geq 0$ then

$$|(P\phi)(t)| < \|\psi\| \prod_{s=0}^{t-1} a(s+r) + |\prod_{u=0}^{t-1} a(u+r)| \|g(\psi(s))\| \sum_{s=-r}^{-1} |a(s+r)|$$

$$+ |L - g(L)| \max_{t \geq 0} \sum_{s=0}^{t-1} |(\prod_{u=s+1}^{t-1} a(u+r))a(s+r)| + g(L) \sum_{s=t-r}^{t-1} |a(s+r)|$$

$$+ g(L) \max_{t \geq 0} \sum_{s=0}^{t-1} |(a(s+r) \prod_{k=s+1}^{t-1} a(k+r)| \sum_{u=s-r}^{s-1} |a(u+r)|$$

$$\leq ||\psi|| \prod_{s=0}^{t-1} |a(s+r)| + |\prod_{u=0}^{t-1} a(u+r)| \, ||g(\psi(s))|| \sum_{s=-r}^{-1} |a(s+r)| + \alpha L$$

$$\leq \prod_{s=0}^{t-1} |a(s+r)| [||\psi|| + K||\psi||] \sum_{s=-r}^{-1} |a(s+r)| + \alpha L. \qquad (3.3.6)$$

If we choose the initial function ψ small enough so that we have

$$\prod_{s=0}^{t-1} |a(s+r)| [||\psi|| + K||\psi||] \sum_{s=-r}^{-1} |a(s+r)| < (1 - \alpha)L,$$

then this yields

$$|(P\phi)(t)| \leq (1 - \alpha)L + \alpha L = L.$$

Thus, $P : S \to S$. This shows that any solution $x(t, 0, \psi)$ of (3.3.2) that is in S, is bounded. Next we show that P defines a contraction map. Using the regular maximum norm will require that the contraction constant to depend on the Lipschitz constant K. Instead, we use the weighted norm $| \cdot |_K$ where for $\phi \in S$, we have

$$|\phi|_K = \sup_{t \geq 0} | \frac{1}{dK} \prod_{s=0}^{t-1} |a(s+r)\phi|, \quad \text{for } d > 0.$$

Proposition 3.2 ([140]). *Let g be odd, increasing on $[0, L]$, satisfy a Lipschitz condition, and let $x - g(x)$ be nondecreasing on $[0, L]$. Suppose that $a(t + r) \neq 0$ for all $t \in \mathbb{Z}^+$ with $|a(t+r)| \leq \frac{1}{2}$. Then P is a contraction with contraction constant $d > 3$.*

Proof. Let $\phi, \varphi \in S$. Then for $t \geq 0$, we have

$$|(P\phi) - (P\varphi)|_K \leq \sum_{s=t-r}^{t-1} |a(s+r)| |g(\phi(s)) - g(\varphi(s))| \frac{1}{dK} \prod_{u=0}^{t-1} |a(u+r)|$$

$$+ \sum_{s=0}^{t-1} |a(s+r) \prod_{k=s+1}^{t-1} a(k+r)| \sum_{u=s-r}^{s-1} |a(u+r)| |g(\phi(s)) - g(\varphi(s))| \frac{1}{dK} \prod_{u=0}^{t-1} |a(u+r)|$$

$$+ \sum_{s=0}^{t-1} (\prod_{u=s+1}^{t-1} |a(u+r)|) |a(s+r)| |\phi(s) - g(\phi(s))$$

$$- (\varphi(s) - g(\varphi(s))| \frac{1}{dK} \prod_{u=0}^{t-1} |a(u+r)|. \qquad (3.3.7)$$

Our aim is to simplify (3.3.7). First we remind the reader that due to the conditions on $g(x)$ and $x - g(x)$, both functions share the same Lipschitz constant K. Moreover, since $|a(t+r)| \leq \frac{1}{2}$, we have $|a(t+r)| \leq 1 - |a(t+r)|$ and $|a(t+r)|^2 \leq 1 - |a(t+r)|^2$. Next, we consider the first term of (3.3.7)

$$\sum_{s=t-r}^{t-1} |a(s+r)| \|g(\phi(s)) - g(\varphi(s))\| \frac{1}{dK} \prod_{u=0}^{t-1} |a(u+r)|$$

$$\leq \sup_{t \geq 0} \frac{K}{dK} \sum_{s=t-r}^{t-1} |a(s+r)| |\phi(s) - \varphi(s)| \prod_{u=0}^{s-1} |a(u+r)| \prod_{u=s}^{t-1} |a(u+r)|$$

$$\leq \frac{1}{d} |\phi - \varphi|_K \sup_{t \geq 0} \sum_{s=t-r}^{t-1} |a(s+r)| \prod_{u=s}^{t-1} |a(u+r)|$$

$$\leq \frac{1}{d} |\phi - \varphi|_K \sup_{t \geq 0} \sum_{s=t-r}^{t-1} |a(s+r)| \prod_{u=s+1}^{t-1} |a(u+r)|$$

$$\leq \frac{1}{d} |\phi - \varphi|_K \sup_{t \geq 0} \sum_{s=t-r}^{t-1} (1 - |a(s+r)|) \prod_{u=s+1}^{t-1} |a(u+r)|$$

$$= \frac{1}{d} |\phi - \varphi|_K \sup_{t \geq 0} \sum_{s=t-r}^{t-1} \triangle_s \left(\prod_{u=s}^{t-1} |a(u+r)| \right)$$

$$= \frac{1}{d} |\phi - \varphi|_K \sup_{t \geq 0} (1 - \prod_{u=t-r}^{t-1} |a(u+r)|)$$

$$\leq \frac{1}{d} |\phi - \varphi|_K.$$

Next we turn our attention to the second term of (3.3.7).

$$\sum_{s=0}^{t-1} |a(s+r)| \prod_{k=s+1}^{t-1} a(k+r)| \sum_{u=s-r}^{s-1} |a(u+r)| \|g(\phi(s)) - g(\varphi(s))\| \frac{1}{dK} \prod_{l=0}^{t-1} |a(l+r)|$$

$$\leq \frac{1}{d} |\phi - \varphi|_K \sup_{t \geq 0} \sum_{s=0}^{t-1} |a(s+r)| \prod_{k=s+1}^{t-1} a(k+r)| \sum_{u=s-r}^{s-1} |a(u+r)| \|g(\phi(s)) - g(\varphi(s))\| \frac{1}{dK} \prod_{l=u+1}^{t-1} |a(l+r)|$$

$$\leq \frac{1}{d} |\phi - \varphi|_K \sup_{t \geq 0} \sum_{s=0}^{t-1} |a(s+r)| \prod_{k=s+1}^{t-1} a(k+r)| \sum_{u=s-r}^{s-1} (1 - |a(u+r)|) \prod_{l=u+1}^{t-1} |a(l+r)|$$

$$= \frac{1}{d} |\phi - \varphi|_K \sup_{t \geq 0} \sum_{s=0}^{t-1} |a(s+r)| \prod_{k=s+1}^{t-1} a(k+r)| \sum_{u=s-r}^{s-1} \triangle_s \left(\prod_{l=u}^{t-1} |a(l+r)| \right)$$

$$\leq \frac{1}{d} |\phi - \varphi|_K \sup_{t \geq 0} \sum_{s=0}^{t-1} |a(s+r)| \prod_{k=s+1}^{t-1} a(k+r)| \prod_{l=s}^{t-1} |a(l+r)|$$

$$= \frac{1}{d} |\phi - \varphi|_K \sup_{t \geq 0} \sum_{s=0}^{t-1} |a(s+r)|^2 (\prod_{k=s+1}^{t-1} |a(k+r)|)^2$$

$$\leq \frac{1}{d} |\phi - \varphi|_K \sup_{t \geq 0} \sum_{s=0}^{t-1} |(1 - |a(s+r)|^2) (\prod_{k=s+1}^{t-1} |a(k+r)|)^2$$

$$\leq \frac{1}{d} |\phi - \varphi|_K \sup_{t \geq 0} \sum_{s=0}^{t-1} |\triangle_s (\prod_{k=s}^{t-1} |a(k+r)|)^2$$

$$\leq \frac{1}{d} |\phi - \varphi|_K.$$

Now we deal with the last term of (3.3.7).

$$\sum_{s=0}^{t-1}(\prod_{u=s+1}^{t-1}|a(u+r))||a(s+r)|||\phi(s)-g(\phi(s))-(\varphi(s)-g(\varphi(s)))||\frac{1}{dK}\prod_{u=0}^{t-1}|a(u+r)|$$

$$\leq \frac{1}{d}|\phi-\varphi|_K\sup_{t\geq0}\sum_{s=0}^{t-1}\prod_{u=s+1}^{t-1}|a(u+r)||a(s+r)|\prod_{u=s}^{t-1}|a(u+r)|$$

$$\leq \frac{1}{d}|\phi-\varphi|_K\sup_{t\geq0}\sum_{s=0}^{t-1}|a(s+r)|^2(\prod_{u=s+1}^{t-1}|a(u+r)|)^2$$

$$\leq \frac{1}{d}|\phi-\varphi|_K\sup_{t\geq0}\sum_{s=0}^{t-1}(1-|a(s+r)|^2)(\prod_{u=s+1}^{t-1}|a(u+r)|)^2$$

$$= \frac{1}{d}|\phi-\varphi|_K\sup_{t\geq0}\sum_{s=0}^{t-1}\triangle_s(\prod_{u=s}^{t-1}|a(u+r)|)^2$$

$$= \frac{1}{d}|\phi-\varphi|_K(1-(\prod_{u=0}^{t-1}|a(u+r)|)^2)$$

$$\leq \frac{1}{d}|\phi-\varphi|_K.$$

Substituting the above three expressions into (3.3.7) yields

$$|(P\phi)-(P\varphi)|_K \leq (\frac{1}{d}+\frac{1}{d}+\frac{1}{d})|\phi-\varphi|_K,$$

which makes P a contraction for $d > 3$. Let $(\mathscr{X},|\cdot|)$ be the Banach space of bounded sequences $\phi : [0,\infty) \to \mathbb{R}$. Since S is a subset of the Banach space $\mathscr{X}$ and S is closed and bounded so S is complete. Thus, $P : S \to S$ has a unique fixed point. This completes the proof.

We have the following corollary.

Corollary 3.1 ([140]). *Let g be odd, increasing on $[0,L]$, satisfy a Lipschitz condition, and let $x - g(x)$ be nondecreasing on $[0,L]$. Suppose that $a(t+r) \neq 0$ for all $t \in \mathbb{Z}^+$. If (3.3.5) hold with $|a(t+r)| \leq \frac{1}{2}$, then the unique solution $x(t,0,\psi)$ of (3.3.2) with small initial function $\psi(t)$ is bounded and its zero solution is stable.*

Proof. Let P be defined by (3.3.4). Then by Theorem 3.3.1, P maps S into S. Moreover, by Proposition 3.2 P is a contraction on S and hence the unique solution of (3.3.2) is bounded by Theorem 3.3.1. Left to show the zero solution is stable. Let L be given by (3.3.6) and set $0 < \varepsilon < L$. Choose $\delta = \dfrac{\varepsilon(1-\alpha)}{(1+K)\prod_{s=0}^{t-1}|a(s+r)|\sum_{s=-r}^{-1}|a(s+r)|}$.
Then for $|\psi| < \delta$, we have by (3.3.6) that

$$|(P\phi)(t)| \leq \prod_{s=0}^{t-1} |a(s+r)| \left[\|\psi\| + K\|\psi\| \right] \sum_{s=-r}^{-1} |a(s+r)| + \alpha L$$

$$\leq \delta(1+K) \prod_{s=0}^{t-1} |a(s+r)| \sum_{s=-r}^{-1} |a(s+r)| + \alpha L$$

$$\leq \delta(1+K) \prod_{s=0}^{t-1} |a(s+r)| \sum_{s=-r}^{-1} |a(s+r)| + \alpha \varepsilon$$

$$\leq \varepsilon(1-\alpha) + \alpha \varepsilon = \varepsilon.$$

Hence the zero solution is stable. This completes the proof.

We mention here that the requirement $|a(t+s)| \leq 1/2$ was necessitated by the use of the norm $|\cdot|_K$. However, in proving that P is a contraction we did not have to involve K in the contraction constant. We have the following application.

Example 3.7 ([140]). Let $a(t+r) \neq 0$ such that $|a(t+r)| \leq \frac{1}{2}$. Consider

$$x(t+1) = -a(t)x^3(t-r). \tag{3.3.8}$$

In view of (3.3.2) we have

$$x(t+1) = a(t+r)x(t) - a(t+r)[x(t) - x^3(t-r)] + \triangle_t \sum_{s=t-r}^{t-1} a(s+r)x^3(s).$$

Let $f(x) = x - x^3$. Then $f(x)$ is increasing on $(0, \frac{1}{\sqrt{3}})$ and has a maximum of $\frac{2}{3\sqrt{3}}$ at $x = \frac{1}{\sqrt{3}}$. For any bounded initial sequence ψ on $[-r, 0]$ with $|\psi(t)| \leq \frac{1}{\sqrt{3}}$ we set

$$S = \left\{ \phi : [-r, \infty) \to \mathbb{R} : \phi_0 = \psi, |\phi(t)| \leq \frac{1}{\sqrt{3}} \right\}.$$

For $\phi \in S$, we define $P : S \to S$ by

$$(P\phi)(t) = \psi(t) \text{ if } -r \leq t \leq 0,$$

and

$$(P\phi)(t) = \psi(0) \prod_{s=0}^{t-1} a(s+r) + \sum_{s=t-r}^{t-1} a(s+r)\phi^3(s) - \prod_{u=0}^{t-1} a(u+r) \sum_{s=-r}^{-1} a(s+r)\psi^3(s)$$

$$- \sum_{s=0}^{t-1} \left(a(s+r) \prod_{k=s+1}^{t-1} a(k+r) \sum_{u=s-r}^{s-1} a(u+r)\phi^3(u) \right)$$

$$- \sum_{s=0}^{t-1} \left(\prod_{u=s+1}^{t-1} a(u+r) \right) a(s+r)[\phi(s) - \phi^3(s)], \ t \geq 0.$$

Let ψ be small enough so that

$$
||\psi||\prod_{s=0}^{t-1}|a(s+r)|+\frac{\sqrt{3}}{9}\sum_{s=t-r}^{t-1}|a(s+r)|+||\psi||\prod_{u=0}^{t-1}|a(u+r)|\sum_{s=-r}^{-1}|a(s+r)|
$$

$$
+\frac{\sqrt{3}}{9}\sum_{s=0}^{t-1}\left(|a(s+r)|\prod_{k=s+1}^{t-1}|a(k+r)|\sum_{u=s-r}^{s-1}|a(u+r)|\right)
$$

$$
+\frac{2}{3\sqrt{3}}\sum_{s=0}^{t-1}(\prod_{u=s+1}^{t-1}|a(u+r)|)|a(s+r)|\leq\frac{1}{\sqrt{3}}.
$$

Then

$$
|(P\phi)(t)|\leq\frac{1}{\sqrt{3}}.
$$

Moreover, it is obvious that the Lipschitz constant $k=1$. Let d be a positive constant such that $d>3$. Using

$$
|\phi|_1=\sup_{t\geq0}|\frac{1}{d}\prod_{s=0}^{t-1}|a(s+r)\phi|,
$$

we have P is a contraction on S and hence all solutions of (3.3.8) are bounded and its zero solution is stable.

3.4 Multiple and Functional Delays

In this section, we consider the multiple and functional delays difference equation

$$
\triangle x(n)=-\sum_{j=1}^{N}a_j(n)(x(n-\tau_j(n))), \tag{3.4.1}
$$

where $a_j:\mathbb{Z}^+\to\mathbb{R}$ and $\tau_j:\mathbb{Z}^+\to\mathbb{Z}^+$ with $n-\tau(n)\to\infty$ as $n\to\infty$. For each n_0, define $m_j(n_0)=\inf\{s-\tau_j(s):s\geq n_0\},m(n_0)=\min\{m_j(n_0):1\leq j\leq N\}$. In [87], Islam and Yankson showed that the zero solution of the equation

$$
x(n+1)=b(n)x(n)+a(n)x(n-\tau(n))
$$

is asymptotically stable with one of the assumptions being that

$$
\prod_{s=0}^{n-1}b(s)\to0\text{ as }n\to\infty. \tag{3.4.2}
$$

However, as pointed out in [136], condition (3.4.2) cannot hold for (3.4.1) since $b(n) = 1$, for all $n \in \mathbb{Z}$. The results we obtain in this section overcome the requirement of (3.4.2). Let $D(n_0)$ denote the set of bounded sequences $\psi : [m(n_0), n_0] \to \mathbb{R}$ with the maximum norm $||\cdot||$. Also, let $(B, ||\cdot||)$ be the Banach space of bounded sequences $\varphi : [m(n_0), \infty) \to \mathbb{R}$ with the maximum norm. Define the inverse of $n - \tau_i(n)$ by $g_i(n)$ if it exists and the set

$$Q(n) = \sum_{j=1}^{N} b(g_j(n)),$$

where

$$\sum_{j=1}^{N} b(g_j(n)) = 1 - \sum_{j=1}^{N} a(g_j(n)).$$

For each $(n_0, \psi) \in \mathbb{Z}^+ \times D(n_0)$, a solution of (3.4.1) through (n_0, ψ) is a function $x : [m(n_0), n_0 + \alpha] \to \mathbb{R}$ for some positive constant $\alpha > 0$ such that $x(n)$ satisfies (3.4.1) on $[n_0, n_0 + \alpha]$ and $x(n) = \psi(n)$ for $n \in [m(n_0), n_0]$. We denote such a solution by $x(n) = x(n, n_0, \psi)$. For a fixed n_0, we define

$$||\psi|| = \max\{|\psi(n)| : m(n_0) \le n \le n_0\}.$$

We begin by rewriting (3.4.1) as

$$\triangle x(n) = -\sum_{j=1}^{N} a_j(g_j(n))x(n) + \triangle_n \sum_{j=1}^{N} \sum_{s=n-\tau_j(n)}^{n-1} a_j(g_j(s))x(s), \qquad (3.4.3)$$

where $\triangle_n$ represents that the difference is with respect to n. But (3.4.3) implies that

$$x(n+1) - x(n) = -\sum_{j=1}^{N} a_j(g_j(n))x(n) + \triangle_n \sum_{j=1}^{N} \sum_{s=n-\tau_j(n)}^{n-1} a_j(g_j(s))x(s)$$

$$x(n+1) = \left(1 - \sum_{j=1}^{N} a_j(g_j(n))\right)x(n) + \triangle_n \sum_{j=1}^{N} \sum_{s=n-\tau_j(n)}^{n-1} a_j(g_j(s))x(s),$$

which is equivalent to

$$x(n+1) = \sum_{j=1}^{N} b_j(g_j(n))x(n) + \triangle_n \sum_{j=1}^{N} \sum_{s=n-\tau_j(n)}^{n-1} a_j(g_j(s))x(s). \qquad (3.4.4)$$

Suppose that $Q(n) \ne 0$ for all $n \in \mathbb{Z}^+$ and the inverse function $g_j(n)$ of $n - \tau_j(n)$ exists. Then $x(n)$ is a solution of (3.4.1) if and only if

$$x(n) = \left(x(n_0) - \sum_{j=1}^{N} \sum_{s=n_0-\tau_j(n_0)}^{n_0-1} a_j(g_j(s))x(s)\right) \prod_{s=n_0}^{n-1} Q(s)$$

$$+ \sum_{j=1}^{N} \sum_{s=n-\tau_j(n)}^{n-1} a_j(g_j(s))x(s)$$

$$- \sum_{s=n_0}^{n-1} \left([1-Q(s)] \prod_{k=s+1}^{n-1} Q(s) \sum_{j=1}^{N} \sum_{u=s-\tau_j(s)}^{s-1} a_j(g_j(u))x(u)\right), \quad n \geq n_0.$$

To see this we have by the variation of parameters formula

$$x(n) = x(n_0) \prod_{s=n_0}^{n-1} Q(s)$$

$$+ \sum_{k=0}^{n-1} \left(\prod_{s=k}^{n-1} Q(s)\triangle_k \sum_{j=1}^{N} \sum_{s=k-\tau_j(k)}^{k-1} a_j(g_j(s))x(s)\right). \qquad (3.4.5)$$

Using the summation by parts formula we obtain

$$\sum_{k=0}^{n-1} \left(\prod_{s=k}^{n-1} Q(s)\triangle_k \sum_{j=1}^{N} \sum_{s=k-\tau_j(k)}^{k-1} a_j(g_j(s))x(s)\right)$$

$$= \sum_{j=1}^{N} \sum_{s=n-\tau_j(n)}^{n-1} a_j(g_j(s))x(s)$$

$$- \prod_{s=n_0}^{n-1} Q(s) \sum_{j=1}^{N} \sum_{s=n_0-\tau_j(n_0)}^{n_0-1} a_j(g_j(s))x(s)$$

$$- \sum_{s=n_0}^{n-1} \left([1-Q(s)] \prod_{k=s+1}^{n-1} Q(k) \sum_{j=1}^{N} \sum_{u=s-\tau_j(s)}^{s-1} a_j(g_j(u))x(u)\right). \qquad (3.4.6)$$

Substituting (3.4.6) into (3.4.5) gives the desired result. We have the following theorem, which is due to Yankosn [166].

Theorem 3.4.1 ([166]). *Suppose that the inverse function $g_j(n)$ of $n - \tau_j(n)$ exists, and assume there exists a constant $\alpha \in (0,1)$ such that*

$$\sum_{j=1}^{N} \sum_{s=n-\tau_j(n)}^{n-1} |a_j(g_j(s))|$$

$$+ \sum_{s=n_0}^{n-1} \left(|[1-Q(s)]| \prod_{k=s+1}^{n-1} Q(k) \left| \sum_{j=1}^{N} \sum_{u=s-\tau_j(s)}^{s-1} |a_j(g_j(u))| \right| \right) \leq \alpha. \qquad (3.4.7)$$

Moreover, assume that there exists a positive constant M such that

$$\left|\prod_{s=n_0}^{n-1} Q(s)\right| \le M.$$

Then the zero solution of (3.4.1) is stable.

Proof. Let $\varepsilon > 0$ be given. Choose $\delta > 0$ such that

$$(M + M\alpha)\delta + \alpha\varepsilon \le \varepsilon.$$

Let $\psi \in D(n_0)$ such that $|\psi(n)| \le \delta$. Define $S = \{\varphi \in B : \varphi(n) = \psi(n)$ if $n \in [m(n_0), n_0], \|\varphi\| \le \varepsilon\}$. Then $(S, \|\cdot\|)$ is a complete metric space, where $\|\cdot\|$ is the maximum norm.
Define the mapping $P : S \to S$ by

$$(P\varphi)(n) = \psi(n) \text{ for } n \in [m(n_0), n_0],$$

and

$$
\begin{aligned}
(P\varphi)(n) = {} & \left(\psi(n_0) - \sum_{j=1}^{N}\sum_{s=n_0-\tau_j(n_0)}^{n_0-1} a_j(g_j(s))\psi(s)\right)\prod_{s=n_0}^{n-1} Q(s) \\
& + \sum_{j=1}^{N}\sum_{s=n-\tau_j(n)}^{n-1} a_j(g_j(s))\varphi(s) \\
& - \sum_{s=n_0}^{n-1}\left([1-Q(s)]\prod_{k-s+1}^{n-1} Q(s)\sum_{j=1}^{N}\sum_{u=s-\tau_j(s)}^{s-1} a_j(g_j(u))\varphi(u)\right).
\end{aligned}
$$

$$(3.4.8)$$

We first show that P maps from S to S.

$$
\begin{aligned}
|(P\varphi)(n)| \le {} & M\delta + M\alpha\delta + \left\{\sum_{j=1}^{N}\sum_{s=n-\tau_j(n)}^{n-1} a_j(g_j(s))\right. \\
& \left. + \sum_{s=n_0}^{n-1}\left([1-Q(s)]\prod_{k=s+1}^{n-1} Q(k)\sum_{j=1}^{N}\sum_{u=s-\tau_j(s)}^{s-1} a_j(g_j(u))\right)\right\}\|\varphi\| \\
\le {} & (M+M\alpha)\delta + \alpha\varepsilon \\
\le {} & \varepsilon.
\end{aligned}
$$

Thus P maps from S into itself. We next show that $P\varphi$ is continuous.
Let $\varphi, \phi \in S$. Given any $\varepsilon > 0$, choose $\delta = \frac{\varepsilon}{\alpha}$ such that $\|\varphi - \phi\| < \delta$. Then,

$$\|(P\varphi) - (P\phi)\| \leq \sum_{j=1}^{N} \sum_{s=n-\tau_j(n)}^{n-1} |a_j(g_j(s))| \|\varphi - \phi\|$$

$$- \sum_{s=n_0}^{n-1} \left([1 - Q(s)] \left| \prod_{k=s+1}^{n-1} Q(s) \right| \sum_{j=1}^{N} \sum_{u=s-\tau_j(s)}^{s-1} |a_j(g_j(u))| \right)$$

$$\times \|\varphi - \phi\|$$

$$\leq \alpha \|\varphi - \phi\|$$

$$\leq \varepsilon.$$

Thus showing that $P\varphi$ is continuous. Finally we show that P is a contraction.

Let $\varphi, \eta \in S$. Then

$$|(P\varphi)(n) - (P\eta)(n)|$$

$$= \left| \left(\psi(n_0) - \sum_{j=1}^{N} \sum_{s=n_0-\tau_j(n_0)}^{n_0-1} a_j(g_j(s))\psi(s) \right) \prod_{s=n_0}^{n-1} Q(s) \right.$$

$$+ \sum_{j=1}^{N} \sum_{s=n-\tau_j(n)}^{n-1} a_j(g_j(s))\varphi(s)$$

$$- \sum_{s=n_0}^{n-1} \left([1 - Q(s)] \prod_{k=s+1}^{n-1} Q(s) \sum_{j=1}^{N} \sum_{u=s-\tau_j(s)}^{s-1} a_j(g_j(u))\varphi(u) \right) \right|$$

$$- \left(\psi(n_0) - \sum_{j=1}^{N} \sum_{s=n_0-\tau_j(n_0)}^{n_0-1} a_j(g_j(s))\psi(s) \right) \prod_{s=n_0}^{n-1} Q(s)$$

$$- \sum_{j=1}^{N} \sum_{s=n-\tau_j(n)}^{n-1} a_j(g_j(s))\eta(s)$$

$$+ \sum_{s=n_0}^{n-1} \left([1 - Q(s)] \prod_{k=s+1}^{n-1} Q(s) \sum_{j=1}^{N} \sum_{u=s-\tau_j(s)}^{s-1} a_j(g_j(u))\eta(u) \right) \right|$$

$$\leq \sum_{j=1}^{N} \sum_{s=n-\tau_j(n)}^{n-1} a_j(g_j(s)) \|\varphi - \eta\|$$

$$+ \sum_{s=n_0}^{n-1} \left(|[1 - Q(s)]| \left| \prod_{k=s+1}^{n-1} Q(s) \right| \sum_{j=1}^{N} \sum_{u=s-\tau_j(s)}^{s-1} |a_j(g_j(u))| \right) \|\varphi - \eta\|$$

$$\leq \left\{ \sum_{j=1}^{N} \sum_{s=n-\tau_j(n)}^{n-1} a_j(g_j(s)) \right.$$

$$+ \sum_{s=n_0}^{n-1} \left(|[1 - Q(s)]| \left| \prod_{k=s+1}^{n-1} Q(s) \right| \sum_{j=1}^{N} \sum_{u=s-\tau_j(s)}^{s-1} |a_j(g_j(u))| \right) \right\} \|\varphi - \eta\|$$

$$\leq \alpha \|\varphi - \eta\|.$$

This shows that P is a contraction. Thus, by the contraction mapping principle, P has a unique fixed point in S which solves (3.4.3) and for any $\varphi \in S$, $\|P\varphi\| \leq \varepsilon$. This proves that the zero solution of (3.4.3) is stable.

In the next theorem we address the asymptotic stability of the zero solution.

Theorem 3.4.2 ([166]). *Assume that the hypotheses of Theorem 3.4.1 hold. Also assume that*

$$\prod_{k=n_0}^{n-1} Q(k) \to 0 \ as \ n \to \infty. \tag{3.4.9}$$

Then the zero solution of (3.4.3) *is asymptotically stable.*

Proof. We have already proved that the zero solution of (3.4.3) is stable. Let $\psi \in D(n_0)$ such that $|\psi(n)| \leq \delta$ and define

$$S^* = \Big\{ \varphi \in B \mid \varphi(n) = \psi(n) \text{ if } n \in [m(n_0), n_0], \ \|\varphi\| \leq \varepsilon \text{ and}$$

$$\varphi(n) \to 0, \text{ as } n \to \infty \Big\}.$$

Define $P : S^* \to S^*$ by (3.4.8). From the proof of Theorem 3.4.1, the map P is a contraction and for every $\varphi \in S^*$, $\|P\varphi\| \leq \varepsilon$.

Next we show that $(P\varphi)(n) \to 0$ as $n \to \infty$. The first term on the right-hand side of (3.4.8) goes to zero because of condition (3.4.9). It is clear from (3.4.7) and the fact that $\varphi(n) \to 0$ as $n \to \infty$ that

$$\sum_{j=1}^{N} \sum_{s=n-\tau_j(n)}^{n-1} \Big| a_j(g_j(s)) \Big| |\varphi(s)| \to 0 \text{ as } n \to \infty.$$

Now we show that the last term on the right-hand side of (3.4.8) goes to zero as $n \to \infty$. Since $\varphi(n) \to 0$ and $n - \tau_j(n) \to \infty$ as $n \to \infty$, for each $\varepsilon_1 > 0$, there exists an $N_1 > n_0$ such that $s \geq N_1$ implies $|\varphi(s - \tau_j(s))| < \varepsilon_1$ for $j = 1, 2, 3, ..., N$. Thus for $n \geq N_1$, the last term I_3 in (3.4.8) satisfies

$$|I_3| = \Bigg| \sum_{s=n_0}^{n-1} \Bigg([1 - Q(s)] \prod_{k=s+1}^{n-1} Q(s) \sum_{j=1}^{N} \sum_{u=s-\tau_j(s)}^{s-1} a_j(g_j(u))\varphi(u) \Bigg) \Bigg|$$

$$\leq \sum_{s=n_0}^{N_1-1} \Bigg(|[1 - Q(s)]| \Big| \prod_{k=s+1}^{n-1} Q(s) \Big| \sum_{j=1}^{N} \sum_{u=s-\tau_j(s)}^{s-1} |a_j(g_j(u))||\varphi(u)| \Bigg)$$

$$+ \sum_{s=N_1}^{n} \Bigg(|[1 - Q(s)]| \Big| \prod_{k=s+1}^{n-1} Q(s) \Big| \sum_{j=1}^{N} \sum_{u=s-\tau_j(s)}^{s-1} |a_j(g_j(u))||\varphi(u)| \Bigg)$$

$$\leq \max_{\sigma \geq m(n_0)} |\varphi(\sigma)| \sum_{s=n_0}^{N_1-1} \Bigg(|[1 - Q(s)]| \Big| \prod_{k=s+1}^{n-1} Q(s) \Big| \sum_{j=1}^{N} \sum_{u=s-\tau_j(s)}^{s-1} |a_j(g_j(u))| \Bigg)$$

$$+ \varepsilon_1 \sum_{s=N_1}^{n} \Bigg(|[1 - Q(s)]| \Big| \prod_{k=s+1}^{n-1} Q(s) \Big| \sum_{j=1}^{N} \sum_{u=s-\tau_j(s)}^{s-1} |a_j(g_j(u))| \Bigg).$$

By (3.4.9), there exists $N_2 > N_1$ such that $n \geq N_2$ implies

$$\max_{\sigma \geq m(n_0)} |\varphi(\sigma)| \sum_{s=n_0}^{N_1-1} \left(|[1-Q(s)]| \prod_{k=s+1}^{n-1} Q(s) \Big| \sum_{j=1}^{N} \sum_{u=s-\tau_j(s)}^{s-1} |a_j(g_j(u))| < \varepsilon_1.$$

Applying (3.4.7) gives $|I_3| \leq \varepsilon_1 + \varepsilon_1 \alpha < 2\varepsilon_1$. Thus, $I_3 \to 0$ as $n \to \infty$. Hence $(P\varphi)(n) \to 0$ as $n \to \infty$, and so $P\varphi \in S^*$.

By the contraction mapping principle, P has a unique fixed point that solves (3.4.3) and goes to zero as n goes to infinity. Therefore the zero solution of (3.4.3) is asymptotically stable.

3.5 Neutral Volterra Equations

The results of this section pertain to asymptotic stability of the zero solution of the neutral type Volterra difference equation

$$x(n+1) = a(n)x(n) + c(n)\triangle x(n-g(n)) + \sum_{s=n-g(n)}^{n-1} k(n,s)h(x(s)) \qquad (3.5.1)$$

where $a, c : \mathbb{Z} \to \mathbb{R}$, $k : \mathbb{Z} \times \mathbb{Z} \to \mathbb{R}$, $h : \mathbb{Z} \to \mathbb{R}$, and $g : \mathbb{Z} \to \mathbb{Z}^+$. Throughout this section we assume that $a(n)$ and $c(n)$ are bounded whereas $0 \leq g(n) \leq g_0$ for some integer g_0. We also assume that $h(0) = 0$ and

$$|h(x) - h(z)| \leq L|x - z|.$$

For any integer $n_0 \geq 0$, we define $\mathbb{Z}_0$ to be the set of integers in $[-g_0, n_0]$. Let $\psi(n) : \mathbb{Z}_0 \to \mathbb{R}$ be an initial discrete bounded function.

Definition 3.5.1. The zero solution of (3.5.1) is Lyapunov stable if for any $\varepsilon > 0$ and any integer $n_0 \geq 0$ there exists a $\delta > 0$ such that $|\psi(n)| \leq \delta$ on $\mathbb{Z}_0$ imply $|x(n, n_0, \psi)| \leq \varepsilon$ for $n \geq n_0$.

Definition 3.5.2. The zero solution of (3.5.1) is asymptotically stable if it is Lyapunov stable and if for any integer $n_0 \geq 0$ there exists $r(n_0) > 0$ such that $|\psi(n)| \leq r(n_0)$ on $\mathbb{Z}_0$ imply $|x(n, n_0, \psi)| \to 0$ as $n \to \infty$.

Suppose that $a(n) \neq 0$ for all $n \in \mathbb{Z}$. Then $x(n)$ is a solution of the equation (3.5.1) if and only if

$$x(n) = [x(n_0) - c(n_0-1)x(n_0 - g(n_0))] \prod_{s=n_0}^{n-1} a(s) + c(n-1)x(n-g(n))$$

$$+ \sum_{r=n_0}^{n-1} [-x(r-g(r))\Phi(r) + \sum_{u=r-g(r)}^{r-1} k(r,u)h(x(u))] \prod_{s=r+1}^{n-1} a(s), n \geq n_0$$

where $\Phi(r) = c(r) - c(r-1)a(r)$.

To see this, we first note that (3.5.1) is equivalent to

$$[\triangle x(n) \prod_{s=n_0}^{n-1} a^{-1}(s)] = \left[c(n)\triangle x(n-g(n)) + \sum_{u=n-g(n)}^{n-1} k(n,u)h(x(u)) \right] \prod_{s=n_0}^{n} a^{-1}(s).$$

Summing the above equation from n_0 to $n-1$ gives

$$\sum_{r=n_0}^{n-1}[\triangle x(r)\prod_{s=n_0}^{r-1} a^{-1}(s)] = \sum_{r=n_0}^{r-1}[c(r)\triangle x(r-g(r)) + \sum_{u=r-g(r)}^{r-1} k(r,u)h(x(u))] \prod_{s=n_0}^{r} a^{-1}(s).$$

Or,

$$x(r)\prod_{s=n_0}^{r-1} a^{-1}(s)\mid_{n_0}^{n} = \sum_{r=n_0}^{n-1}[\sum_{r=n_0}^{r-1} k(r,u)h(x(u)) + c(r)\triangle x(r-g(r))]\prod_{s=n_0}^{r} a^{-1}(s).$$

Thus,

$$x(n) = x(n_0)\prod_{s=n_0}^{n-1} a(s) + \sum_{r=n_0}^{n-1}[\sum_{r=n_0}^{r-1} k(r,u)h(x(u)) + c(r)\triangle x(r-g(r))]\prod_{s=n_0}^{r} a^{-1}(s)\prod_{s=n_0}^{n-1} a(s).$$

Performing a summation by parts yields,

$$\sum_{r=n_0}^{n-1}[c(r)\triangle x(r-g(r))\prod_{s=r+1}^{n-1} a(s)] = [c(r-1)x(r-g(r))\prod_{s=r}^{n-1} a(s)\mid_{n_0}^{n}$$

$$-\sum_{r=n_0}^{n-1} x(r-g(r))\triangle[c(r-1)\prod_{s=r}^{n-1} a(s)]$$

$$= c(n-1)x(n-g(n))\prod_{s=n}^{n-1} a(s) - c(n_0-1)x(n_0-g(n_0))\prod_{s=n_0}^{n-1} a(s)$$

$$-\sum_{r=n_0}^{n-1} x(r-g(r))\triangle[c(r-1)\prod_{s=r}^{n-1} a(s)].$$

Also,

$$\sum_{r=n_0}^{n-1}[c(r)\triangle x(r-g(r))\prod_{s=r+1}^{n-1} a(s)] = c(n-1)x(n-g(n)) - c(n_0-1)x(n_0-g(n_0))\prod_{s=n_0}^{n-1} a(s)$$

$$-\sum_{r=n_0}^{n-1} x(r-g(r))\triangle[c(r-1)\prod_{s=r}^{n-1} a(s)].$$

A substitution into the above expression gives,

$$x(n) = x(n_0) \prod_{s=n_0}^{n-1} a(s) + \sum_{r=n_0}^{n-1} [\sum_{u=r-g(r)}^{r-1} k(r,u)h(x(u))] \prod_{s=r+1}^{n-1} a(s) + c(n-1)x(n-g(n))$$

$$- c(n_0-1)x(n_0-g(n_0)) \prod_{s=n_0}^{n-1} a(s) - \sum_{r=n_0}^{n-1} x(r-g(r)) \triangle [c(r-1) \prod_{s=r}^{n-1} a(s)]$$

$$= [xn_0 - c(n_0-1)x(n_0-g(n_0)] \prod_{s=r+1}^{n-1} a(s) + c(n-1)x(n-g(n))$$

$$+ \sum_{r=n_0}^{n-1} (-x(r-g(r)) \triangle [c(r-1) \prod_{s=r}^{n-1} a(s) \sum_{u=r-g(r)}^{r-1} k(r,u)h(x(u))] \prod_{s=r+1}^{n-1} a(s).$$

Combining all expressions, we arrive at

$$x(n) = [xn_0 - c(n_0-1)x(n_0-g(n_0)] \prod_{s=r+1}^{n-1} a(s) + c(n-1)x(n-g(n))$$

$$+ \sum_{r=n_0}^{n-1} [-x(r-g(r))\Phi(r) + \sum_{u=r-g(r)}^{r-1} k(r,u)h(x(u))] \prod_{s=r+1}^{n-1} a(s)$$

$$= [xn_0 - c(n_0-1)x(n_0-g(n_0)] \prod_{s=r+1}^{n-1} a(s) + c(n-1)x(n-g(n))$$

$$+ \sum_{r=n_0}^{n-1} (-x(r-g(r))\Phi(r) + \sum_{u=r-g(r)}^{r-1} k(r,u)h(x(u))] \prod_{s=r+1}^{n-1} a(s), n \geq n_0.$$

This completes the process.
Define
$$S = \{\varphi : \mathbb{Z} \to \mathbb{R} \mid \|\varphi\| \to 0 \ as \ n \to \infty\},$$
where
$$\|\varphi\| = \max\{|\varphi(n)|, \ n \geq n_0\}.$$
Then $(S, \|\cdot\|)$ is a Banach space. Let $\psi : (-\infty, n_0] \to \mathbb{R}$ be a given initial bounded sequence. Define mapping $H : S \to S$ by
$$(H\varphi)(n) = \psi(n) \text{ for } n \leq n_0,$$
and

$$(H\varphi)(n) = [\psi(n_0) - c(n_0-1)\psi(n_0-g(n_0))] \prod_{s=n_0}^{n-1} a(s) + c(n-1)\varphi(n-g(n))$$

$$+ \sum_{r=n_0}^{n-1} (-\varphi(r-g(r))\Phi(r) + \sum_{u=r-g(r)}^{r-1} k(r,u)h(\varphi(u))] \prod_{s=r+1}^{n-1} a(s), n \geq n_0.$$

$$(3.5.2)$$

It should cause no confusion to write

$$\|\psi\| = \max\{|\psi(n)|, \ n \le n_0\}.$$

We state Krasnoselskii's fixed point theorem which will be used to prove the zero solution of (3.5.1) is asymptotically stable. We emphasize that it is the only appropriate theorem to use for such equation since the inversion of a neutral equation results in two mappings.

Theorem 3.5.1 (Krasnoselskii [97]). *Let* $\mathbb{M}$ *be a closed convex nonempty subset of a Banach space* $(\mathbb{B}, \|\cdot\|)$. *Suppose that C and B map* $\mathbb{M}$ *into* $\mathbb{B}$ *such that*
    [(iii)]
 (i) C is continuous and C$\mathbb{M}$ is contained in a compact set,

 (ii) B is a contraction mapping.

(iii) $x, y \in \mathbb{M}$ implies $Cx + By \in \mathbb{M}$.

Then there exists $z \in \mathbb{M}$ with $z = Cz + Bz$.

We are now ready to prove our main results. According to Theorem 3.5.1 we need to construct two mappings, one is a contraction and the other is compact. Hence we write the mapping H that is given by (3.5.2) as

$$(H\varphi)(n) = (Q\varphi)(n) + (A\varphi)(n),$$

where $A, Q : S \to S$ are given by

$$(Q\varphi)(n) = [\psi(n_0) - c(n_0 - 1)\psi(n_0 - g(n_0))] \prod_{s=n_0}^{n-1} a(s) + c(n-1)\varphi(n - g(n)) \tag{3.5.3}$$

and

$$(A\varphi)(n) = \sum_{r=n_0}^{n-1} \left[-\varphi(r - g(r))\Psi(r) + \sum_{u=r-g(r)}^{r-1} k(r,u)h(\varphi(u)) \right] \prod_{s=r+1}^{n-1} a(s). \tag{3.5.4}$$

Theorem 3.5.2. *Assume the Lipschitz condition on h. Suppose that*

$$\prod_{s=n_0}^{n-1} a(s) \to 0 \text{ as } n \to \infty, \tag{3.5.5}$$

$$n - g(n) \to \infty \text{ as } n \to \infty, \tag{3.5.6}$$

and there exist $\alpha \in (0,1)$ such that,

$$|c(n-1)| + \sum_{r=n_0}^{n-1} \left[|\Phi(r)| + L \sum_{u=r-g(r)}^{r-1} k(r,u) \right] \left| \prod_{s=r+1}^{n-1} a(s) \right| \le \alpha, n \ge n_0. \tag{3.5.7}$$

Then the zero solution of (3.5.1) is asymptotically stable.

Proof. First we show the mapping H defined by (3.5.2) $\to 0$ as $n \to \infty$. The first term on the right of (3.5.2) goes to zero because of condition (3.5.5). The second term on the right goes to zero because of condition (3.5.6) and the fact that $\varphi \in S$. Left to show that the last term

$$\sum_{r=n_0}^{n-1} \left[(-\Phi(r)\varphi(r-g(r)) + \sum_{u=r-g(r)}^{r-1} k(r,u)h(\varphi(u))) \right] \prod_{s=r+1}^{n-1} a(s)$$

on the right of (3.5.2) goes to zero as $n \to \infty$. Let $m > 0$ such that for $\varphi \in S$, $|\varphi(n-g(n))| < \sigma$ for $\sigma > 0$. Also, since $\varphi(n-g(n)) \to 0$ as $n-g(n) \to \infty$, there exists an $n_2 > m$ such that for $n > n_2$, $|\varphi(n-g(n))| < \varepsilon_2$ for $\varepsilon_2 > 0$. Due to condition (3.5.5) there exists an $n_3 > n_2$ such that for $n > n_3$ implies that

$$\left| \prod_{s=n_2}^{n-1} a(s) \right| < \frac{\varepsilon_2}{\alpha \sigma}.$$

Thus for $n > n_3$, we have

$$\left| \sum_{r=n_0}^{n-1} [(-\varphi(r-g(r))\Phi(r) + \sum_{u=r-g(r)}^{r-1} k(r,u)h(\varphi(u))] \prod_{s=r+1}^{n-1} a(s) \right|$$

$$\leq \sum_{r=n_0}^{n-1} \left| [(-\varphi(r-g(r))\Phi(r) + \sum_{u=r-g(r)}^{r-1} k(r,u)h(\varphi(u))] \prod_{s=r+1}^{n-1} a(s) \right|$$

$$\leq \sum_{r=n_0}^{n_2-1} \left| [(\varphi(r-g(r))\Phi(r) + L \sum_{u=r-g(r)}^{r-1} k(r,u)\varphi(u)] \prod_{s=r+1}^{n-1} a(s) \right|$$

$$+ \sum_{r=n_0}^{n_2-1} \left| [(\varphi(r-g(r))\Phi(r) + L \sum_{u=r-g(r)}^{r-1} k(r,u)h(\varphi(u))] \prod_{s=r+1}^{n-1} a(s) \right|$$

$$\leq \sigma \left[\sum_{r=n_0}^{n_2-1} |\Phi(r)| + L \sum_{u=r-g(r)}^{r-1} k(r,u) \right] \left| \prod_{s=r+1}^{n-1} a(s) \right| + \varepsilon_2 \alpha$$

$$\leq \sigma \left[\sum_{r=n_0}^{n_2-1} |\Phi(r)| + L \sum_{u=r-g(r)}^{r-1} k(r,u) \right] \left| \prod_{s=r+1}^{n_2-1} a(s) \prod_{s=n_2}^{n-1} a(s) \right| + \varepsilon_2 \alpha$$

$$\leq \sigma \alpha \left| \prod_{s=n_2}^{n-1} a(s) \right| + \varepsilon_2 \alpha \leq \varepsilon_2 + \varepsilon_2 \alpha.$$

Hence, $(Q\varphi)(n) + (A\varphi)(n) : S \to S$. Next we show that Q is a contraction. Let Q be given by (3.5.3). Then $\varphi, \zeta \in S$, we have from (3.5.7) that

$$\|(Q\varphi) - (Q\zeta)\| \leq |c(n-1)| \|\varphi - \zeta\|$$
$$\leq \eta \|\varphi - \zeta\|, \text{ for some } \eta \in (0,1).$$

Now we are ready to prove the map A is compact. We note that the proof that is given in [167] for the compactness of A is not correct since our map is defined on an unbounded interval which rules out the use of Ascoli-Arzelà's theorem. First we show A is continuous. Let $\{\varphi^l\}$ be a sequence in S such that

$$\lim_{l\to\infty}||\varphi^l - \varphi|| = 0.$$

Since S is closed, we have $\varphi \in S$. Then by the definition of A

$$||A(\varphi^l) - A(\varphi)|| = \max_{n\in\mathbb{Z}}|A(\varphi^l) - A(\varphi)|.$$

Thus, for $\varphi \in S$, we have by (3.5.4) that

$$
\begin{aligned}
|(A\varphi^l)(n) - (A\varphi)(n)| &\leq \sum_{r=n_0}^{n-1}|\Phi(r)|\left|\varphi^l(r-g(r)) - \varphi(r-g(r))\right|\left|\prod_{s=r+1}^{n-1}a(s)\right| \\
&+ \sum_{r=n_0}^{n-1}\left|\sum_{u=r-g(r)}^{r-1}k(r,u)h(\varphi^l(u)) - \sum_{u=r-g(r)}^{r-1}k(r,u)h(\varphi(u))\right|\left|\prod_{s=r+1}^{n-1}a(s)\right| \\
&= \sum_{r=n_0}^{n-1}|\Phi(r)|\left|\varphi^l(r-g(r)) - \varphi(r-g(r))\right|\left|\prod_{s=r+1}^{n-1}a(s)\right| \\
&+ \sum_{r=n_0}^{n-1}\sum_{u=r-g(r)}^{r-1}|k(r,u)||(h(\varphi^l(u)) - h(\varphi(u))||\prod_{s=r+1}^{n-1}a(s)|.
\end{aligned}
$$

The continuity of φ and h along with Lebesgue dominated convergence theorem imply that

$$\lim_{l\to\infty}\max|A(\varphi^l)(n) - A(\varphi)(n)| = 0, \; n \in \mathbb{Z}.$$

This shows A is continuous. Finally, we have to show that AS is precompact. Let φ^l be a sequence in S. Then for each $n \in \mathbb{Z}$, φ^l is a bounded sequence of real numbers. This shows that $\{\varphi^l\}$ has a convergent subsequence. By the diagonal process, we can construct a convergent subsequence $\{\varphi^{l_k}\}$ of $\{\varphi^l\}$ in S. Since A is continuous, we know that $\{A\varphi^l\}$ has a convergent subsequence in AS. This means AS is precompact. This completes the proof for compactness. Left to show the zero solution is stable.

Due to condition (3.5.5) there exists a positive constant ρ such that $\left|\prod_{s=n_0}^{n-1}a(s)\right| \leq \rho$.

Let $\varepsilon > 0$ be given. Choose $\delta > 0$ such that

$$|1 - c(n_0 - 1)|\delta\rho + \alpha\varepsilon < \varepsilon.$$

Let $\psi(n)$ be any given initial function such that $|\psi(n)| < \delta$.

Define $\mathbb{M} = \{\varphi \in S : \|\varphi\| < \varepsilon\}$. Let $\varphi, \zeta \in \mathbb{M}$, then

$$
\|(Q\zeta) - (A\varphi)\| \leq \left| [\psi(n_0) - c(n_0 - 1)\psi(n_0 - g(n_0))] \prod_{s=n_0}^{n-1} a(s) \right| + |c(n-1)\zeta(n - g(n))|
$$

$$
+ \sum_{r=n_0}^{n-1} \left| \varphi(r - g(r))\Phi(r) + \sum_{r=n_0}^{n-1} k(r,u)h(\varphi(u)) \prod_{s=r+1}^{n-1} a(s) \right|
$$

$$
\leq |1 - c(n_0 - 1)|\,\delta\rho + |c(n-1)| + \sum_{r=n_0}^{n-1} \left| \Phi(r) + L \sum_{u=r-g(r)}^{r-1} k(r,u) \right| \left| \prod_{s=r+1}^{n-1} a(s) \right| \varepsilon
$$

$$
\leq |1 - c(n_0 - 1)|\,\delta\rho + \left\{ |c(n-1)| + \sum_{r=n_0}^{n-1} \left| \Phi(r) + L \sum_{u=r-g(r)}^{r-1} k(r,u) \right| \left| \prod_{s=r+1}^{n-1} a(s) \right| \right\} \varepsilon
$$

$$
\leq |1 - c(n_0 - 1)|\,\delta\rho + \alpha\varepsilon
$$

$$
\leq \varepsilon.
$$

It follows from the above work that all the conditions of the Krasnoselskii's fixed point theorem are satisfied on $\mathbb{M}$. Thus there exists a fixed point z in $\mathbb{M}$ such that $z = Az + Qz$. This completes the proof.

We end this section with the following example.

Example 3.8 ([167]). Consider the difference equation

$$
x(n+1) = \frac{1}{1+n}x(n) + \frac{2^{n+1}}{16(n+1)!} \triangle x(n-2) + \sum_{s=n-2}^{n-1} \frac{2^n}{8(1-n)!(s+2)} x(s), n \geq 0.
$$

$$(3.5.8)$$

In this example we take $n_0 = 0$. We observe that

$$
\prod_{s=0}^{n-1} \frac{1}{1+s} = \frac{1}{n!} \to 0 \ as \ n \to \infty,
$$

and hence condition (3.5.5) is satisfied. Condition (3.5.6) also satisfied since

$$
n - 2 \to \infty \ as \ n \to \infty.
$$

Next we verify condition (3.5.7).

$$
\left| \frac{2^n}{16n!} \right| + \sum_{r=0}^{n-1} \left[\frac{2^{r+1}}{16(r+1)!} + \frac{2^r}{16r!(r+1)!} \right] \prod_{s=r+1}^{n-1} \frac{1}{1+s}
$$

$$
+ \sum_{r=0}^{n-1} \sum_{u=r-2}^{r-1} \frac{2^r}{8(1-r)!(u+2)} \prod_{s=r+1}^{n-1} \frac{1}{1+s}
$$

$$
= \left| \frac{2^n}{16n!} \right| + \frac{1}{8n!} \sum_{r=0}^{n-1} 2r - \frac{1}{16n!} \sum_{r=0}^{n-1} 2r + \sum_{r=0}^{n-1} \frac{2^r}{8n!}
$$

$$\leq \left| \frac{2^n}{16n!} \right| + \frac{1}{8n!}(2^n - 1) + \frac{1}{8n!}(2^n - 1)$$

$$\leq \left| \frac{2^n}{16n!} \right| + \frac{1}{8n!}2^n + \frac{1}{8n!}2^n$$

$$\leq \frac{1}{8} + \frac{1}{4} + \frac{1}{4}$$

$$= \frac{5}{8} < 1.$$

Hence condition (3.5.7) is satisfied. All the conditions of Theorem 3.5.2 are satisfied and the zero solution of (3.5.1) is asymptotically stable.

3.6 Almost-Linear Volterra Equations

We consider the scalar Volterra difference equation

$$\triangle x(n) = a(n)h(x(n)) + \sum_{k=0}^{n-1} c(n,k)g(x(k)), \; x(0) = x_0, \; n \geq 0. \tag{3.6.1}$$

We assume that the functions h and g are continuous and that there exist positive constants $H, H^*, G,$ and G^* such that

$$| h(x) - Hx | \leq H^*, \tag{3.6.2}$$

and

$$| g(x) - Gx | \leq G^*. \tag{3.6.3}$$

Equation (3.6.1) will be called Almost-Linear if (3.6.2) and (3.6.3) hold. In [53] Burton introduced this concept of Almost-Linear equations for the continuous case and studied certain important properties of the resolvent of a linear Volterra equation. The work of this section is found in [150]. Our objective here is to apply the concept of Almost-Linear equations to Volterra difference equations and prove that the solutions of these Volterra difference equations are also bounded if they satisfy (3.6.2) and (3.6.3). Due to (3.6.2) and (3.6.3) contraction mapping principle cannot be used since our mapping cannot be made into a contraction. Therefore, we resort to the use of Krasnoselskii's fixed point theorem. At the end of the section we will construct a suitable Lyapunov functional and refer to Chapter 2 to deduce that all solutions of (3.6.1) are bounded. It turns out that either method has advantages and disadvantages.

We begin with the following lemma which is essential to the construction of our mappings. Consider the general difference equation

$$\triangle x(n) - Ha(n)x(n) = f(n), \ x(0) = x_0, \ n \geq 0. \tag{3.6.4}$$

Lemma 3.3. *Suppose* $1 + Ha(n) \neq 0$ *for all* $n \in [0, \infty) \cap \mathbb{Z}$. *Then* $x(n)$ *is a solution of equation* (3.6.4) *if and only if*

$$x(n) = x(0) \prod_{s=0}^{n-1}(1 + Ha(s)) + \sum_{u=0}^{n-1} f(u) \prod_{s=u+1}^{n-1}(1 + Ha(s)). \tag{3.6.5}$$

Proof. First we note that (3.6.4) is equivalent to

$$\triangle \left[\prod_{s=0}^{n-1}(1 + Ha(s))^{-1}x(n) \right] = f(n) \prod_{s=0}^{n}(1 + Ha(s))^{-1} \tag{3.6.6}$$

Summing equation (3.6.6) from 0 to $n-1$ and dividing both sides by

$$\prod_{s=0}^{n-1}(1 + Ha(s))^{-1}$$

gives (3.6.5).

Lemma 3.4. *Suppose* $1 + Ha(n) \neq 0$ *for all* $n \in [0, \infty) \cap \mathbb{Z}$. *Then* $x(n)$ *is a solution of equation* (3.6.1) *if and only if*

$$x(n) = x(0) \prod_{s=0}^{n-1}(1 + Ha(s)) + \sum_{u=0}^{n-1} \left[a(u) \left(-Hx(u) + h(x(u)) \right) \right] \prod_{s=u+1}^{n-1}(1 + Ha(s))$$

$$+ \sum_{u=0}^{n-1}\sum_{k=0}^{u-1} c(u,k) \left[g(x(k)) - Gx(k) \right] \prod_{s=u+1}^{n-1}(1 + Ha(s))$$

$$+ \sum_{u=0}^{n-1}\sum_{k=0}^{u-1} c(u,k)Gx(k) \prod_{s=u+1}^{n-1}(1 + Ha(s)). \tag{3.6.7}$$

Proof. Rewrite equation (3.6.1) as

$$\triangle x(n) - Ha(n)x(n) = -Ha(n)x(n) + a(n)h(x(n))$$

$$+ \sum_{k=0}^{n-1} c(n,k) \left[g(x(k)) - Gx(k) \right] + \sum_{k=0}^{n-1} c(n,k)Gx(k).$$

If we let

$$f(n) = -Ha(n)x(n) + a(n)h(x(n)) + \sum_{k=0}^{n-1} c(n,k)\Big[g(x(k)) - Gx(k)\Big]$$

$$+ \sum_{k=0}^{n-1} c(n,k)Gx(k),$$

then the results follow from Lemma 3.3.

We rely on the following theorem for the relative compactness criterion since the Ascolli-Arzelà's theorem cannot be utilized here due to the unbounded domain.

Theorem 3.6.1 ([7]). *Let M be the space of all bounded continuous (vector-valued) functions on $[0,\infty)$ and $S \subset M$. Then S is relatively compact in M if the following conditions hold:*

(i) S is bounded in M;
(ii) the functions in S are equicontinuous on any compact interval of $[0,\infty)$;
(iii) the functions in S are equiconvergent, that is, given $\varepsilon > 0$, there exists a $T = T(\varepsilon) > 0$ such that $\| \phi(t) - \phi(\infty) \|_{\mathbb{R}^n} < \varepsilon$, for all $t > T$ and all $\phi \in S$.

We assume that

$$\lim_{n \to \infty} a(n) = 0, \tag{3.6.8}$$

and for some positive constant L,

$$0 \le \sum_{k=0}^{u-1} |c(u,k)| \le L|a(u)| \text{ for all } u \in [0,\infty) \cap \mathbb{Z}, \tag{3.6.9}$$

and

$$H|a(n)| \le 1 - \big|1 + Ha(n)\big| \text{ for all } n \in [0,\infty) \cap \mathbb{Z}. \tag{3.6.10}$$

Moreover, we assume

$$\sum_{u=0}^{n-1} \Big| \prod_{s=u+1}^{n-1} (1 + Ha(s)) \Big| \sum_{k=0}^{u-1} G|c(u,k)| \le \alpha < 1, \tag{3.6.11}$$

and

$$\sum_{u=0}^{n-1} \Big| \prod_{s=u+1}^{n-1} (1 + Ha(s)) \Big| \Big[|a(u)|H^* + \sum_{k=0}^{u-1} G^*|c(u,k)| \Big] \le \beta < \infty. \tag{3.6.12}$$

Finally, choose a constant $\rho > 0$ such that

$$|x_0| \Big| \prod_{s=0}^{n-1} (1 + Ha(s)) \Big| + \alpha\rho + \beta \le \rho \tag{3.6.13}$$

for all $n \geq 0$. Let S be the Banach space of bounded sequences with the maximum norm. Let

$$M = \{\psi \in S, \ \psi(0) = x_0 \ : \ ||\psi|| \leq \rho\}. \tag{3.6.14}$$

Then M is a closed convex subset of S.
Define mappings $\mathscr{A} : M \to S$ and $\mathscr{B} : M \to M$ as follows.

$$(\mathscr{A}\phi)(n) = \sum_{u=0}^{n-1} \left[a(u)\left(-H\phi(u) + h(\phi(u)) \right) \right] \prod_{s=u+1}^{n-1} (1 + Ha(s)) \tag{3.6.15}$$

$$+ \sum_{u=0}^{n-1}\sum_{k=0}^{u-1} c(u,k)\left[g(\phi(k)) - G\phi(k) \right] \prod_{s=u+1}^{n-1} (1 + Ha(s)),$$

and

$$(\mathscr{B}\phi)(n) = x(0)\prod_{s=0}^{n-1}(1 + Ha(s))$$

$$+ \sum_{u=0}^{n-1}\sum_{k=0}^{u-1} c(u,k)G\phi(k) \prod_{s=u+1}^{n-1} (1 + Ha(s)). \tag{3.6.16}$$

We have the following lemma.

Lemma 3.5. *Suppose* (3.6.11) *and* (3.6.13) *hold. The map* $\mathscr{B}$ *is a contraction from M into M.*

Proof. Let $\phi \in M$. It follows from (3.6.11) and (3.6.13) that

$$|(\mathscr{B}\phi)(n)| \leq |x_0|\left| \prod_{s=0}^{n-1}(1 + Ha(s)) \right| + \alpha\rho \leq \rho. \tag{3.6.17}$$

Also, for $\phi, \psi \in M$, we obtain

$$|(\mathscr{B}\phi)(n) - (\mathscr{B}\psi)(n)| \leq \sum_{u=0}^{n-1}\left| \prod_{s=u+1}^{n-1} (1 + Ha(s)) \right| \sum_{k=0}^{u-1} G|c(u,k)| ||\phi - \psi||$$

$$\leq \alpha||\phi - \psi||.$$

Therefore proving that $\mathscr{B}$ is a contraction from M into M.

Lemma 3.6. *The mapping* $\mathscr{A}$ *is a continuous mapping on M.*

Proof. Let $\{\phi_n\}$ be any sequence of functions in M with $|| \phi_n - \phi || \to 0$ as $n \to \infty$. Then one can easily verify that

$$|| \mathscr{A}\phi_n - \mathscr{A}\phi || \to 0 \text{ as } n \to \infty.$$

Lemma 3.7. *Suppose* (3.6.2), (3.6.3), (3.6.8), (3.6.9), *and* (3.6.10) *hold. Then* $\mathscr{A}(M)$ *is relatively compact.*

Proof. We use Theorem 3.6.1 to prove the relative compactness of $\mathscr{A}(M)$ by showing that all three conditions of Theorem 3.6.1 hold. Thus to see that $\mathscr{A}(M)$ is uniformly bounded, we use conditions (3.6.2), (3.6.3), (3.6.9), and (3.6.10) to obtain

$$
\begin{aligned}
|(\mathscr{A}\phi)(n)| &\leq \frac{H^* + LG^*}{H} \sum_{u=0}^{n-1} H|a(u)| \Big| \prod_{s=u+1}^{n-1} (1 + Ha(s)) \Big| \\
&\leq \frac{H^* + LG^*}{H} \sum_{u=0}^{n-1} \big(1 - |1 + Ha(u)|\big) \Big| \prod_{s=u+1}^{n-1} (1 + Ha(s)) \Big| \\
&= \frac{H^* + LG^*}{H} \sum_{u=0}^{n-1} \triangle_u \Big[\prod_{s=u}^{n-1} |(1 + Ha(s))| \Big] \\
&\leq \frac{H^* + LG^*}{H} \Big[1 - \prod_{s=0}^{n-1} |(1 + Ha(s))| \Big] := \sigma \text{ for all } n \in [0, \infty) \cap \mathbb{Z}.
\end{aligned}
$$

This shows that $\mathscr{A}(M)$ is uniformly bounded.

To show equicontinuity of $\mathscr{A}(M)$, without loss of generality, we let $n_1 > n_2$ for $n_1, n_2 \in [0, \infty) \cap \mathbb{Z}$ and use the notations

$$
F(\phi(u)) = a(u)[H\phi(u) - h(\phi(u))],
$$

and

$$
J(\phi(u)) = \sum_{k=0}^{u-1} c(u,k) \Big[g(\phi(k)) - G\phi(k) \Big].
$$

Then, we may write

$$
(\mathscr{A}\phi)(n) = \sum_{u=0}^{n-1} \prod_{s=u+1}^{n-1} (1 + Ha(s)) \Big[F(\phi(u)) + J(\phi(u)) \Big]. \tag{3.6.18}
$$

Hence we have

$$
\begin{aligned}
|(\mathscr{A}\phi)(n_1) - (\mathscr{A}\phi)(n_2)| &= \Big| \sum_{u=0}^{n_1-1} \prod_{s=u+1}^{n_1-1} (1 + Ha(s)) \Big[F(\phi(u)) + J(\phi(u)) \Big] \\
&\quad - \sum_{u=0}^{n_2-1} \prod_{s=u+1}^{n_2-1} (1 + Ha(s)) \Big[F(\phi(u)) + J(\phi(u)) \Big] \Big| \\
&= \Big| \sum_{u=0}^{n_2-1} \Big[\prod_{s=u+1}^{n_1-1} (1 + Ha(s)) \\
&\quad - \prod_{s=u+1}^{n_2-1} (1 + Ha(s)) \Big] \Big[F(\phi(u)) + J(\phi(u)) \Big] \Big| \\
&\quad + \Big| \sum_{u=n_2}^{n_1-1} \prod_{s=u+1}^{n_1-1} (1 + Ha(s)) \Big[F(\phi(u)) + J(\phi(u)) \Big] \Big|
\end{aligned}
$$

$$= \sum_{u=0}^{n_2-1} \left| \prod_{s=u+1}^{n_2-1} (1+Ha(s)) \right.$$

$$- \prod_{s=u+1}^{n_1-1} (1+Ha(s)) \right| \left| F(\phi(u)) + J(\phi(u)) \right|$$

$$+ \sum_{u=n_2}^{n_1-1} \prod_{s=u+1}^{n_1-1} |(1+Ha(s))| \left| F(\phi(u)) + J(\phi(u)) \right|$$

$$\leq \sigma \sum_{u=0}^{n_2-1} H|a(u)| \left| \prod_{s=u+1}^{n_2-1} |(1+Ha(s))| - \prod_{s=u+1}^{n_1-1} |(1+Ha(s))| \right|$$

$$+ \sigma \sum_{u=n_2}^{n_1-1} H|a(u)| \prod_{s=u+1}^{n_1-1} |(1+Ha(s))|$$

$$\leq \sigma \sum_{u=0}^{n_2-1} [1 - |1+Ha(u)|] \left| \prod_{s=u+1}^{n_2-1} |(1+Ha(s))| - \prod_{s=u+1}^{n_1-1} |(1+Ha(s))| \right|$$

$$+ \sigma \sum_{u=n_2}^{n_1-1} [1 - |1+Ha(u)|] \left| \prod_{s=u+1}^{n_1-1} |(1+Ha(s))| \right|$$

$$\leq \sigma \sum_{u=0}^{n_2-1} \left| \triangle_u \left[\prod_{s=u}^{n_2-1} |(1+Ha(s))| - \prod_{s=u}^{n_1-1} |(1+Ha(s))| \right] \right|$$

$$+ \sigma \sum_{u=n_2}^{n_1-1} \triangle_u \left[\prod_{s=u}^{n_1-1} |(1+Ha(s))| \right]$$

$$\leq \sigma \left[2 - 2 \prod_{s=n_2}^{n_1-1} |(1+Ha(s))| - \prod_{s=0}^{n_2-1} |(1+Ha(s))| \right.$$

$$+ \left. \prod_{s=0}^{n_1-1} |(1+Ha(s))| \right] \to 0 \text{ as } n_2 \to n_1.$$

This shows that $\mathscr{A}$ is equicontinuous.
To see that $\mathscr{A}$ is equiconvergent, we let

$$\lim_{n \to \infty} \sum_{u=0}^{n-1} \prod_{s=u}^{n-1} (1+Ha(s)) \left[F(\phi(u)) + J(\phi(u)) \right] =$$

$$\sum_{u=0}^{\infty} \prod_{s=u}^{\infty} (1+Ha(s)) \left[F(\phi(u)) + J(\phi(u)) \right].$$

Then we have

$$
\begin{aligned}
|(\mathscr{A}\phi)(\infty) - (\mathscr{A}\phi)(n)| &= \Big| \sum_{u=0}^{\infty} \prod_{s=u+1}^{\infty} (1+Ha(s)) \Big[F(\phi(u)) + J(\phi(u)) \Big] \\
&\quad - \sum_{u=0}^{n-1} \prod_{s=u+1}^{n-1} (1+Ha(s)) \Big[F(\phi(u)) + J(\phi(u)) \Big] \Big| \\
&= \Big| \sum_{u=0}^{n-1} \Big[\prod_{s=u+1}^{\infty} (1+Ha(s)) \\
&\quad - \prod_{s=u+1}^{n-1} (1+Ha(s)) \Big] \Big[F(\phi(u)) + J(\phi(u)) \Big] \Big| \\
&\quad + \Big| \sum_{u=n}^{\infty} \prod_{s=u+1}^{\infty} (1+Ha(s)) \Big[F(\phi(u)) + J(\phi(u)) \Big] \Big| \\
&= \sum_{u=0}^{n-1} \Big| \prod_{s=u+1}^{n-1} (1+Ha(s)) \\
&\quad - \prod_{s=u+1}^{\infty} (1+Ha(s)) \Big| \Big| F(\phi(u)) + J(\phi(u)) \Big| \\
&\quad + \sigma \sum_{u=n}^{\infty} \triangle_u \Big[\prod_{s=u}^{\infty} |(1+Ha(s))| \Big] \\
&\le \sigma \sum_{u=0}^{n-1} \Big| \triangle_u \Big[\prod_{s=u}^{n-1} |(1+Ha(s))| - \prod_{s=u}^{\infty} |(1+Ha(s))| \Big] \Big| \\
&\quad + \sigma [1 - \prod_{s=n}^{\infty} |(1+Ha(s))|] \\
&\le \sigma \Big[2 - 2 \prod_{s=n}^{\infty} |(1+Ha(s))| - \prod_{s=0}^{n-1} |(1+Ha(s))| \\
&\quad + \prod_{s=0}^{\infty} |(1+Ha(s))| \Big] \to 0 \text{ as } n \to \infty,
\end{aligned}
$$

where we used (3.6.8) which yields $\lim_{n\to\infty} \prod_{s=n}^{\infty} (1+Ha(s)) = 1$.

Theorem 3.6.2. *Assume* (3.6.2), (3.6.3), *and* (3.6.8)–(3.6.13) *hold. Then* (3.6.1) *has a bounded solution.*

Proof. For $\phi, \psi \in M$, we obtain

$$
|(\mathscr{A}\phi)(n) + (\mathscr{B}\psi)(n)| \le |x_0| \Big| \prod_{s=0}^{n-1} (1+Ha(s)) \Big| + \alpha\rho + \beta \le \rho.
$$

Thus, $\mathscr{A}\phi + \mathscr{B}\psi \in M$. Moreover, Lemmas 3.5–3.7 satisfy the requirements of Krasnoselskii's fixed point theorem and hence there exists a function $x(n) \in M$ such that

$$x(n) = \mathscr{A}x(n) + \mathscr{B}x(n).$$

This proves that (3.6.1) has a bounded solution $x(n)$.

3.6.1 Application to Nonlinear Volterra Difference Equations

Consider the Volterra difference equation

$$\triangle x(n) = -\frac{1}{2^n}h(x(n)) + \sum_{k=0}^{n-1} \frac{4^k}{4(2^n)n!}g(x(k)), \ x(0) = x_0, \ n \geq 0, \quad (3.6.19)$$

where the functions h and g satisfy conditions (3.6.2) and (3.6.3), respectively. Let H, G, H^*, and G^* be positive constants with $G < 1$ and $H = 1$. We choose $\rho > 0$ such that for any initial point x_0, the inequality

$$\left| x_0 \right| \left| \prod_{s=0}^{n-1}(1 - 2^{-s}) \right| + G\rho + (H^* + G^*) \leq \rho$$

holds. Then (3.6.19) has a bounded solution $x(n)$ satisfying $||x|| \leq \rho$.
We let $a(n) = -\frac{1}{2^n}$ and $c(n,k) = \frac{4^k}{4(2^n)n!}$.
Thus,

$$\sum_{u=0}^{n-1}|c(n,u)| = \sum_{u=0}^{n-1}\frac{4^u}{4(2^n)n!}$$
$$\leq \frac{1}{4(2^n)n!}\left(4^n - 1\right)$$
$$\leq \frac{1}{2^n}.$$

This shows that condition (3.6.9) is satisfied with $L = 1$. Condition (3.6.8) can be easily verified. Moreover,

$$H|a(n)| = 2^{-n} = 1 - (1 - 2^{-n}) \leq 1 - |1 + Ha(n)|,$$

thus showing that condition (3.6.10) is satisfied. Next, we verify (3.6.11) as follows.

$$\sum_{u=0}^{n-1}|\prod_{s=u+1}^{n-1}(1 - 2^{-s})|G\sum_{k=0}^{u-1}\frac{4^k}{4(2^n)n!}$$
$$\leq G\sum_{u=0}^{n-1}\frac{1}{2^u} = G(1 - \frac{1}{2^n})$$
$$\leq G < 1.$$

Finally, we verify (3.6.12).

$$\sum_{u=0}^{n-1} \Big| \prod_{s=u+1}^{n-1} (1-2^{-s}) \Big| \Big[2^{-u}H^* + \sum_{k=0}^{u-1} G^* \frac{4^k}{4(2^n)n!} \Big]$$

$$\leq \sum_{u=0}^{n-1} \Big[2^{-u}H^* + G^* \frac{1}{2^u} \Big]$$

$$= (H^* + G^*) \sum_{u=0}^{n-1} \frac{1}{2^u}$$

$$\leq (H^* + G^*)(1 - \frac{1}{2^n}) < (H^* + G^*).$$

Thus, by Theorem 3.6.2, Equation (3.6.19) has a bounded solution.

3.7 Lyapunov Functionals or Fixed Points

In this section, we construct a Lyapunov functional and then refer to Theorem 2.1.1 to deduce boundedness on all solutions of (3.6.1). Then we will compare the results via an example with Theorem 3.6.2. First we rewrite (3.6.1) as

$$x(n+1) = b(n)h(x(n)) + \sum_{s=0}^{n-1} C(n,s)g(x(s)), \; x(0) = x_0, \; n \geq 0, \qquad (3.7.1)$$

where $b(n) = 1 - a(n)$. Before we state the next theorem we note that as a consequence of (3.6.2) and (3.6.3) we have, respectively, that

$$| h(x) | \leq H | x | + H^*, \qquad (3.7.2)$$

and

$$|g(x)| \leq G | x | + G^*. \qquad (3.7.3)$$

Theorem 3.7.1. *Suppose (3.7.2) and (3.7.3) hold and for some $\alpha \in (0,1)$, we have that*

$$H|b(n)| + G \sum_{j=n+1}^{\infty} |C(j,n)| - 1 \leq -\alpha. \qquad (3.7.4)$$

Also, assume that

$$\sum_{s=0}^{n} \sum_{j=n}^{\infty} |C(j,s)| < \infty \qquad (3.7.5)$$

and

$$\triangle_s |C(j,s)| \geq 0 \qquad (3.7.6)$$

then solutions of (3.7.1) are bounded.

Proof. Define

$$V(n,x(\cdot)) = |x(n)| + \sum_{s=0}^{n-1} \sum_{j=n}^{\infty} |C(j,s)||g(x(s))|. \qquad (3.7.7)$$

Then along solutions of (3.7.1), we have

$$\triangle V(n,x(\cdot)) = |x(n+1)| - |x(n)| + \sum_{s=0}^{n} \sum_{j=n+1}^{\infty} |C(j,s)||g(x(s))|$$

$$- \sum_{s=0}^{n-1} \sum_{j=n}^{\infty} |C(j,s)||g(x(s))|$$

$$= \left| b(n)h(x(n)) + \sum_{s=0}^{n-1} C(n,s)g(x(s)) \right|$$

$$- |x(n)| + \sum_{s=0}^{n} \sum_{j=n+1}^{\infty} |C(j,s)||g(x(s))|$$

$$- \sum_{s=0}^{n-1} \sum_{j=n}^{\infty} |C(j,s)||g(x(s))|$$

$$\leq \left[H|b(n)| + G \sum_{j=n+1}^{\infty} |C(j,n)| - 1 \right] |x(n)| + M$$

$$\leq -\alpha|x(n)| + M,$$

where $M = H^*|b(n)| + G^* \sum_{j=n+1}^{\infty} |C(j,n)|$.

Let $\varphi(n,s) = \sum_{j=n}^{\infty} |C(j,s)|$. Then, all the conditions of Theorem 2.1.1 are satisfied which implies that all solutions of (3.7.1) are bounded.

We note that Theorem 2.1.1 gives conditions under which all solutions of (3.7.1) are bounded, unlike Theorem 3.6.2 from which one can only conclude the existence of a bounded solution.

Next, we use the results of Section 3.6.1 to compare the conditions of Theorem 2.1.1 to those of Theorem 3.6.2. Let $a(n), G$, and H be given as in Theorem 3.7.1 and consider condition (3.7.4) for $n \geq 0$. Then,

$$H|b(n)| + G \sum_{j=n+1}^{\infty} |C(j,n)| - 1 = |1 - \frac{1}{2^n}| - 1 + \sum_{j=n+1}^{\infty} \frac{4^n}{4(2^j)j!}$$

$$= -\frac{1}{2^n} + 4^{n-1} \Big[\sum_{n=0}^{\infty} \frac{1}{(2^n)n!} - \sum_{j=0}^{n} \frac{1}{(2^j)j!} \Big]$$

$$= -\frac{1}{2^n} + 4^{n-1} \Big[\sqrt{e} - \sum_{j=0}^{n} \frac{1}{(2^j)j!} \Big]. \qquad (3.7.8)$$

Next we perform the following calculations by using $n! > 2^n$ for $n \geq 4$.

$$-\sum_{j=0}^{n} \frac{1}{(2^j)j!} = -\frac{3}{2} - \frac{1}{8} - \frac{1}{48} - \sum_{j=4}^{n} \frac{1}{(2^j)j!}$$

$$\geq -\frac{3}{2} - \frac{1}{8} - \frac{1}{48} - \sum_{j=4}^{n} \frac{1}{(4^j)}$$

$$= -\frac{3}{2} - \frac{1}{8} - \frac{1}{48} + \frac{1}{4} + \frac{1}{4^2} + \frac{1}{4^3} - \sum_{j=1}^{n} \frac{1}{(4^j)}$$

$$= -\frac{78}{48} + \frac{21}{4^3} - \frac{1}{4} \Big(\frac{1 - (1/4)^n}{1 - 1/4} \Big). \qquad (3.7.9)$$

Thus, substitution of (3.7.9) into (3.7.8) yields,

$$H|b(n)| + G \sum_{j=n+1}^{\infty} |C(j,n)| - 1 \geq -\frac{1}{2^n} + 4^{n-1} \Big[\sqrt{e} - \frac{78}{48} + \frac{21}{4^3} - \frac{1}{4} \Big(\frac{1 - (1/4)^n}{1 - 1/4} \Big) \Big]$$

$$> 0, \text{ for } n = 3.$$

This shows that condition (3.7.4) does not hold for all $n \geq 0$. Hence, Theorem 2.1.1 gives no information regarding the solutions and yet Theorem 3.6.2 implies the existence of at least one bounded solution.

3.8 Delay Functional Difference Equations

We consider a functional infinite delay difference equation and use fixed point theory to obtain necessary and sufficient conditions for the asymptotic stability of its zero solution. We will apply the results to nonlinear Volterra difference equations. Let $\mathbb{R} = (-\infty, \infty)$, $\mathbb{Z}^+ = [0, \infty)$ and $\mathbb{Z}^- = (-\infty, 0]$, respectively. We concentrate on the delay functional difference equation

$$x(t+1) = a(t)x(t) + g(t, x_t), \qquad (3.8.1)$$

where $a : \mathbb{Z}^+ \to \mathbb{R}$, and $g : \mathbb{Z}^+ \times \mathscr{C}$, is continuous with $\mathscr{C}$ being the Banach space of bounded functions $\phi : \mathbb{Z}^- \to \mathbb{R}$ with the maximum norm $||\cdot||$. If $x_t \in \mathscr{C}$, then $x_t(s) = x(t+s)$ for $s \in \mathbb{Z}^-$.

We will use fixed point theory to obtain necessary and sufficient conditions for the asymptotic stability of the zero solution of (3.8.1). Throughout this section we assume $g(t,0) = 0$ so that $x = 0$ is a solution of (3.8.1). For every positive $\beta > 0$, we define the set

$$\mathscr{C}(\beta) = \{\phi \in \mathscr{C} : ||\phi|| \le \beta\}.$$

Given a function $\psi : \mathbb{Z} \to \mathbb{Z}$, we define $||\psi||^{[s,t]} = \max\{|\psi(u)| : s \le u \le t\}$. Moreover, for $D > 0$ a sequence $x : (-\infty, D] \to \mathbb{R}$ is called a solution of (3.8.1) through $(t_0, \phi) \in \mathbb{Z}^+ \times \mathscr{C}$ if $x_{t_0} = \phi$ and x satisfies (3.8.1) on $[t_0, D]$. Due to the importance of the next result, we summarize it in the following lemma.

Lemma 3.8. *Suppose that $a(t) \ne 0$ for all $t \in \mathbb{Z}^+$. Then $x(t)$ is a solution of equation (3.8.1) if and only if*

$$x(t) = \phi(t_0) \prod_{s=t_0}^{t-1} a(s) + \sum_{s=t_0}^{t-1} \prod_{u=s+1}^{t-1} a(u)\, g(s, x_s)\ for\ t \ge t_0. \qquad (3.8.2)$$

The proof of lemma 3.8 follows easily from the variation of parameters formula given in Chapter 1, and hence we omit.

In the preparation for our next theorem we let $L > 0$ be a constant, $\delta_0 \ge 0$ and $t_0 \ge 0$ Let $\phi \in \mathscr{C}(\delta_0)$ be fixed and set

$$S = \left\{x : \mathbb{Z} \to \mathbb{R} : x_{t_0} = \phi,\ x_t \in \mathscr{C}(L)\ \text{for}\ t \ge t_0, x(t) \to 0,\ \text{as}\ t \to \infty\right\}.$$

Then, S is a complete metric space with metric

$$\rho(x,y) = \max_{t \ge t_0} |x(t) - y(t)|.$$

Define the mapping $P : S \to S$ by

$$(Px)(t) = \phi(t)\ \text{if}\ t \le t_0$$

and

$$(Px)(t) = \phi(t_0) \prod_{s=t_0}^{t-1} a(s) + \sum_{s=t_0}^{t-1} \prod_{u=s+1}^{t-1} a(u)\, g(s, x_s)\ \text{for}\ t \ge t_0.$$

It is clear that for $\varphi \in S$, $P\varphi$ is continuous.

Theorem 3.8.1 ([146]). *Assume the existence of positive constants* α, L, *and a sequence* $b : \mathbb{Z}^+ \to [0, \infty)$ *such that the following conditions hold:*

(i) $a(t) \neq 0$ *for all* $t \in \mathbb{Z}^+$.

(ii) $\sum_{s=0}^{t-1} \Big| \prod_{u=s+1}^{t-1} a(u) \Big| b(s) \leq \alpha < 1$ *for all* $t \in \mathbb{Z}^+$.

(iii) $|g(t, \phi) - g(t, \psi)| \leq b(t) \|\phi - \psi\|$ *for all* $\phi, \psi \in \mathscr{C}(L)$.

(iv) For each $\varepsilon > 0$ *and* $t_1 \geq 0$, *there exists a* $t_2 > t_1$ *such that for* $t > t_2, x_t \in \mathscr{C}(L)$
imply

$$|g(t, x_t)| \leq b(t) \Big(\varepsilon + \|x\|^{[t_1, t-1]} \Big).$$

Then the zero solution of (3.8.1) *is asymptotically stable if and only if*

(v) $\Big| \prod_{s=0}^{t-1} a(s) \Big| \to 0$ *as* $t \to \infty$.

Proof. Suppose *(v)* hold and let $K = \max_{t \geq t_0} |\prod_{s=t_0}^{t-1} a(s)|$. Then $K > 0$ due to (i). Choose $\delta_0 > 0$ such that $\delta_0 K + \alpha L \leq L$. Then for $x(t) \in S$ and for fixed $\phi \in \mathscr{C}(\delta_0)$ we have

$$|(Px)(t)| \leq |\phi(t_0)| \, |\prod_{s=t_0}^{t-1} a(s)| + \sum_{s=t_0}^{t-1} | \prod_{u=s+1}^{t-1} a(u)| b(s) \|x_s\|$$

$$\leq \delta_0 K + \alpha L \leq L, \text{ for } t \geq t_0.$$

Hence, $(Px) \in \mathscr{C}(L)$. Next we show that $(Px)(t) \to 0$ as $t \to \infty$. Let $x \in S$. As a consequence of $x(t) \to 0$ as $t \to \infty$, there exists $t_1 > t_0$ such that $|x(t)| < \varepsilon$ for all $t \geq t_1$. Moreover, since $|x(t)| \leq L$, for all $t \in \mathbb{Z}$, by *(iv)* there is a $t_2 > t_1$ such that for $t > t_2$ we have

$$|g(t, x_t)| \leq b(t) \Big(\varepsilon + \|x\|^{[t_1, t-1]} \Big).$$

Thus, for $t \geq t_2$, we have

$$\Big| \sum_{s=t_0}^{t-1} \prod_{u=s+1}^{t-1} a(u) \, g(s, x_s) \Big| \leq \sum_{s=t_0}^{t_2-1} | \prod_{u=s+1}^{t-1} a(u)| \, |g(s, x_s)|$$

$$+ \sum_{s=t_2}^{t-1} | \prod_{u=s+1}^{t-1} a(u)| \, |g(s, x_s)|$$

$$\leq \sum_{s=t_0}^{t_2-1} | \prod_{u=s+1}^{t-1} a(u)| \, \|x_s\|$$

$$+ \sum_{s=t_2}^{t-1} | \prod_{u=s+1}^{t-1} a(u)| \, b(s) \Big(\varepsilon + \|x\|^{[t_1, s-1]} \Big)$$

$$\leq \sum_{s=t_0}^{t_2-1} |\prod_{u=s+1}^{t_2-1} a(u)| |\prod_{u=t_2}^{t-1} a(u)| \,||x_s|| + 2\alpha\varepsilon$$

$$\leq \alpha L |\prod_{u=t_2}^{t-1} a(u)| + 2\alpha\varepsilon.$$

By *(v)*, there exists $t_3 > t_2$ such that

$$\delta_0 |\prod_{u=s+1}^{t-1} a(u)| + L |\prod_{u=t_2}^{t-1} a(u)| < \varepsilon.$$

Thus, for $t \geq t_3$, we have

$$|(Px)(t)| \leq \delta_0 |\prod_{u=s+1}^{t-1} a(u)| + \alpha L |\prod_{u=t_2}^{t-1} a(u)| + 2\alpha\varepsilon < 3\varepsilon.$$

Hence, $(Px)(t) \to 0$ as $t \to \infty$. Left to show that $(P\varphi)(t)$ is a contraction under the maximum norm. Let $\zeta, \eta \in S$. Then

$$\left|(P\zeta)(t) - (P\eta)(t)\right| \leq \sum_{s=t_0}^{t-1} |\prod_{u=s+1}^{t-1} a(u)| \,|g(s,\zeta_s) - g(s,\eta_s)|$$

$$\leq \sum_{s=t_0}^{t-1} |\prod_{u=s+1}^{t-1} a(u)| b(s) \,|\zeta_s - \eta_s|$$

$$\leq \alpha\rho(\zeta,\eta).$$

Or,

$$\rho(P\zeta, P\eta) \leq \alpha\rho(\zeta,\eta).$$

Thus, by the contraction mapping principle P has a unique fixed point in S which solves (3.8.1) with $\phi \in \mathscr{C}(\delta_0)$ and $x(t) = x(t, t_0, \phi) \to 0$ as $t \to \infty$. We are left with showing that the zero solution of (3.8.1) is stable. Let $\varepsilon > 0, \varepsilon < L$ be given and chose $0 < \delta < \varepsilon$ so that $\delta K + \alpha\varepsilon < \varepsilon$. By the choice of δ we have $|x(t_0)| < \varepsilon$. Let $t^* \geq t_0 + 1$ be such that $|x(t^*)| \geq \varepsilon$ and $|x(s)| < \varepsilon$ for $t_0 \leq s \leq t^* - 1$. If $x(t) = x(t, t_0, \phi)$ is a solution for (3.8.1) with $||\phi|| < \delta$, then

$$|x(t^*)| \leq \delta |\prod_{s=t_0}^{t^*-1} a(s)| + \sum_{s=t_0}^{t^*-1} |\prod_{u=s+1}^{t^*-1} a(u)| b(s)||x_s||$$

$$\leq \delta K + \alpha\varepsilon < \varepsilon,$$

which contradict the definition of t^*. Thus $|x(t)| < \varepsilon$ for all $t \geq t_0$ and hence the zero solution of (3.8.1) is asymptotically stable.

Conversely, suppose *(v)* does not hold. Then by *(i)* there exists a sequence $\{t_n\}$ such that for positive constant q,

$$\left(\left| \prod_{u=0}^{t_n-1} a(u) \right| \right)^{-1} = q, \text{ for } n = 1, 2, 3, \cdots .$$

Now by *(ii)* we have that

$$\sum_{s=0}^{t_n-1} \left| \prod_{u=s+1}^{t_n-1} a(u) \right| b(s) \leq \alpha,$$

from which we get that

$$\left(\left| \prod_{u=0}^{t_n-1} a(u) \right| \right)^{-1} \sum_{s=0}^{t_n-1} \left| \prod_{u=s+1}^{t_n-1} a(u) \right| b(s) \leq \alpha \left(\left| \prod_{u=0}^{t_n-1} a(u) \right| \right)^{-1}.$$

This simplifies to

$$\sum_{s=0}^{t_n-1} \left(\left| \prod_{u=0}^{s} a(u) \right| \right)^{-1} b(s) \leq \alpha q.$$

Thus the sequence $\left\{ \sum_{s=0}^{t_n-1} \left(\left| \prod_{u=0}^{s} a(u) \right| \right)^{-1} b(s) \right\}$ is bounded and hence there is a convergent subsequence. Thus, for the sake of keeping a simple notation we may assume

$$\lim_{n\to\infty} \sum_{s=0}^{t_n-1} \left(\left| \prod_{u=0}^{s} a(u) \right| \right)^{-1} b(s) = \omega$$

for some positive constant ω. Next we may choose a positive integer $\tilde{n}$ large enough so that

$$\sum_{s=t_{\tilde{n}}}^{t_n-1} \left(\left| \prod_{u=0}^{s} a(u) \right| \right)^{-1} b(s) < \frac{1-\alpha}{2K^2}$$

for all $n \geq \tilde{n}$.

Consider the solution $x(t, t_{\tilde{n}}, \phi)$ with $\phi(s) = \delta_0$ for $s \leq \tilde{n}$. Then, $|x(t)| \leq L$ for all $n \geq \tilde{n}$ and

$$|x(t)| \leq \delta_0 \left| \prod_{s=t_{\tilde{n}}}^{t-1} a(s) \right| + \sum_{s=t_{\tilde{n}}}^{t-1} \left| \prod_{u=s+1}^{t-1} a(u) \right| b(s) \|x_s\|$$

$$\leq \delta_0 K + \alpha \|x_t\|.$$

This implies

$$|x(t)| \leq \frac{\delta_0 K}{1-\alpha}, \text{ for all } t \geq t_{\tilde{n}}.$$

On the other hand, for $n \geq \widetilde{n}$, we also have

$$
|x(t)| \geq \delta_0 \, | \prod_{s=t_{\widetilde{n}}}^{t_n-1} a(s)| - \sum_{s=t_{\widetilde{n}}}^{t-1} | \prod_{u=s+1}^{t_n-1} a(u)| \, |b(s)| \, \|x_s\|
$$

$$
\geq \delta_0 \, | \prod_{s=t_{\widetilde{n}}}^{t_n-1} a(s)| - \frac{\delta_0 K}{1-\alpha} | \prod_{u=0}^{t_n-1} a(u)| \sum_{s=t_{\widetilde{n}}}^{t-1} | \Big(\prod_{u=0}^{s} a(u)| \Big)^{-1} b(s)
$$

$$
= \delta_0 \, | \prod_{s=t_{\widetilde{n}}}^{t_n-1} a(s)| - \frac{\delta_0 K}{1-\alpha} | \prod_{u=0}^{t_{\widetilde{n}}-1} a(s)| \prod_{u=t_{\widetilde{n}}}^{t_n-1} a(s)| \sum_{s=t_{\widetilde{n}}}^{t-1} | \Big(\prod_{u=0}^{s} a(u)| \Big)^{-1} b(s)
$$

$$
\geq | \prod_{s=t_{\widetilde{n}}}^{t_n-1} a(s)| \Big(\delta_0 - \frac{\delta_0 K}{1-\alpha} K \sum_{s=t_{\widetilde{n}}}^{t-1} \big(| \prod_{u=0}^{s} a(u)| \big)^{-1} b(s) \Big)
$$

$$
\geq | \prod_{s=t_{\widetilde{n}}}^{t_n-1} a(s)| (\delta_0 - \frac{\delta_0 K}{1-\alpha} K \frac{1-\alpha}{2K^2}) = \frac{\delta_0}{2} | \prod_{s=t_{\widetilde{n}}}^{t_n-1} a(s)|
$$

$$
= \frac{\delta_0}{2} | \prod_{u=0}^{t_n-1} a(s)| \Big(| \prod_{u=0}^{t_{\widetilde{n}}-1} a(s)| \Big)^{-1} \to \frac{\delta_0}{2} q/q \neq 0 \text{ as } n \to \infty.
$$

Hence, condition (v) is necessary. This completes the proof.

Now we apply the results of Theorem 3.8.1 to the nonlinear Volterra infinite delay equation

$$
x(t+1) = a(t)x(t) + \sum_{s=-\infty}^{t-1} G(t,s,x(s)) \tag{3.8.3}
$$

where $a : \mathbb{Z}^+ \to \mathbb{R}$ and $G : \Omega \times \mathbb{R} \to \mathbb{R}, \Omega = \{(t,s) \in \mathbb{Z}^2 : t \geq s\}$ and G is continuous in x. The next theorem gives necessary and sufficient conditions for the stability of the zero solution of (3.8.3).

Theorem 3.8.2. *Assume the existence of positive constants α, L, and a sequence $p : \Omega \to \mathbb{R}^+$ such that the following conditions hold:*

(I) $a(t) \neq 0$ for all $t \in \mathbb{Z}^+$,

(II) $\max\limits_{t \in \mathbb{Z}^+} \sum\limits_{s=0}^{t-1} | \prod\limits_{u=s+1}^{t-1} a(u)| \sum\limits_{\tau=0}^{s-1} p(s,\tau) \leq \alpha < 1$ for all $t \in \mathbb{Z}^+$,

(III) If $|x|, |y| \leq L$, then

$$
|G(t,s,x) - G(t,s,y)| \leq p(t,s)|x-y|
$$

and $G(t,s,0) = 0$ for all $(t,s) \in \Omega$,

(IV) For each $\varepsilon > 0$ and $t_1 \geq 0$, there exists a $t_2 > t_1$ such that for $t \geq t_2$, implies

$$\sum_{s=-\infty}^{t_1-1} p(t,s) \leq \varepsilon \sum_{s=-\infty}^{t-1} p(t,s).$$

Then the zero solution of (3.8.3) *is asymptotically stable if and only if*

$$(V) \ |\prod_{s=0}^{t-1} a(s)| \to 0 \quad as \ t \to \infty.$$

Proof. We only need to verify that *(iii)* and *(iv)* of Theorem 3.8.1 hold. First we remark that due to condition *(III)* we have that $|G(t,s,x)| \leq p(t,s)L$. Equation (3.8.3) can be put in the form of Equation (3.8.1) by letting

$$g(t,\phi) = \sum_{s=-\infty}^{-1} G(t,t+s,\phi(s)).$$

To verify *(iii)* we let $b(t) = \sum_{s=-\infty}^{t-1} p(t,s)$ and then for any functions $\phi, \varphi \in \mathscr{C}(L)$, we have

$$|g(t,\phi) - g(t,\varphi)| \leq \left| \sum_{s=-\infty}^{-1} G(t,t+s,\phi(s)) - \sum_{s=-\infty}^{-1} G(t,t+s,\varphi(s)) \right|$$

$$\leq \sum_{s=-\infty}^{-1} p(t,t+s) \, \|\phi - \varphi\|$$

$$= b(t)\|\phi - \varphi\|.$$

Next we verify *(iv)*. Let $\varepsilon > 0$ and $t_1 \geq 0$ be given. By *(IV)* there exists a $t_2 > t_1$ such that

$$L \sum_{s=-\infty}^{t_1-1} p(t,s) < \varepsilon \sum_{s=-\infty}^{t-1} p(t,s) \text{ for all } t > t_2.$$

Let $x_t \in \mathscr{C}(L)$ and for $t > t_2$ we have

$$|g(t,x_t)| \leq \sum_{s=-\infty}^{t_1-1} |G(t,s,x(s))| + \sum_{s=t_1}^{t-1} |G(t,s,x(s))|$$

$$\leq \sum_{s=-\infty}^{t_1-1} Lp(t,s) + \sum_{s=t_1}^{t-1} p(t,s)|x(s)|$$

$$\leq \varepsilon \sum_{s=-\infty}^{t-1} p(t,s) + \sum_{s=t_1}^{t-1} p(t,s)\|x\|^{[t_1,t-1]}$$

$$\leq b(t)\left(\varepsilon + \|x\|^{[t_1,t-1]}\right).$$

This implies that *(iv)* is satisfied, and hence by Theorem 3.8.1, the zero solution of (3.8.3) is asymptotically stable if and only if *(V)* holds.

We end the paper with the following example.

Example 3.9. Consider the difference equation

$$x(t+1) = \frac{1}{2^t}x(t) + \sum_{s=-\infty}^{t-1} 2^{s-t}x(s), n \geq 0. \tag{3.8.4}$$

In this example we take $t_0 = 0$. We make sure all conditions of Theorem 3.8.2 are satisfied. We observe that $a(t) = \frac{1}{2^t}$, and $G(t,s,x) = 2^{s-t}x(s)$. Thus,

$$\prod_{s=0}^{t-1} \frac{1}{2^s} \to 0 \text{ as } t \to \infty,$$

and hence condition(V) is satisfied. It is clear that $p(t,s) = 2^{s-t}$. Next we make sure condition (II) is satisfied.

$$\max_{t \in \mathbf{Z}^+} \sum_{s=0}^{t-1} \left| \prod_{u=s+1}^{t-1} a(u) \right| \sum_{\tau=0}^{s-1} p(s,\tau)$$

$$= \max_{t \in \mathbf{Z}^+} \sum_{s=0}^{t-1} \left| \prod_{u=s+1}^{t-1} 2^{-u} \right| \sum_{\tau=0}^{s-1} 2^{s-\tau}$$

$$< \max_{t \in \mathbf{Z}^+} \sum_{s=0}^{t-1} \left| \prod_{u=s+1}^{t-1} 2^{-u}(1-2^{-s}) \right.$$

$$\leq \max_{t \in \mathbf{Z}^+} \sum_{s=0}^{t-1} 2^{1-t}(1-2^{-s})$$

$$\leq 2^{1-t}[-2^{1-t} + 2 + \frac{4^{1-t}}{3} - 4/3]$$

$$\leq 2/3, \text{ for all } t \in \mathbf{Z}^+.$$

Hence (II) is satisfied. Left to show (IV) is satisfied. Let $t_1 \geq 0$ be given. Then

$$\sum_{s=-\infty}^{t_1-1} p(t,s) = \sum_{s=-\infty}^{t_1-1} 2^{-t+s}$$

$$= 2^{-t}[2^{t_1} - 2^{-\infty}]$$

$$\leq 2^{t-t_2}$$

$$= 2^{-t_2} \sum_{s=-\infty}^{t-1} 2^{-t+s}$$

$$\leq \varepsilon \sum_{s=-\infty}^{t-1} p(t,s), t \geq t_2 \geq t_1.$$

Thus all the conditions of Theorem 3.8.2 are satisfied and the zero solution of (3.8.4) is asymptotically stable.

3.9 Volterra Summation Equations

We shift our focus to different types of Volterra difference equations, which we call Volterra summation equations. Volterra integral equations were first studied by Miller [123], in which he proposed the extension of the use of Lyapunov functionals in Volterra integro-differential equations to integral equations. Years later, Burton took upon himself such a tedious task and successfully used Lyapunov functionals in the qualitative analysis of Integral equations. For such a reference we mention the papers [28, 29, 30]. Since then, the study of integral equations has been fully developed, unlike its counterpart, Volterra summation equations. All the results of this section are new and not published anywhere else. Volterra summation equations play major role in the qualitative analysis of neutral difference equations. To see this we consider the neutral difference equation

$$\triangle\big(D(n,x_n)\big) = f(n,x_n), \ \ n \in \mathbb{Z}^+. \tag{3.9.1}$$

If

$$D(n,0) = f(n,0) = 0,$$

then one would have to ask that the zero solution of $D(n,0) = 0$ be stable in order for the zero solution of (3.9.1) to be stable. On the other hand, if we are interested in studying boundedness of solutions of (3.9.1), one would have to require that the solutions of

$$D(n,x_n) = h(n),$$

be bounded for a suitable function $h(n)$ with suitable conditions. Equation (3.9.1) is typified by neutral equations of the form

$$\triangle\left(x(n) + \sum_{s=n-h}^{n-1} b(s+h)x(s)\right) = f(n,x_n). \tag{3.9.2}$$

Equation (3.9.2) will be studied in detail in Chapter 6. For more on neutral difference equations, we refer to [183]. Most of this section's materials can be found in [142]. We consider the vector Volterra summation equation

$$x(t) = a(t) - \sum_{s=0}^{t-1} C(t,s)x(s), \ t \in \mathbb{Z}^+ \tag{3.9.3}$$

where x and a are k-vectors, $k \geq 1$, while C is an $k \times k$ matrix. To clear any confusion, we note that the summation term in (3.9.3) could have been started at any initial time $t_0 \geq 0$. We will use the resolvent equation that was established on time scales in [2], combined with Lyapunov functionals and fixed point theory to obtain boundedness of solutions and their asymptotic behaviors. One of the major difficulties when using a suitable Lyapunov functional on Volterra summation equation is relating the solution back to that Lyapunov functional. For $x \in \mathbb{R}$, $|x|$ denotes the Euclidean norm of x. For any $k \times k$ matrix A, define the norm of A by $|A| = \sup\{|Ax| : |x| \leq 1\}$. Let X

denote the set of functions $\phi : [0,n] \to \mathbb{R}$ and $\|\phi\| = \max\{|\phi(s)| : 0 \le s \le n\}$. We have the following theorem regarding the existence of solutions of (3.9.3).

Theorem 3.9.1. *Assume the existence of two positive constants* K *and* $\alpha \in (0,1)$ *such that*

$$|a(t)| \le K, \text{ and } \max_{t \ge 0} \sum_{s=0}^{t-1} |C(t,s)| \le \alpha, \qquad (3.9.4)$$

then there is a unique bounded solution of (3.9.3).

Proof. Define a mapping $D : X \to X$, by

$$(D\phi)(t) = a(t) - \sum_{s=t_0}^{t-1} C(t,s)\phi(s).$$

It is clear that $(X, \|\cdot\|)$ is a Banach space. Now for $\phi \in X$, with $\|\phi\| \le q$ for positive constant q we have that

$$\|(D\phi)\| \le K + \alpha q.$$

Thus $D : X \to X$. Left to show that D defines a contraction mapping on X. Let $\phi, \varphi \in X$. Then

$$\|(D\phi) - (D\varphi)\| \le \max_{t \ge 0} \sum_{s=0}^{t-1} |C(t,s)| \|\phi - \varphi\|$$

$$\le \alpha \|\phi - \varphi\|.$$

Hence, D is a contraction, and by the contraction mapping principle it has a unique solution in X that solves (3.9.3). This completes the proof.

We have the following theorem in which we use a Lyapunov functional to drive solutions to zero.

Theorem 3.9.2. *Assume the existence of two positive constants* K_1 *and* $\alpha \in (0,1)$ *such that*

$$\sum_{t=0}^{\infty} |a(t)| \le K_1, \text{ and } \sum_{u=1}^{\infty} |C(u+t,t)| \le \alpha, \qquad (3.9.5)$$

then every solution $x(t)$ *of* (3.9.3) *satisfies* $x \in l_1[0,\infty)$ *and* $x(t) \to 0$, *as* $t \to \infty$.

Proof. Using (3.9.3) we obtain

$$|x(t)| - |a(t)| \le \sum_{s=0}^{t-1} |C(t,s)||x(s)|. \qquad (3.9.6)$$

Define the Lyapunov functional V by

$$V(t) = \sum_{s=0}^{t-1} \sum_{u=t-s}^{\infty} |C(u+s,s)||x(s)|.$$

Moreover,

$$\triangle V(t) = |x(t)| \sum_{u=1}^{\infty} |C(u+s,s)| - \sum_{s=0}^{t-1} |C(t,s)||x(s)|.$$

A substitution of (3.9.6) in the above expression yields

$$\triangle V(t) \le |x(t)| \sum_{u=1}^{\infty} |C(u+s,s)| - |x(t)| + |a(t)|$$
$$= (\alpha - 1)|x(t)| + |a(t)|.$$

Summing the above inequality from 0 to $t-1$ gives

$$0 \le V(t) \le V(0) + (\alpha - 1) \sum_{s=0}^{t-1} |x(s)| + \sum_{s=0}^{t-1} |a(s)|.$$

Since $V(0) = 0$ and $\alpha \in (0,1)$ we arrive at

$$\sum_{s=0}^{t-1} |x(s)| \le \frac{1}{1-\alpha} \sum_{s=0}^{t-1} |a(s)|.$$

Letting $t \to \infty$ we have

$$\sum_{t=0}^{\infty} |x(t)| \le \frac{1}{1-\alpha} \sum_{t=0}^{\infty} |a(t)| \le \frac{1}{1-\alpha} K_1,$$

which is automatically implicd that $x(t) \;\rightarrow\; 0$, as $t \;\rightarrow\; \infty$. This completes the proof.

We note that the use of Lyapunov functional has an advantage here due to the absence of a linear term in our original equation (3.9.3) which is necessary for the use of variation of parameters. The next result is about the existence of a unique periodic solution for the Volterra summation equation

$$x(t) = a(t) - \sum_{s=-\infty}^{t-1} C(t,s)x(s), \; t \in \mathbb{Z} \tag{3.9.7}$$

where x and a are k-vectors, $k \ge 1$, while C is a $k \times k$ matrix.

Theorem 3.9.3. *Assume the existence of a constant* $\alpha \in (0,1)$ *such that*

$$\max_{t \ge 0} \sum_{s=-\infty}^{t-1} |C(t,s)| \le \alpha,$$

and there is a positive integer T such that

$$a(t+T) = a(t) \text{ and } C(t+T,s+T) = C(t,s),$$

then there is a unique periodic solution of (3.9.7).

Proof. Let X be the space of periodic sequences of period T. Then, it is clear that $(X, \|\cdot\|)$ is a Banach space. Now for $\phi \in X$, we define $D : X \to X$ by

$$(D\phi)(t) = a(t) - \sum_{s=-\infty}^{t-1} C(t,s)\phi(s).$$

It is clear that D is periodic of period T. That is $(D\phi)(t+T) = (D\phi)(t)$. Left to show that D defines a contraction mapping on X. Let $\phi, \varphi \in X$. Then

$$\|(D\phi) - (D\varphi)\| \leq \max_{t \in \mathbb{Z}} \sum_{s=-\infty}^{t-1} |C(t,s)| \|\phi - \varphi\|$$

$$\leq \alpha \|\phi - \varphi\|.$$

Hence, D is a contraction and by the contraction mapping principle it has a unique solution in X that solves (3.9.7). This completes the proof.

In the next theorem we return to (3.9.3) and rewrite it so we can show it has an asymptotically periodic solution. Thus we rewrite (3.9.3) in the form

$$x(t) = a(t) - \sum_{s=-\infty}^{t-1} C(t,s)x(s) + \sum_{s=-\infty}^{-1} C(t,s)x(s). \qquad (3.9.8)$$

Note that the term $a(t) - \sum_{s=-\infty}^{t-1} C(t,s)x(s)$ produced a unique periodic solution as we have seen in Theorem 3.9.3 and this indicates that for any bounded x, the term $\sum_{s=-\infty}^{-1} C(t,s)x(s) \to 0$, $t \to \infty$. Hence it is intuitive to expect a solution x of (3.9.8) to be written as $x = y + z$ where y is periodic and $z \to 0$, as $t \to \infty$. We need to properly define our spaces. Let

$$P_T = \{\phi : \mathbb{Z} \to \mathbb{R}^k \mid \phi(t+T) = \phi(t)\}$$

and

$$Q = \{q : \mathbb{Z}^+ \to \mathbb{R}^k \mid q(t) \to 0, \text{ as } t \to \infty\}.$$

We have the following theorem.

Theorem 3.9.4. *Suppose for $\phi \in P_T$,*

$$\sum_{s=-\infty}^{-1} C(t,s)\phi(s) \to 0, \text{ as } t \to \infty$$

and for each $z \in Q$,

$$\sum_{s=0}^{t-1} C(t,s)z(s) \to 0, \text{ as } t \to \infty.$$

Assume the existence of a constant $\alpha \in (0,1)$ such that

$$\max_{t \geq 0} \sum_{s=0}^{t-1} |C(t,s)| \leq \alpha,$$

and there is a positive constant integer T such that $a(t+T) = a(t)$ and $C(t+T, s+T) = C(t,s)$. Then (3.9.3) has a solution $x(t) = y(t) + z(t)$ where $\phi \in P_T$ and $z \in Q$.

Proof. Let X be the space of sequences $\phi : \mathbb{Z}^+ \to \mathbb{R}^k$ such that $\phi \in X$ implies there is a $y \in P_T$ and $z \in Q$ with $\phi = y + z$. We claim that $(X, \|\cdot\|)$ is a Banach space, where $\|\cdot\|$ is the maximum norm. To see this, we let $\{y_n + z_n\}$ be a Cauchy sequence in $(X, \|\cdot\|)$. Given an $\varepsilon > 0$ there is an N such that for $n, m \geq N$ we have

$$|y_n(t) + z_n(t) - y_m(t) - z_m(t)| < \frac{\varepsilon}{2}.$$

Since $z \in Q$ we have for each $\varepsilon > 0$ and each $z \in Q$ there is an $L > 0$ such that $t \geq L$, implies that $|z(t)| \leq \varepsilon/4$. Fix $n, m \geq N$ and for the $\varepsilon/4$ find $L > 0$ such that $t \geq L$ implies that

$$|y_n(t) - z_m(t)| - |z_n(t) - y_m(t)| \leq |y_n(t) + z_n(t) - y_m(t) - z_m(t)| < \frac{\varepsilon}{2}$$

so that $t \geq L$ implies that

$$|y_n(t) - y_m(t)| \leq \frac{\varepsilon}{2} + |z_n(t)| + |z_m(t)| < \varepsilon,$$

for all t since $y_n, y_m \in P_T$. Since this holds for every pair with $n \geq N$ and $m \geq N$, it follows that $\{y_n\}$ is a Cauchy sequence. The same argument can be repeated for the sequence $\{y_n\}$. This completes the proof of the claim.

Let $\phi = y + z$ where $y \in P_T$ and $z \in Q$. Define the mapping $H : X \to X$ by

$$(H\phi)(t) = a(t) - \sum_{s=0}^{t-1} C(t,s)x(s).$$

Then H is a contraction mapping by Theorem 3.9.3. Now we observe that

$$
\begin{aligned}
(H\phi)(t) &= a(t) - \sum_{s=0}^{t-1} C(t,s)\phi(s) \\
&= a(t) - \sum_{s=0}^{t-1} C(t,s)(y(s) + z(s)) \\
&= \left[a(t) - \sum_{s=-\infty}^{t-1} C(t,s)y(s) \right] \\
&\quad + \left[\sum_{s=-\infty}^{-1} C(t,s)y(s) - \sum_{s=0}^{t-1} C(t,s)z(s) \right] \\
&:= B\phi + A\phi.
\end{aligned}
$$

This defines mappings A and B on X. Note that $B : X \to P_T \subset X$ and $A : X \to Q \subset X$. Since H defines a contraction mapping, it has a unique fixed point in X that solves (3.9.3). The proof is complete.

Next we discuss the equation for Volterra summation equation (3.9.3) and use variation of parameters to obtain variety of different results concerning the solutions. Adivar and Raffoul [2] were the first to establish the existence of the resolvent of an equation that is similar to (3.9.3) on time scales. Due to the importance of such results, we will state them as they were presented in [2] on time scales. We then set the time scale equal to $\mathbb{Z}$ to suit our equation (3.9.3). For a good reference on time scales we refer the reader to the famous book by Martin Bohner and Al Peterson [18]. A time scale, denoted $\mathbb{T}$, is a nonempty closed subset of real numbers. The set $\mathbb{T}^\kappa$ is derived from the time scale $\mathbb{T}$ as follows: if $\mathbb{T}$ has a left-scattered maximum M, then $\mathbb{T}^\kappa = \mathbb{T} - \{M\}$, otherwise $\mathbb{T}^\kappa = \mathbb{T}$. The delta derivative f^Δ of a function $f : \mathbb{T} \to \mathbb{R}$, defined at a point $t \in \mathbb{T}^\kappa$ by

$$f^\Delta(t) := \lim_{s \to t} \frac{f(\sigma(t)) - f(s)}{\sigma(t) - s}, \quad \text{where } s \to t, \ s \in \mathbb{T} \backslash \{\sigma(t)\}, \qquad (3.9.9)$$

In (3.9.9), $\sigma : \mathbb{T} \to \mathbb{T}$ is the forward jump operator defined by $\sigma(t) := \inf\{s \in \mathbb{T} : s > t\}$. Hereafter, we denote by $\mu(t)$ the step size function $\mu : \mathbb{T} \to \mathbb{R}$ defined by $\mu(t) := \sigma(t) - t$. A point $t \in \mathbb{T}$ is said to be right dense (right scattered) if $\mu(t) = 0$ ($\mu(t) > 0$). A point is said to be left dense if $\sup\{s \in \mathbb{T} : s < t\} = t$. We note that when the time scale is the set on integers, $\mathbb{T} = \mathbb{Z}$, then $\sigma(t) = t + 1$ and $\mu(t) = 1$. where $t_0 \in \mathbb{T}^\kappa$ is fixed and the functions $a : I_\mathbb{T} \to \mathbb{R}$, $C : I_\mathbb{T} \times I_\mathbb{T} \to \mathbb{R}$. Based on the results of [2], we have the following. Given a linear system of integral equations of the form

$$x(t) = a(t) - \int_{t_0}^t C(t, s) x(s) \Delta s, \quad t_0 \in \mathbb{T}^\kappa \qquad (3.9.10)$$

the corresponding resolvent equation associated with $C(t, s)$ is given by

$$R(t, s) = C(t, s) - \int_{\sigma(s)}^t R(t, u) C(u, s) \Delta u. \qquad (3.9.11)$$

If C is scalar valued, then so is R. If C is $n \times n$ matrix, then so is R. Moreover, the solution of (3.9.10) in terms of R is given by the variation of parameters formula

$$x(t) = a(t) - \int_{t_0}^t R(t, u) a(u) \Delta u. \qquad (3.9.12)$$

It should cause no difficulties to take the initial time $t_0 = 0$. With this in mind, if we set $\mathbb{T} = \mathbb{Z}$, then equations (3.9.10) and (3.9.11) become

$$x(t) = a(t) - \sum_{s=0}^{t-1} C(t, s) x(s), \qquad (3.9.13)$$

and

$$R(t,s) = C(t,s) - \sum_{u=s+1}^{t-1} R(t,u)C(u,s), \qquad (3.9.14)$$

respectively. If C is scalar valued, then so is R. If C is $n \times n$ matrix, then so is R. Moreover, the solution of (3.9.13) in terms of R is given by

$$x(t) = a(t) - \sum_{s=0}^{t-1} R(t,s)a(s). \qquad (3.9.15)$$

For the remainder of this section we denote the vector space of bounded sequences $\phi : \mathbb{Z}^+ \to \mathbb{R}^k$ by $\mathscr{BC}$. The next theorem is an extension of Perron's Theorem for integral equation over the reals to an arbitrary time scale. Its proof can be found in [2]. Only for the next theorem we define $\mathscr{B}$ to be the space of bounded continuous functions on $I_{\mathbb{T}} = [t_0, \infty)_{\mathbb{T}}$ endowed with the supremum norm, $||\cdot||_{\mathscr{B}}$, given by

$$||f||_{\mathscr{B}} := \sup_{t \in I_{\mathbb{T}}} |f(t)|.$$

One can easily see that $(\mathscr{B}, ||\cdot||_{\mathscr{B}})$ is a Banach space.

Theorem 3.9.5 ([2]). *Let $C : I_{\mathbb{T}} \times I_{\mathbb{T}} \to \mathbb{R}$ be continuous real valued function on $\leq t_0 \leq s \leq t < \infty$. If $\int_{t_0}^{t} R(t,s)f(s)\Delta s \in \mathscr{BC}$ for each $f \in \mathscr{BC}$, then there exists a positive constant K such that $\int_{t_0}^{t} |R(t,s)|\, \Delta s < K$, for all $t \in I_{\mathbb{T}}$.*

Just for the record, we restate Theorem 3.9.5 for $\mathbb{T} = \mathbb{Z}$

Theorem 3.9.6. *Let $C : \mathbb{Z}^+ \times \mathbb{Z}^+ \to \mathbb{R}$ be real valued sequence on $0 \leq t_0 \leq s \leq t < \infty$. If $\sum_{s=0}^{t-1} R(t,s)f(s) \in \mathscr{BC}$ for each $f \in \mathscr{BC}$, then there exists a positive constant K such that $\sum_{s=0}^{t-1} |R(t,s)| < K$, for all $t \in \mathbb{Z}^+$.*

Theorem 3.9.7. *Suppose $R(t,s)$ satisfies (3.9.14) and that $a \in \mathscr{BC}$. Then every solution $x(t)$ of (3.9.13) is bounded if and only if*

$$\max_{t \in \mathbb{Z}^+} \sum_{s=0}^{t-1} |R(t,s)| < \infty \qquad (3.9.16)$$

holds.

Proof. Suppose (3.9.16) holds. Then, using (3.9.15), it is trivial to show that $x(t)$ is bounded. If $x(t)$ and $a(t)$ are bounded, then from (3.9.15), we have

$$\sum_{s=0}^{t-1} |R(t,s)||a(s)| \leq \gamma$$

for some positive constant γ and the proof follows from Theorem 3.9.6.

The intuitive idea here is that for $C(t,s)$ well behaved then the solution of (3.9.13) follows $a(t)$.

Theorem 3.9.8. *Let C be a $k \times k$ matrix. Assume the existence of a constant $\alpha \in (0, 1)$ such that*

$$\max_{t \in \mathbb{Z}^+} \sum_{s=0}^{t-1} |C(t,s)| \le \alpha, \tag{3.9.17}$$

(i) If $a \in \mathcal{BC}$ so is the solution x of (3.9.13); hence, (3.9.17) holds.

(ii) Suppose, in addition, that for each $T > 0$ we have $\sum_{s=0}^{T} |C(t,s)| \to 0$ as $t \to \infty$. If $a(t) \to 0$ as $t \to \infty$, then so does $x(t)$ and $\sum_{s=0}^{t-1} R(t,s)a(s)$.

(iii) $\sum_{s=0}^{t-1} |R(t,s)| \le \dfrac{\alpha}{1-\alpha}$.

Proof. The proof of (i) is the same as the proof of Theorem 3.9.1. For the proof of (ii) we define the set

$$M = \{\phi : \mathbb{Z}^+ \to \mathbb{R}^k \mid |\phi(t)| \to 0, \text{ as } t \to \infty\}.$$

For $\phi \in M$, define the mapping Q by

$$(Q\phi)(t) = a(t) - \sum_{s=0}^{t-1} C(t,s)\phi(s).$$

Then

$$|(Q\phi)(t)| \le |a(t)| + \sum_{s=0}^{t-1} |C(t,s)\phi(s)|.$$

We already know that $a(t) \to 0$ as $t \to \infty$. Given an $\varepsilon > 0$ and $\phi \in M$, find T such that $|\phi(t)| < \varepsilon$ if $t \ge T$ and find d with $|\phi(t)| \le d$ for all $t \ge T$. For this fixed T, find $\eta > T$ such that $t \ge \eta$ implies that $\sum_{s=0}^{T-1} |C(t,s)| \le \dfrac{\varepsilon}{d}$. Then $t \ge \eta$ implies that

$$\sum_{s=0}^{t-1} |C(t,s)| \le \sum_{s=0}^{T-1} |C(t,s)| \sum_{s=T}^{t-1} |C(t,s)|$$
$$\le (d\varepsilon)/d + \alpha\varepsilon < 2\varepsilon.$$

Thus, $Q : M \to M$ and the fixed point satisfies $x(t) \to 0$, as $t \to \infty$, for every vector sequence $a \in M$. Using (3.9.15) we have

$$\sum_{s=0}^{t-1} R(t,s)a(s) = a(t) - x(t) \to 0, \text{ as } t \to \infty.$$

This completes the proof of (ii). Using (3.9.14) and (3.9.17) we have by changing of order of summations

$$\sum_{s=0}^{t-1} |R(t,s)| \leq \sum_{s=0}^{t-1} |C(t,s)| + \sum_{s=0}^{t-1} \sum_{u=s+1}^{t-1} |R(t,u)||C(u,s)|$$

$$= \sum_{s=0}^{t-1} |C(t,s)| + \sum_{u=0}^{t-1} |R(t,u)| \sum_{s=0}^{u-1} |C(u,s)|$$

$$\leq \alpha + \alpha \sum_{u=0}^{t-1} |R(t,u)|.$$

Therefore,

$$(1-\alpha)\sum_{s=0}^{t-1} |R(t,s)| \leq \alpha.$$

That is,

$$\max_{t\in\mathbb{Z}^+} \sum_{s=0}^{t-1} |R(t,s)| \leq \frac{\alpha}{1-\alpha}.$$

Example 3.10. Suppose there is a sequence $r : \mathbb{Z}^+ \to (0,1]$, with $r(t) \downarrow 0$ with

$$\max_{t\in\mathbb{Z}^+} \sum_{s=0}^{t-1} |C(t,s)|(r(s)/r(t)) \leq \alpha, \ \alpha \in (0,1) \qquad (3.9.18)$$

and

$$|a(t)| \leq kr(t) \qquad (3.9.19)$$

for some positive constant k. Then the unique solution $x(t)$ of (3.9.13) is bounded and goes to zero as t approaches infinity. Moreover, $\sum_{s=0}^{t-1} R(t,s)a(s) \to 0$, as $t \to \infty$.

Proof. Let

$$\mathscr{M} = \{\phi : [0,\infty) \to \mathbb{R}^k \mid |\phi|_r \leq \max_{t\in\mathbb{Z}^+} \frac{|\phi(t)|}{|r(t)|} < \infty\}.$$

Then $(\mathscr{M}, |\cdot|_r)$ is a Banach space. For $\phi \in \mathscr{M}$, define the mapping Q by

$$(Q\phi)(t) = a(t) - \sum_{s=0}^{t-1} C(t,s)\phi(s).$$

Then,

$$|(Q\phi)(t)|/r(t) \leq |a(t)|/r(t) + \sum_{s=0}^{t-1} |C(t,s)|(r(s)/r(t))|\phi(s)|/r(s)$$

$$\leq k + |\phi|_r \sum_{s=0}^{t-1} |C(t,s)|(r(s)/r(t))$$

$$\leq k + \alpha|\phi|_r,$$

which shows that $Q\phi \in \mathcal{M}$. Let $\phi, \eta \in \mathcal{M}$, then we readily have that

$$\left|(Q\phi)(t)) - (Q\eta)(t)\right|/r(t) \leq \alpha|\phi - \eta|_r$$

and so we have Q is a contraction on $\mathcal{M}$ and therefore it has a unique fixed point $x(t)$ in $\mathcal{M}$ that solves (3.9.13). Moreover, $\max_{t \in \mathbb{Z}^+} \frac{|x(t)|}{|r(t)|} < \infty$, implies that $|x(t)| \leq k^*r(t) \to 0$, as $t \to \infty$. Also by (3.9.19) we have $|a(t)| \to 0$, as $t \to \infty$ and hence using (3.9.15) we have

$$\sum_{s=0}^{t-1} |R(t,s)a(s)| \leq |a(t)| + |x(t)| \to 0, \text{ as } t \to \infty.$$

This completes the proof.

The next theorem relates the kernel to a kernel of convolution type. Also, unlike Theorem 3.9.2 we only ask for boundedness on $a(t)$.

Theorem 3.9.9. *Assume the existence of a constant $\alpha \in (0,1)$ such that*

$$\sum_{u=1}^{\infty} |C(u+t,t)| \leq \alpha, \tag{3.9.20}$$

and let $a(t)$ be a bounded sequence. Suppose there is a decreasing sequence Φ : $[0,\infty) \to (0,\infty)$ with $\Phi \in l_1[0,\infty)$, and

$$\Phi(t-s) \geq \sum_{u=t-s}^{\infty} |C(u+s,s)|. \tag{3.9.21}$$

If in addition there exists a positive constant K with

$$\sum_{u=t-s}^{\infty} |C(u+s,s)| \geq K|C(t,s)|, \tag{3.9.22}$$

then the unique solution $x(t)$ of (3.9.13) is bounded and

$$\max_{t \in \mathbb{Z}^+} \sum_{s=0}^{t-1} |R(t,s)| < \infty.$$

Proof. Define the Lyapunov functional V by

$$V(t) = \sum_{s=0}^{t-1} \sum_{u=t-s}^{\infty} |C(u+s,s)||x(s)|.$$

Then along the solutions of (3.9.13) we have

$$\triangle V(t) = |x(t)| \sum_{u=1}^{\infty} |C(u+s,s)| - \sum_{s=0}^{t-1} |C(t,s)||x(s)|.$$

Using (3.9.13) we arrive at

$$|x(t)| - |a(t)| \le \sum_{s=0}^{t-1} |C(t,s)||x(s)|.$$

Hence,

$$\triangle V(t) \le |x(t)| \sum_{u=1}^{\infty} |C(u+s,s)| - |x(t)| + |a(t)|$$
$$= (\alpha - 1)|x(t)| + |a(t)| := -\delta|x(t)| + |a(t)|,$$

for $\delta > 0$. Replace t with s in the above expression and then multiply both sides by $\Phi(t-s)$ for $0 \le s \le t < \infty$.

$$\triangle_s V(s)\Phi(t-s) \le -\delta|x(s)|\Phi(t-s) + |a(s)|\Phi(t-s). \tag{3.9.23}$$

Suppose there is a $t > 0$ satisfying

$$V(t) = \max_{0 \le s \le t-1} V(s+1).$$

Then summing from 0 to $t-1$ followed with summation by parts and by noting that $V(0) = 0$ we arrive at

$$\sum_{s=0}^{t-1} \triangle_s V(s)\Phi(t-s) = V(s)\Phi(t-s)\Big|_{s=0}^{t} - \sum_{s=0}^{t-1} V(s+1)\triangle_s\Phi(t-s)$$

$$\ge V(t)\Phi(0) - V(0)\Phi(t) - V(t)\sum_{s=0}^{t-1}\triangle_s\Phi(t-s)$$

$$= V(t)\Phi(0) - V(t)[\Phi(0) - \Phi(t)]$$

$$= V(t)\Phi(t). \tag{3.9.24}$$

Hence (3.9.24) combined with (3.9.23) and making use of (3.9.22) yield

$$V(t)\Phi(t) \le \sum_{s=0}^{t-1} \triangle_s V(s)\Phi(t-s)$$

$$\le -\delta \sum_{s=0}^{t-1} \Phi(t-s)|x(s)| + \sum_{s=0}^{t-1} \Phi(t-s)|a(s)|$$

$$\le -\delta \sum_{s=0}^{t-1} \sum_{u=t-s}^{\infty} |C(u+s,s)||x(s)| + \sum_{s=0}^{t-1} \Phi(t-s)|a(s)|$$

$$\le -\delta V(t) + ||a||q,$$

where $||a||$ is the maximum norm of a and q is a positive constant that we get from $|\Phi| \in l_1[0,\infty)$. The above inequality implies that

$$V(t) \le \frac{||a||q}{\Phi(t)+\delta},$$

and $V(t)$ is bounded. Using (3.9.22) in $V(t)$ gives

$$V(t) \ge K \sum_{s=0}^{t-1} |C(t,s)||x(s)| \ge K[|x(t)| - |a(t)|],$$

from which we conclude $x(t)$ is bounded since both $V(t)$ and $a(t)$ are bounded. Now from (3.9.15) we have that $\sum_{s=0}^{t-1} |R(t,s)a(s)| \le |x(t)| + |a(t)|$ and hence $\sum_{s=0}^{t-1} |R(t,s)a(s)|$ is bounded and by Theorem 3.9.6, we have that $\max_{t \in \mathbb{Z}^+} \sum_{s=0}^{t-1} |R(t,s)| < \infty$. This completes the proof.

For Theorem 3.9.10 we assume (3.9.13) is scalar.

Theorem 3.9.10. *Assume the existence of constants* $\alpha, \beta \in (0,1)$ *such that*

$$\sum_{u=1}^{\infty} |C(u+t,t)| < \alpha, \tag{3.9.25}$$

and

$$\max_{t \in \mathbb{Z}^+} \sum_{s=0}^{t-1} |C(t,s)| \le \beta. \tag{3.9.26}$$

If $a \in l_2[0,\infty)$ *so is the solution* x *of* (3.9.13).

Proof. Define the Lyapunov functional V by

$$V(t) = \sum_{s=0}^{t-1} \sum_{u=t-s}^{\infty} |C(u+s,s)|x^2(s).$$

Then along the solutions of (3.9.13) we have that

$$\Delta V(t) = x^2(t) \sum_{u=1}^{\infty} |C(u+s,s)| - \sum_{s=0}^{t-1} |C(t,s)|x^2(s).$$

Squaring both sides of (3.9.13) gives

$$x^2(t) = a^2(t) - 2a(t) \sum_{s=0}^{t-1} C(t,s)x(s) + \left(\sum_{s=0}^{t-1} C(t,s)x(s) \right)^2$$

$$\le 2\left(a^2(t) + \left(\sum_{s=0}^{t-1} C(t,s)x(s) \right)^2 \right)$$

$$\leq 2a^2(t) + 2 \sum_{s=0}^{t-1} |C(t,s)| \sum_{s=0}^{t-1} |C(t,s)| x^2(s)$$

$$\leq 2a^2(t) + 2\beta \sum_{s=0}^{t-1} |C(t,s)| x^2(s).$$

This implies that

$$-\sum_{s=0}^{t-1} |C(t,s)| x^2(s) \leq a^2(t)/\beta - x^2(t)/(2\beta).$$

Substituting into $\triangle V$ gives

$$\triangle V(t) = x^2(t) \sum_{u=1}^{\infty} |C(u+s,s)| - \sum_{s=0}^{t-1} |C(t,s)| x^2(s)$$

$$\leq a^2(t)/\beta - (1/(2\beta) - \alpha) x^2(t).$$

Summing the above inequality for 0 to $n-1$ yields

$$0 \leq V(t) - V(0) \leq 1/\beta \sum_{s=0}^{n-1} a^2(s) - (1/(2\beta) - \alpha) \sum_{s=0}^{n-1} x^2(s),$$

and hence the results. This completes the proof.

3.10 The Need for Large Contraction

So far, we have been successful in using fixed point theorems including the contraction mapping principles in obtaining different results concerning functional difference equations. It is naive to believe that every map can be defined so that it is a contraction, even with the strictest conditions. For example, consider

$$f(x) = x - x^3$$

then for $x, y \in \mathbb{R}$ we have that

$$|f(x) - f(y)| = |x - x^3 - y + y^3| \leq |x - y| \left(1 - \frac{x^2 + y^2}{2} \right)$$

and the contraction constant tends to one as $x^2 + y^2 \to 0$. As a consequence, the regular contraction mapping principle failed to produce any results. This forces us to look for other alternative, namely the concept of Large Contraction. We will restate the contraction mapping principle and Krasnoselskii's fixed point theorems in which the regular contraction is replaced with Large Contraction. Then based on the notion of Large Contraction, we introduce two theorems to obtain boundedness and periodicity results in which Large Contraction is substituted for regular contraction.

Definition 3.10.1. *Let $(\mathcal{M},d)$ be a metric space and $B: \mathcal{M} \to \mathcal{M}$. The map B is said to be large contraction if $\phi, \varphi \in \mathcal{M}$, with $\phi \neq \varphi$ then $d(B\phi, B\varphi) \leq d(\phi, \varphi)$ and if for all $\varepsilon > 0$, there exists a $\delta \in (0,1)$ such that*

$$[\phi, \varphi \in \mathcal{M}, d(\phi, \varphi) \geq \varepsilon] \Rightarrow d(B\phi, B\varphi) \leq \delta d(\phi, \varphi).$$

The next theorems are alternative to the regular Contraction Mapping Principle and Krasnoselskii's fixed point theorem in which we substitute Large Contraction for regular contraction. The proofs of the two theorems and the statement of Definition 3.10.1 can be found in [24].

Theorem 3.10.1. *Let $(\mathcal{M}, \rho)$ be a complete metric space and B be a large contraction. Suppose there are an $x \in \mathcal{M}$ and an $L > 0$ such that $\rho(x, B^n x) \leq L$ for all $n \geq$. Then B has a unique fixed point in $\mathcal{M}$.*

Theorem 3.10.2. *Let $\mathcal{M}$ be a bounded convex nonempty subset of a Banach space $(\mathbb{B}, \|\cdot\|)$. Suppose that A and B map $\mathcal{M}$ into $\mathbb{B}$ such that*

i. $x, y \in \mathcal{M}$ implies $Ax + By \in \mathcal{M}$;
ii. A is compact and continuous;
iii. B is a large contraction mapping.

Then there exists $z \in \mathcal{M}$ with $z = Az + Bz$.

Next, we consider the completely nonlinear difference equation

$$x(t+1) = a(t)x(t)^5 + p(t), \tag{3.10.1}$$

where $a, p : \mathbb{Z} \to \mathbb{R}$. To invert our equation, we create a linear term by letting

$$H(x) = -x + x^5. \tag{3.10.2}$$

It would become clearer later on that $H(x)$ is not a contraction and, as a consequence, the Contraction Mapping Principle cannot be used. Instead, we will show that H is a Large Contraction and hence our mapping, to be constructed, will define a Large Contraction. Then we use Theorem 3.10.2 and show that solutions of (3.10.1) are bounded. This allows us to put (3.10.1) in the form

$$x(t+1) - a(t)x(t) = a(t)H(x(t)) + p(t). \tag{3.10.3}$$

Let $x(0) = x_0$, then by the variation of parameters formula, one can easily show that for $t \geq 0$, $x(t)$ is a solution of (3.10.3) if and only if

$$x(t) = x_0 \prod_{s=0}^{t-1} a(s) + \sum_{s=0}^{t-1} \left(a(s)H(x(s)) \prod_{u=s+1}^{t-1} a(u) \right) + \sum_{s=0}^{t-1} \left(p(s) \prod_{u=s+1}^{t-1} a(u) \right). \tag{3.10.4}$$

We begin with the following lemma.

Lemma 3.9. *Let* $\| \cdot \|$ *denote the maximum norm. If*

$$\mathbb{M} = \left\{ \phi : \mathbb{Z} \to \mathbb{R} \mid \phi(0) = \phi_0, \ \text{and} \ \|\phi\| \leq 5^{-1/4} \right\},$$

then the mapping H defined by (3.10.2) is a large contraction on the set $\mathbb{M}$.

Proof. For any reals a and b we have the following inequalities

$$0 \leq (a+b)^4 = a^4 + b^4 + ab(4a^2 + 6ab + 4b^2),$$

and

$$-ab(a^2 + ab + b^2) \leq \frac{a^4 + b^4}{4} + \frac{a^2 b^2}{2} \leq \frac{a^4 + b^4}{2}.$$

If $x, y \in \mathbb{M}$ with $x \neq y$, then $x(t)^4 + y(t)^4 < 1$. Hence, we arrive at

$$
\begin{aligned}
|H(u) - H(v)| &\leq |u-v| \left| 1 - \left(\frac{u^5 - v^5}{u - v} \right) \right| \\
&= |u-v| \left\{ 1 - u^4 - v^4 - uv(u^2 + uv + v^2) \right\} \\
&\leq |u-v| \left\{ 1 - \frac{(u^4 + v^4)}{2} \right\} \leq |u-v|,
\end{aligned}
\tag{3.10.5}
$$

where we use the notations $u = x(t)$ and $v = y(t)$ for brevity. Now, we are ready to show that H is a large contraction on $\mathbb{M}$. For a given $\varepsilon \in (0,1)$, suppose $x, y \in \mathbb{M}$ with $\|x - y\| \geq \varepsilon$. There are two cases:

a.
$$\frac{\varepsilon}{2} \leq |x(t) - y(t)| \ \text{for some} \ t \in \mathbb{Z},$$

   or
b.
$$|x(t) - y(t)| \leq \frac{\varepsilon}{2} \ \text{for some} \ t \in \mathbb{Z}.$$

If $\varepsilon/2 \leq |x(t) - y(t)|$ for some $t \in \mathbb{Z}$, then

$$(\varepsilon/2)^4 \leq |x(t) - y(t)|^4 \leq 8(x(t)^4 + y(t)^4),$$

or

$$x(t)^4 + y(t)^4 \geq \frac{\varepsilon^4}{2^7}.$$

For all such t, we get by (3.10.5) that

$$|H(x(t)) - H(y(t))| \leq |x(t) - y(t)| \left(1 - \frac{\varepsilon^4}{2^7} \right).$$

On the other hand, if $|x(t) - y(t)| \leq \varepsilon/2$ for some $t \in \mathbb{Z}$, then along with (3.10.5) we find

$$|H(x(t)) - H(y(t))| \leq |x(t) - y(t)| \leq \frac{1}{2}\|x - y\|.$$

Hence, in both cases we have

$$|H(x(t)) - H(y(t))| \leq \min\left\{1 - \frac{\varepsilon^4}{2^7}, \frac{1}{2}\right\}\|x - y\|.$$

Thus, H is a large contraction on the set $\mathbb{M}$ with $\delta = \min\{1 - \varepsilon^4/2^7, 1/2\}$. The proof is complete.

Remark 3.6. It is clear from inequality (3.10.5) that $(u^4 + v^4)/2 \to 0$, the contraction constant approaches one. Hence, $H(x)$ does not define a contraction mapping as we have claimed before.

For $\psi \in \mathbb{M}$, we define the map $B : \mathbb{M} \to \mathbb{M}$ by

$$(B\psi)(t) = \psi_0 \prod_{s=0}^{t-1} a(s) + \sum_{s=0}^{t-1}\left(a(s)H(\psi(s)) \prod_{u=s+1}^{t-1} a(u)\right) + \sum_{s=0}^{t-1}\left(p(s) \prod_{u=s+1}^{t-1} a(u)\right).$$

$$(3.10.6)$$

Lemma 3.10. *Assume for all $t \in \mathbb{Z}$*

$$|\psi_0|\left|\prod_{s=0}^{t-1} a(s)\right| + 4(5^{-5/4})\sum_{s=0}^{t-1}\left|\prod_{u=s}^{t-1} a(u)\right| + \sum_{s=0}^{t-1}\left(\left|p(s) \prod_{u=s+1}^{t-1} a(u)\right|\right) \leq 5^{-1/4}. \quad (3.10.7)$$

If H is a large contraction on $\mathbb{M}$, then so is the mapping B.

Proof. It is easy to see that

$$|H(x(t))| = |x(t) - x(t)^5| \leq 4(5^{-5/4}) \text{ for all } x \in \mathbb{M}.$$

By Lemma 3.9 H is a large contraction on $\mathbb{M}$. Hence, for $x, y \in \mathbb{M}$ with $x \neq y$, we have $\|Hx - Hy\| \leq \|x - y\|$. Hence,

$$|Bx(t) - By(t)| \leq \sum_{s=0}^{t-1}|H(x(s)) - H(y(s))|\left|\prod_{u=s}^{t-1} a(u)\right|$$

$$\leq 4(5^{-5/4})\sum_{s=0}^{t-1}\left|\prod_{u=s}^{t-1} a(u)\right|\|x - y\|$$

$$= \|x - y\|.$$

Taking maximum norm over the set $[0, \infty)$, we get that $\|Bx - By\| \leq \|x - y\|$. Now, from the proof of Lemma 3.9, for a given $\varepsilon \in (0, 1)$, suppose $x, y \in \mathbb{M}$ with $\|x - y\| \geq$

ε. Then $\delta = \min\{1 - \varepsilon^4/2^7, 1/2\}$, which implies that $0 < \delta < 1$. Hence, for all such $\varepsilon > 0$ we know that

$$[x, y \in \mathbb{M}, \|x - y\| \geq \varepsilon] \Rightarrow \|Hx - Hy\| \leq \delta \|x - y\|.$$

Therefore, using (3.10.7), one easily verify that

$$\|Bx - By\| \leq \delta \|x - y\|.$$

The proof is complete.

We arrive at the following theorem in which we prove boundedness.

Theorem 3.10.3. *Assume* (3.10.7). *Then* (3.10.3) *has a unique solution in* $\mathbb{M}$ *which is bounded.*

Proof. $(\mathbb{M}, \|\cdot\|)$ is a complete metric space of bounded sequences. For $\psi \in \mathbb{M}$ we must show that $(B\psi)(t) \in \mathbb{M}$. From (3.10.6) and the fact that

$$|H(x(t))| = |x(t) - x(t)^5| \leq 4(5^{-5/4}) \text{ for all } x \in \mathbb{M},$$

we have

$$|(B\psi)(t)| \leq |\psi_0| \left| \prod_{s=0}^{t-1} a(s) \right| + 4(5^{-5/4}) \sum_{s=0}^{t-1} \left| \prod_{u=s}^{t-1} a(u) \right| + \sum_{s=0}^{t-1} \left(\left| p(s) \prod_{u=s+1}^{t-1} a(u) \right| \right)$$

$$\leq 5^{-1/4}.$$

This shows that $(B\psi)(t) \in \mathbb{M}$. Lemma 3.9 implies the map B is a large contraction and hence by Theorem 3.10.1, the map B has a unique fixed point in $\mathbb{M}$ which is a solution of (3.10.3). This completes the proof.

Next we use Theorem 3.10.2 and prove the existence of a periodic solution of the nonlinear delay difference equation

$$x(t+1) = a(t)x(t)^5 + G(t, x(t-r)) + p(t), \ t \in \mathbb{Z}, \tag{3.10.8}$$

where r is a positive integer and

$$a(t+T) = a(t), \ p(t+T) = p(t), \text{ and } G(t+T, \cdot) = G(t, \cdot) \tag{3.10.9}$$

and T is the least positive integer for which these hold. As before, for the sake of inversion, we rewrite (3.10.8) as

$$x(t+1) - a(t)x(t) = a(t)H(x(t)) + G(t, x(t-r)) + p(t), \tag{3.10.10}$$

where

$$H(x) = -x + x^5. \tag{3.10.11}$$

We begin with the following lemma which we omit its proof.

Lemma 3.11. *Suppose that* $1 - \prod_{s=t-T}^{t-1} a(s) \neq 0$ *for all* $t \in \mathbb{Z}$. *Then* $x(t)$ *is a solution of* (3.10.10) *if and only if*

$$x(t) = \left(1 - \prod_{s=t-T}^{t-1} a(s)\right)^{-1} \sum_{u=t-T}^{t-1} \left(a(u)H(x(u)) + G(t, x(u-r)) + p(u)\right) \prod_{s=u+1}^{t-1} a(s).$$

Let P_T be the set of all sequences $x(t)$, periodic in t of period T. Then $(P_T, \|\cdot\|)$ is a Banach space when it is endowed with the maximum norm

$$\|x\| = \max_{t \in \mathbb{Z}} |x(t)| = \max_{t \in [0, T-1]} |x(t)|.$$

Set

$$\mathbb{M} = \{\varphi \in P_T : \|\varphi\| \leq 5^{-1/4}\}. \tag{3.10.12}$$

Obviously, $\mathbb{M}$ is bounded and convex subset of the Banach space P_T. Let the map $A : \mathbb{M} \to P_T$ be defined by

$$(A\varphi)(t) = \left(1 - \prod_{s=t-T}^{t-1} a(s)\right)^{-1} \sum_{u=t-T}^{t-1} \left(G(t, \varphi(u-r)) + p(u)\right) \prod_{s=u+1}^{t-1} a(s), \; t \in \mathbb{Z}.$$
$$\tag{3.10.13}$$

In a similar way, we set the map $B : \mathbb{M} \to P_T$ by

$$(B\psi)(t) = \left(1 - \prod_{s=t-T}^{t-1} a(s)\right)^{-1} \sum_{u=t-T}^{t-1} \left(a(u)H(\psi(u))\right) \prod_{s=u+1}^{t-1} a(s), \; t \in \mathbb{Z}. \tag{3.10.14}$$

It is clear from (3.10.13) and (3.10.14) that $A\varphi$ and $B\psi$ are T-periodic in t. For simplicity we let

$$\eta := \left|\left(1 - \prod_{s=t-T}^{t-1} a(s)\right)^{-1}\right|.$$

Let

$$G(u, \psi(u-r)) = b(u)\psi(u-r)^5. \tag{3.10.15}$$

For $x \in \mathbb{M}$, we have

$$|x(t)|^5 \leq 5^{-5/4},$$

and therefore,

$$\begin{aligned} G(u, x(u-r)) + p(u) &= b(u)x(u-r)^5 + p(u) \\ &\leq 5^{-5/4}|b(u)| + |p(u)| \end{aligned} \tag{3.10.16}$$

and

$$|H(x(t))| = |x(t) - x(t)^5| \leq 4(5^{-5/4}) \quad \text{for all } x \in \mathbb{M}.$$

We have the following theorem.

Theorem 3.10.4. *Suppose* $G(u, \psi(u-r))$ *is given by* (3.10.15). *Assume for all* $t \in \mathbb{Z}$

$$\eta \sum_{u=t-T}^{t-1} \left(5^{-5/4}|b(u)| + |p(u)| + 4(5^{-5/4})|a(u)|\right) \Big| \prod_{u=s+1}^{t-1} a(u) \Big| \leq 5^{-1/4}. \quad (3.10.17)$$

Then (3.10.8) has a periodic solution.

Proof. Using condition (3.10.17) and by a similar argument as in Lemma 3.9, one can easily show that B is a large contraction since H is a large contraction. Also, the map A is continuous and maps bounded sets into compact sets and hence it is compact. Moreover, for $\varphi, \psi \in \mathbb{M}$, we have by (3.10.17) that

$$A\varphi + B\psi : \mathbb{M} \to \mathbb{M}.$$

Hence an application of Theorem 3.10.2 implies the existence of a periodic solution in $\mathbb{M}$. This completes the proof.

It is evident from Lemma 3.9 that proving large contraction is tedious, long, and not very practical to consider case by case function. Consider the mapping H be defined by

$$H(x(u)) = x(u) - h(x(u)). \quad (3.10.18)$$

We have observed from Lemma 3.9 that the properties of the function h in (3.10.18) plays a substantial role in obtaining a large contraction on a convenient set. Next we state and prove a remarkable theorem that generalizes the concept of Large Contraction. The theorem provides easily checked sufficient conditions under which a mapping is a Large Contraction. The next theorem is due to Adivar, Raffoul, and Islam. Several authors have published it in their work without the proper citations. Let $\alpha \in (0, 1]$ be a fixed real number. Define the set $\mathbb{M}_\alpha$ by

$$\mathbb{M}_\alpha = \{\phi : \phi \in C(\mathbb{R}, \mathbb{R}) \text{ and } \|\phi\| \leq \alpha\}. \quad (3.10.19)$$

H.1. $h : \mathbb{R} \to \mathbb{R}$ is continuous on $[-\alpha, \alpha]$ and differentiable on $(-\alpha, \alpha)$,
H.2. The function h is strictly increasing on $[-\alpha, \alpha]$,
H.3. $\sup\limits_{t \in (-\alpha, \alpha)} h'(t) \leq 1$.

Theorem 3.10.5. *[Adivar-Raffoul-Islam [4] (Classifications of Large Contraction Theorem)] Let $h : \mathbb{R} \to \mathbb{R}$ be a function satisfying (H.1-H.3). Then the mapping H in (3.10.18) is a large contraction on the set $\mathbb{M}_\alpha$.*

Proof. Let $\phi, \varphi \in \mathbb{M}_\alpha$ with $\phi \neq \varphi$. Then $\phi(t) \neq \varphi(t)$ for some $t \in \mathbb{R}$. Let us denote this set by $D(\phi, \varphi)$, i.e.,

$$D(\phi, \varphi) = \{t \in \mathbb{R} : \phi(t) \neq \varphi(t)\}.$$

For all $t \in D(\phi, \varphi)$, we have

$$|H\phi(t) - H\varphi(t)| = |\phi(t) - h(\phi(t)) - \varphi(t) + h(\varphi(t))|$$

$$= |\phi(t) - \varphi(t)| \left| 1 - \left(\frac{h(\phi(t)) - h(\varphi(t))}{\phi(t) - \varphi(t)}\right) \right|. \quad (3.10.20)$$

Since h is a strictly increasing function we have

$$\frac{h(\phi(t)) - h(\varphi(t))}{\phi(t) - \varphi(t)} > 0 \text{ for all } t \in D(\phi, \varphi). \qquad (3.10.21)$$

For each fixed $t \in D(\phi, \varphi)$ define the interval $U_t \subset [-\alpha, \alpha]$ by

$$U_t = \begin{cases} (\varphi(t), \phi(t)) \text{ if } \phi(t) > \varphi(t) \\ (\phi(t), \varphi(t)) \text{ if } \phi(t) < \varphi(t) \end{cases}.$$

Mean Value Theorem implies that for each fixed $t \in D(\phi, \varphi)$ there exists a real number $c_t \in U_t$ such that

$$\frac{h(\phi(t)) - h(\varphi(t))}{\phi(t) - \varphi(t)} = h'(c_t).$$

By (H.2-H.3) we have

$$0 \le \inf_{u \in (-\alpha, \alpha)} h'(u) \le \inf_{u \in U_t} h'(u) \le h'(c_t) \le \sup_{u \subset U_t} h'(u) \le \sup_{u \in (-\alpha, \alpha)} h'(u) \le 1. \quad (3.10.22)$$

Hence, by (3.10.20)–(3.10.22) we obtain

$$|H\phi(t) - H\varphi(t)| \le \left|1 - \inf_{u \in (-\alpha, \alpha)} h'(u)\right| |\phi(t) - \varphi(t)|. \qquad (3.10.23)$$

for all $t \in D(\phi, \varphi)$. This implies a large contraction in the supremum norm. To see this, choose a fixed $\varepsilon \in (0, 1)$ and assume that ϕ and φ are two functions in $\mathbb{M}_\alpha$ satisfying

$$\varepsilon \le \sup_{t \in D(\phi, \varphi)} |\phi(t) - \varphi(t)| = \|\phi - \varphi\|.$$

If $|\phi(t) - \varphi(t)| \le \frac{\varepsilon}{2}$ for some $t \in D(\phi, \varphi)$, then we get by (3.10.22) and (3.10.23) that

$$|H(\phi(t)) - H(\varphi(t))| \le |\phi(t) - \varphi(t)| \le \frac{1}{2} \|\phi - \varphi\|. \qquad (3.10.24)$$

Since h is continuous and strictly increasing, the function $h\left(u + \frac{\varepsilon}{2}\right) - h(u)$ attains its minimum on the closed and bounded interval $[-\alpha, \alpha]$. Thus, if $\frac{\varepsilon}{2} < |\phi(t) - \varphi(t)|$ for some $t \in D(\phi, \varphi)$, then by (H.2) and (H.3) we conclude that

$$1 \ge \frac{h(\phi(t)) - h(\varphi(t))}{\phi(t) - \varphi(t)} > \lambda,$$

where

$$\lambda := \frac{1}{2\alpha} \min \{h(u + \varepsilon/2) - h(u) : u \in [-\alpha, \alpha]\} > 0.$$

Hence, (3.10.20) implies

$$|H\phi(t) - H\varphi(t)| \leq (1 - \lambda) \|\phi(t) - \varphi(t)\|. \tag{3.10.25}$$

Consequently, combining (3.10.24) and (3.10.25) we obtain

$$|H\phi(t) - H\varphi(t)| \leq \delta \|\phi - \varphi\|,$$

where

$$\delta = \max \left\{ \frac{1}{2}, 1 - \lambda \right\} < 1.$$

The proof is complete.

Example 3.11. Let $\alpha \in (0,1)$ and $k \in \mathbb{N}$ be fixed elements and $u \in (-1,1)$.

1. The condition (H.2) is not satisfied for the function $h_1(u) = \frac{1}{2k} u^{2k}$.
2. The function $h_2(u) = \frac{1}{2k+1} u^{2k+1}$ satisfies (H.1-H.3).

Proof. Since $h_1'(u) = u^{2k-1} < 0$ for $-1 < u < 0$, the condition (H.2) is not satisfied for h_1. Evidently, (H.1-H.2) hold for h_2. (H.3) follows from the fact that $h_2'(u) \leq \alpha^{2k}$ and $\alpha \in (0,1)$.

3.11 Open Problems

Open Problem 1.
Consider the neutral delay functional difference equation

$$x(n+1) = \alpha x(n+1-h) + ax(n) - q(n,x(n),x(n-h)), \, n \in \mathbb{Z}^+ \tag{3.11.1}$$

where the function $q : \mathbb{Z}^+ \times \mathbb{R} \times \mathbb{R} \to \mathbb{R}$ is continuous and α and a are constants. For $a > 1$, (3.11.1) is equivalent to

$$\triangle \left[(x(n) - \alpha x(n-h)a^{1-n}) \right] = \left[a\alpha x(n-h) - q(n,x(n),x(n-h)) \right] a^{-n}. \tag{3.11.2}$$

We search for a solution of (3.11.1) having the property

$$(x(n) - \alpha x(n-h))a^{-n} \to 0, \text{ as } n \to \infty.$$

Hence, by summing (3.11.2) from 0 to ∞ we get the following advanced type Volterra difference equation,

$$x(n+1) = \alpha x(n-h) - \sum_{s=n}^{\infty} \left[a\alpha x(s-h) - q(s,x(s),x(s-h)) \right] a^{n-s} \tag{3.11.3}$$

which is an indication to study the general advanced type Volterra difference equation

$$x(n) = f(n,x(n),x(n-h)) - \sum_{s=n}^{\infty} Q(s,x(s),x(s-h))C(n-s). \qquad (3.11.4)$$

Under suitable conditions, one might explore the boundedness of solutions and the existence of periodic solutions. Finding suitable Lyapunov functionals to imply the results is almost impossible. It is my suggestion that the use of fixed point theory would be fruitful. For instance, if the right spaces are set up, one have the choice to use the contraction mapping principle; Theorem 3.5.1 or the following Krasnoselskii-Schaefer Theorem.

Theorem 3.11.1 (Krasnosselskii-Schaefer Theorem [25]). *Let* $(S, \|\cdot\|)$ *be a Banach space. Suppose* $B : S \to S$ *is a contraction map, and* $A : S \to S$ *is continuous and maps bounded sets into compact sets. Then either*

(i) $x = \lambda B(\frac{x}{\lambda}) + \lambda Ax$ *has a solution in* S *for* $\lambda = 1$, *or*
(ii) the set of all such solutions, $0 < \lambda < 1$, *is unbounded.*

It is noted that Krasnoselskii-Schaefer's theorem requires *a priori* bounds on solutions of a corresponding auxiliary equation. To obtain such bounds, one would have to construct a Lyapunov functional that is suitable for Krasnoselskii-Schaefer's theorem. For problem (3.11.4), the auxiliary equation is given by

$$x(n) = \lambda f(n,x(n)/\lambda,x(n-h)/\lambda) - \lambda \sum_{s=n}^{\infty} Q(s,x(n),x(n-h))C(n-s).$$

Open Problem 2.
Extend the results of Section 3.9 to Volterra summation equations of the form

$$x(t) = a(t) - \sum_{s=0}^{n-1} C(n,s)g(s,x(s)).$$

This is an unexplored area of research.

Open Problem 3.
We have seen in Chapter 3 that fixed point theory was successfully used to obtain asymptotic stability, while as in Chapter 2 we obtained uniform asymptotic stability using Lyapunov functionals. To my knowledge, no one has been able to use fixed point theory to obtain uniform asymptotic stability. Such results would mainly rest on how the set S is defined. Being able to do so would revolutionize the concept of fixed point theory and open the door for new research in differential/difference equations and even in dynamical systems on time scales.

Open Problem 4.
Consider the following system of two neurons:

$$\begin{cases} x_1(n+1) = h_1 x_1(n) + \beta_1 f(x_1(n-\tau)) + a_1 f(x_2(n-\tau_1)) \\ x_2(n+1) = h_2 x_2(n) + \beta_2 f(x_2(n-\tau)) + a_2 f(x_1(n-\tau_2)) \end{cases} \qquad (3.11.5)$$

where $x_1(n)$ and $x_1(n)$ denote the activations of the two neurons, $\tau_i, (i=1,2)$ and τ denote the synaptic transmission delays, a_1 and a_2 are the synaptic coupling weights, $f : \mathbb{R} \to \mathbb{R}$ is the activation function with $f(0) = 0$.

Use either Lyapunov functionals (Chapter 2) or fixed point theory (Chapter 3) to analyze the system and compare both methods.

Chapter 4
Periodic Solutions

This chapter is devoted to the study of periodic solutions of functional difference systems with finite and infinite delay. We will obtain different results concerning Volterra difference equations with finite and infinite delays, using fixed point theory. Fixed point theory will enable us to obtain results concerning stability, classification of solutions, existence of positive solutions, and the existence of periodic solutions and positive periodic solutions. In the analysis, we make use of Schaefer fixed point theorem, [159], Krasnoselskii's fixed point theorem, [97], and Schauder fixed point theorem. We apply our results to infinite delay Volterra difference equations, by constructing suitable Lyapunov functionals to obtain the a priori bound on all possible solutions. We transition to systems or coupled Volterra infinite delay difference equations and show the existence of a periodic solution and asymptotically periodic solution. For some classes of nonlinear systems with delay, it is shown that the presence of the time delay results in the existence of periodic solutions. We end the chapter by considering functional difference equation that has the characteristic that every constant is a solution. Then by means of fixed point theory we show that the unique solution converges to a pre-determined constant or a periodic solution. In addition we show the solution is stable and that its limit function serves as a global attractor. Most of the results of this chapter can be found in [1, 52, 127, 131, 135], and [137].

4.1 Periodic Solutions in Finite and Infinite Delays Equations

This chapter is entirely devoted to the study of existence of periodic solutions of functional difference equations and in particular Volterra infinite delay difference equations. We begin by discussing some results from the celebrated paper of Elaydi [52], in which the existence of a periodic solution is directly tied up to (UAS). We consider the following systems of difference equations of non-convolution type

© Springer Nature Switzerland AG 2018
Y. N. Raffoul, *Qualitative Theory of Volterra Difference Equations*,
https://doi.org/10.1007/978-3-319-97190-2_4

$$x(n+1) = A(n)x(n) + \sum_{r=0}^{n} B(n,r)x(r) \qquad (4.1.1)$$

and its corresponding perturbed system

$$x(n+1) = A(n)x(n) + \sum_{r=0}^{n} B(n,r)x(r) + g(n) \qquad (4.1.2)$$

where a, B are $k \times k$ matrix functions on $\mathbb{Z}^+$ and $\mathbb{Z}^+ \times \mathbb{Z}^+$, respectively, and g is a vector function on $\mathbb{Z}^+$. As before, we let $R(n,m)$ be the resolvent matrix of (4.1.1). Our objective is to find a periodic solution for the difference system with infinite delay

$$z(n+1) = A(n)x(n) + \sum_{r=-\infty}^{n} B(n,r)z(r) + g(n), \qquad (4.1.3)$$

where

$$A(n+N) = A(n), \; B(n+N, m+N) = B(n,m), \; g(n+N) = g(n). \qquad (4.1.4)$$

It can be easily shown, see [52], that

$$R(n+N, m+N) = R(n,m).$$

Hence we have the following theorem.

Theorem 4.1.1 ([52]). *Suppose that the zero solution of Equation* (4.1.1) *is (UAS). Then Equation* (4.1.3) *has the unique N-periodic solution*

$$z(n) = \sum_{m=-\infty}^{n-1} R(n, m+1)g(m).$$

The next theorem provides criteria for the (UAS) of Equation (4.1.1).

Theorem 4.1.2 ([52]). *Let*

$$|x| = \sum_{i=1}^{k} |x_i|, \; \beta_{jn}(n) = \sum_{s=n}^{\infty} b_{ji}(s,n)| < \infty.$$

Assume that

$$\sum_{j=1}^{k} [|a_{ji}(n)| + |\beta_{ji}(n)| \le 1 - c, 1 \le i \le k, \, n \ge n_0, \textit{ for some } c \in (0,1).$$

Then the zero solution of Equation (4.1.1) *is (UAS).*

As in the case of Chapter 2, we feel the need for the development of a more general theory for the existence of periodic solutions that will accommodate a wider range of equations. Thus, in this section we consider the functional nonlinear system of difference equations with either finite or infinite delay,

$$\triangle x(n) = F(n, x_n), n \in \mathbb{Z} \tag{4.1.5}$$

where $F : \mathbb{Z} \times BC \to \mathbb{R}^k$ is continuous in x and T-periodic in n. Here BC is the space of bounded sequences $\phi : (-\infty, 0] \to \mathbb{R}^k$ with the maximum norm $||\cdot||$. By x_n we mean that $x_n(s) = x(n+s)$ for $s \leq 0$. Let $(P_T, ||\cdot||)$ be the Banach space of T-periodic sequences $\phi : \mathbb{Z} \to \mathbb{R}^k$ with the maximum norm

$$||x|| = \max_{n \in [0, T-1]} |x(n)|.$$

Also, we let

$$P_T^0 = \Big\{ \phi \in P_T : \sum_{s=0}^{T-1} \phi(s) = 0 \Big\}.$$

Proving the existence of a periodic solution of (4.1.5) rest on the following Schaefer fixed point theorem.

Theorem 4.1.3 ([159]). *Let $(\mathbb{B}, |\cdot|)$ be a normed linear space, H a continuous mapping of $\mathbb{B}$ into $\mathbb{B}$ which is compact on each bounded subset of $\mathbb{B}$. Then either (i) the equation $x = \lambda H x$ has a solution for $\lambda = 1$, or (ii) the set of all solutions x, for $0 < \lambda < 1$, is unbounded.*

We make the following assumptions.
(a) For every $\phi \in P_T^0$, there exists a constant $d_\phi \in \mathbb{R}$ such that $\sum_{s=0}^{T-1} F(s, \psi_s) = 0$ where

$$\begin{cases} \psi(n) = d_\phi + \sum_{s=0}^{n-1} \phi(s), \text{ for } n \geq 1, \\ \psi(n) = d_\phi + \sum_{s=0}^{j-1} \phi(s), \text{ for } n \leq 0, \, n = j \mod T, \, 1 \leq j \leq T. \end{cases} \tag{4.1.6}$$

(b) Let $E(\phi)(n) = \psi(n)$ be continuous in ϕ with $E : P_T^0 \to P_T$ such that for each $\alpha > 0$, there exists a constant $L_\alpha > 0$ such that $|d_\phi| \leq L_\alpha$ whenever $||\phi|| \leq \alpha$.
The following proposition assures that ψ is well defined. Its proof is straightforward and therefore omitted.

Proposition 4.1. *Let n and $T \geq 1$ be any given integers. Then there exist unique integers K and j, $1 \leq j \leq T$, such that $n = KT + j$.*

Theorem 4.1.4 ([135]). *Suppose conditions (i) and (ii) hold. For $0 < \lambda < 1$, define $G_\lambda : P_T^0 \to P_T^0$ by*

$$G_\lambda(\phi)(n) = \lambda F(n, \psi_n).$$

If there is a constant $D > 0$ such that $||\phi|| < D$ whenever ϕ is a fixed point of G_λ, then Equation (4.1.5) has a T-periodic solution .

Proof. First we note that P_T is equivalent to $\mathbb{R}^{Tk}$. Let $n \in \mathbb{Z}$. By the continuity of F and condition (i), we can easily see that

$$\sum_{s=0}^{T-1} G_\lambda(\phi)(s) = \lambda \sum_{s=0}^{T-1} F(s, \psi_s) = 0.$$

Hence, we have that $G_\lambda(\phi) \in P_T^0$. For each $\alpha > 0$, the set $S = \{E(\phi) : \phi \in P_T^0, ||\phi|| \leq \alpha\}$ is closed and bounded by (ii). Let $Q = \{G_\lambda(\phi)(n) : \phi \in S\}$. Then S is a subset of $\mathbb{R}^{Tk}$ which is closed and bounded and thus compact. As G_λ is continuous in ϕ, it maps compact sets into compact sets. Therefore, $Q = G_\lambda(S)$ is compact. The hypothesis $||\phi|| < D$ rules out Condition (i) of Theorem 4.1.3 and thus applying Schaefer's theorem to $\phi = G_\lambda(\phi)$ we conclude that G_λ has a fixed point for $\lambda = 1$. That is $\phi = G_1\phi = F(n, \psi_n)$. On the other hand, $\phi(n) = \triangle\psi(n) = F(n, \psi_n)$. Thus, ψ is a T-periodic solution of (4.1.5). This completes the proof.

Corollary 4.1. *Suppose conditions (a) and (b) hold. Assume the functional F maps bounded sets into compact sets. If there exists a positive constant J such that any T-periodic solution $x(n)$ of*

$$\triangle x(n) = \lambda F(n, x_n), \ \lambda \in (0, 1) \tag{4.1.7}$$

satisfies $||x|| < J$, then (4.1.5) has a T-periodic solution .

Proof. Since $G_\lambda(\phi)(n) = \lambda F(n, \psi_n)$, then any fixed solution ϕ of G_λ implies the existence of a T-periodic solution of (4.1.5). As the functional F maps bounded sets into compact subsets, we have, whenever $||\psi|| \leq J$, that $|F(n, \psi_n)| \leq R$, where R depends on the a priori bound J. Let ϕ be a fixed solution of G_λ. Then $\phi(n) = \triangle\psi(n) = \lambda F(n, \psi_n)$. Since all T-periodic solutions of (4.1.7) have a priori bound J, by Theorem 4.1.4, Equation (4.1.5) has a T-periodic solution. This completes the proof.

Corollary 4.2. *Suppose conditions (i) and (ii) hold. If there exist constants $M, r, \ 0 < r < 1$ such that*

$$|F(n, \psi_n)| \leq r||\phi|| + M, \text{ for all } \phi \in P_T^0$$

where ψ is given by (4.1.6), then Equation (4.1.5) has a T-periodic solution.

Proof. The proof is straightforward. To see this, let ϕ be a fixed solution of G_λ. Then for $\phi \in P_T^0$

$$|\phi(n)| = \lambda|F(n, \psi_n)| \leq r||\phi|| + M,$$

from which we arrive at

$$||\phi|| \leq \frac{M}{1 - r}.$$

Hence, Equation (4.1.5) has a T-periodic solution by Theorem 4.1.4.

For the next theorem we consider the functional delay equation

$$\triangle x(n) = L(n, x_n) + p(n), n \in \mathbb{Z} \tag{4.1.8}$$

where $L : \mathbb{Z} \times BC \to \mathbb{R}^k$ is continuous and linear in x, T-periodic in n and $p \in P_T$.

Theorem 4.1.5 ([135]). *Suppose that for every d in $\mathbb{R}^k$, the $k \times k$ matrix $L(n, \cdot)$ satisfies the relation*

$$L(n,\cdot)d = L(n,d) \text{ and } \sum_{n=0}^{T-1} L(n,\cdot) \text{ is invertible.}$$

If there is a priori bound on all possible T-periodic solutions of

$$\triangle x(n) = \lambda \Big[L(n, x_n) + p(n)\Big], \; \lambda \in (0,1) \tag{4.1.9}$$

then Equation (4.1.8) has a T-periodic solution .

Proof. First we note that

$$\sum_{n=0}^{T-1} L(n,\cdot) \text{ is invertible if and only if the matrix } \Big(\sum_{n=0}^{T-1} L(n,\cdot) \Big)^{-1}$$

exists. In view of Corollary 4.1, we only need to verify *(i)* and *(ii)*. Set $F(n,\psi_n) = L(n,\psi_n) + p(n)$ and

$$d_\phi = -\Big(\sum_{n=0}^{T-1} L(n,\cdot) \Big)^{-1} \Big[\sum_{n=0}^{T-1} L\Big(n, (\sum_{s=0}^{n-1} \phi(s))_n\Big) + \sum_{n=0}^{T-1} p(n) \Big]. \tag{4.1.10}$$

For $\phi \in P_T^0$, $d_\phi \in \mathbb{R}^k$ is uniquely determined by (4.1.10). Since $L(n,\cdot)d = L(n,d)$ we have

$$\sum_{n=0}^{T-1} L\Big(n, (d_\phi + (\sum_{s=0}^{n-1} \phi(s))_n\Big) + \sum_{n=0}^{T-1} p(n) = 0.$$

Thus, $\sum_{s=0}^{T-1} F(s,\psi_s) = 0$. Let E be defined as in *(ii)*, then it is readily verified that $E : P_T^0 \to P_T$ and continuous in ϕ. Now, since L is linear and continuous in the second argument, there exists a $\beta > 0$ such that for any $\psi \in BC$, $|L(n,\psi_n)| \le \beta \|\psi\|$. This yields

$$\Big| L\Big(n, (\sum_{s=0}^{n-1} \phi(s))_n\Big) \Big| \le \beta T \|\phi\|.$$

Thus, from (4.1.10) one obtains for $\|\phi\| \le \alpha$ that

$$d_\phi \le \Big| \Big(\sum_{n=0}^{T-1} L(n,\cdot) \Big)^{-1} \Big| \Big(\beta T \alpha + \|p\| \Big) T =: L\alpha.$$

Thus, by Corollary 4.1 Equation (4.1.8) has a T-periodic solution and the proof is complete.

4.2 Application to Functional Difference Equations

It is obvious that Theorem 4.1.4 is of general nature and hence we will apply it to different types of functional difference equations. Namely, we will obtain existence of periodic solutions of scalar Volterra difference equations with finite or infinite delay.

4.2.1 Finite Delay Difference Equations

We will use Theorem 4.1.4 to prove the existence of a periodic solution for a scalar difference equation with finite delay.

Theorem 4.2.1 ([135]). *Consider the scalar delay difference equation*

$$\triangle x(n) = a(n)x(n) + b(n)x(n-h) + p(n), \, n \in \mathbb{Z}, \qquad (4.2.1)$$

where the sequences $a(n), b(n)$, and $p(n)$ are T-periodic sequences, and $h \in \mathbb{Z}$ with $h \geq 0$.
Suppose that either $a(n) > 0$ or $a(n) < 0$ for all $n \in \mathbb{Z}$. Suppose there exists a constant $N > 1$ such that

$$|a(n)| - N|b(n+h)| \geq 0.$$

If
(i) $\rho - ||b|| - \rho T(||a|| + ||b||) > 0$
where $\rho = \min_{n \in [0,T-1]} |a(n)|$, then Equation (4.2.1) has a T-periodic solution.

Proof. First we note that since either $a(n) > 0$ or $a(n) < 0$ for all $n \in \mathbb{Z}$, we have $\sum_{n=0}^{T-1} a(n) \neq 0$. Define L by $L(n, x_n) = a(n)x(n) + b(n)x(n-h)$. Then L is linear and $L(n, \cdot) = a(n) + b(n)$. In view of Theorem 4.1.5, we need to show that $\sum_{n=0}^{T-1} L(n, \cdot) \neq 0$ and all T-periodic solutions of

$$\triangle x(n) = \lambda \Big[a(n)x(n) + b(n)x(n-h) + p(n) \Big], \, \lambda \in (0,1) \qquad (4.2.2)$$

have a priori bound. By noting that $b(n+h)$ is also T-periodic, we have

$$\sum_{n=0}^{T-1} |b(n)| = \sum_{s=-h}^{T-h-1} |b(s+h)| = \sum_{s=0}^{T-1} |b(s+h)|.$$

Thus for $a(n) > 0$,

$$\sum_{n=0}^{T-1} (a(n) + b(n)) \geq \sum_{n=0}^{T-1} (|a(n)| - |b(n)|) = \sum_{n=0}^{T-1} (|a(n)| - |b(n+h)|).$$

By making use of $|a(n)| - N|b(n+h)| \geq 0$ in the above inequality, we get

$$\sum_{n=0}^{T-1} (a(n) + b(n)) \geq \sum_{n=0}^{T-1} (|a(n)| - |b(n+h)|)$$

$$= \frac{N-1}{N} \sum_{n=0}^{T-1} |a(n)| + \frac{1}{N} \sum_{n=0}^{T-1} (|a(n)| - N|b(n+h)|)$$

$$\geq \frac{N-1}{N} \sum_{n=0}^{T-1} a(n) > 0.$$

Next, suppose $a(n) < 0$ for all $n \in \mathbb{Z}$. Then

$$\sum_{n=0}^{T-1} (a(n) + b(n)) \leq \sum_{n=0}^{T-1} (-|a(n)| + |b(n+h)|)$$

$$= \frac{1-N}{N} \sum_{n=0}^{T-1} |a(n)| - \frac{1}{N} \sum_{n=0}^{T-1} (|a(n)| - N|b(n+h)|)$$

$$\leq \frac{1-N}{N} \sum_{n=0}^{T-1} |a(n)| < 0.$$

Hence, we have shown that $\sum_{n=0}^{T-1} L(n, \cdot) \neq 0$ for all $n \in \mathbb{Z}$. Now we turn our attention to finding the a priori bound. Let $x(n)$ be a T-periodic solution of (4.2.2). By summing equation (4.2.2) from n to $n+T-1$ we get

$$0 = x(n+T) - x(n) = \lambda \sum_{s=n}^{n+T-1} \left[a(s)x(s) + b(s)x(s-h) + p(s) \right].$$

Thus,

$$\sum_{s=n}^{n+T-1} a(s)x(s) = - \sum_{s=n}^{n+T-1} (b(s)x(s-h) + p(s)).$$

Since there exists an $n^* \in [n, n+T-1]$ such that

$$T|a(n^*)||x(n^*)| \leq \sum_{s=n}^{n+T-1} |a(s)||x(s)|,$$

we arrive from the above relation that

$$T|a(n^*)||x(n^*)| \leq \sum_{s=n}^{n+T-1} (|b(s)||x(s-h)| + |p(s)|)$$

$$\leq T||b|| \, ||x|| + T||p||.$$

As a consequence, we get

$$|x(n^*)| \leq \frac{||b||}{\rho} ||x|| + \frac{||p||}{\rho}.$$

Using Equation (4.2.2) we have

$$|\triangle x| \leq |a(n)||x(n)| + |b(n)||x(n-h)| + |p(n)|$$

$$\leq ||a|| \, ||x|| + ||b|| \, ||x|| + ||p||$$

$$= (||a|| + ||b||)||x|| + ||p||.$$

For all $n \in \mathbb{Z}$, we write $x(n) \in P_T$ as

$$x(n) = x(n^*) + \sum_{s=n^*}^{n+T-1} \triangle x(s).$$ (4.2.3)

Using (4.2.3) and then the norms of $x(n^*), \triangle x$ and x we get

$$|x(n)| \leq |x(n^*)| + \sum_{s=n}^{n+T-1} |\triangle x(s)|$$

$$\leq ||x(n^*)|| + T||\triangle x||$$

$$\leq \frac{||b||}{\rho}||x|| + \frac{||p||}{\rho} + T\Big((||a|| + ||b||)||x|| + ||p||\Big).$$

The above inequality yields

$$||x|| \leq \frac{T\rho||p||}{\rho - ||b|| - \rho T(||a|| + ||b||)}.$$

This defines a priori bound on all possible T-periodic solutions of Equation (4.2.2). Hence, Equation (4.2.1) has a T-periodic solution by Theorem 4.1.4.

In the next corollary, we relax condition *(i)* of Theorem 4.2.1.

Corollary 4.3 ([135]). *Suppose the hypothesis of Theorem 4.2.1 holds with (i) being replaced by*

$$|a(n) + b(n)|\Big(|d^{-1}|T^2(||a|| + ||b||) + T\Big) = \varsigma < 1,$$

where

$$d^{-1} = \Big[\sum_{n=0}^{T-1}(a(n) + b(n))\Big]^{-1}.$$

Then Equation (4.2.1) has a T-periodic solution.

Proof. Take ϕ, ψ, and $L(n, \cdot)$ to be as in Theorem 4.2.1. In view of Corollary 4.2 we only need to show that $|F(n, \psi_n)| \leq r||\phi|| + M$, $M > 0$ is a constant and $0 < r < 1$. By a similar argument as in Theorem 4.1.5, one may easily find that

$$d_\phi = -\Big[\sum_{n=0}^{T-1}(a(n) + b(n))\Big]^{-1}\Big\{\sum_{n=0}^{T-1}(a(n) + b(n))\sum_{s=0}^{n-1}\phi(s) + \sum_{n=0}^{T-1}p(n)\Big\}.$$

Now,

$$|d_\phi| \leq |d^{-1}|T(|a| + |b|)T|\phi| + T|P|$$

$$\leq |d^{-1}|T^2(|a| + |b|)T||\phi|| + T||P||.$$

This yields to

$$|F(n,\psi_n)| \le |a(n)+b(n)||d_\phi| + |a(n)+b(n)|T||\phi|| + ||p||$$
$$\le |a(n)+b(n)| \left(|d^{-1}|T^2(||a||+||b||) + T \right) = \varsigma < 1.$$

Thus, by Corollary 4.2, Equation (4.2.1) has a T-periodic solution. This completes the proof.

4.2.2 Infinite Delay Volterra Difference Systems

In this section, we apply Corollary 4.2 and Theorem 4.1.5 to show that the Volterra difference system with infinite delay given by

$$\triangle x(n) = A(n)x(n) + \sum_{s=-\infty}^{n} B(n,s)x(s) + g(n), \quad -\infty < s \le t < \infty \qquad (4.2.4)$$

where A, B are $k \times k$ T-periodic matrices and g is a given $k \times 1$ T-periodic sequence, has a T-periodic solution. We begin with the following theorem.

Theorem 4.2.2 ([135]). *Suppose that*

$$D = \sum_{s=0}^{T-1} \left(A(n) + \sum_{s=-\infty}^{n} B(n,s) \right) \text{ is invertible,} \qquad (4.2.5)$$

$$\max_{n\in[0,T-1]} \left[\left| A(n) + \sum_{s=-\infty}^{n} B(n,s) \right| MT + (|A(n)| + \sum_{s=-\infty}^{n} |B(n,s)|)T \right] =: \varsigma < 1, \quad (4.2.6)$$

where

$$M = |D^{-1}| \sum_{u=0}^{T-1} \left(|A(u)| + \sum_{s=-\infty}^{u} |B(u,s)| \right).$$

Then Equation (4.2.4) has a T-periodic solution.

Proof. Set $F(n,x_n) = A(n)x(n) + \sum_{s=-\infty}^{n} B(n,s)x(s) + p(n)$.
Then (4.2.5) and Theorem 4.1.4 imply that for each $\phi \in P_T^0$, there exists a unique $d_\phi \in \mathbb{R}$ such that $\sum_{n=0}^{T-1} F(n,\psi_n) = 0$, where $\psi(n)$ is defined by (4.1.6). In fact for $\sum_{n=0}^{T-1} F(n,\psi_n) = 0$ gives

$$\sum_{n=0}^{T-1} (A(n)(d + \sum_{s=0}^{n-1} \phi(s)) + \sum_{n=0}^{T-1} \left[\sum_{s=-\infty}^{n} B(n,s)(d + \sum_{u=0}^{s-1} \phi(u)) \right] + \sum_{n=0}^{T-1} p(n) = 0.$$

This yields

$$d_\phi = -D^{-1} \left[\sum_{n=0}^{T-1} \left(A(n) \sum_{s=0}^{n-1} \phi(s) + \sum_{s=-\infty}^{n} B(n,s) \sum_{u=0}^{s-1} \phi(u) + p(n) \right) \right].$$

Thus,

$$|d_\phi| \leq MT||\phi|| + |D^{-1}|||p||T.$$

On the other hand,

$$|F(n, \psi_n)| \leq \left| A(n)\psi_n + \sum_{s=-\infty}^{n} B(n,s)\psi_s + p(n) \right|$$

$$\leq \left| A(n)(d_\phi + \sum_{s=0}^{n-1} \phi(s)) \right.$$

$$\left. + \sum_{s=-\infty}^{n} B(n,s)(d_\phi + \sum_{u=0}^{s-1} \phi(u)) + \sum_{n=0}^{T-1} p(n) \right|$$

$$\leq \left| A(n) + \sum_{s=-\infty}^{n} B(n,s) \right| |d_\phi|$$

$$+ \left(|A(n)| + \sum_{s=-\infty}^{n} |B(n,s)| \right) T||\phi|| + ||p||.$$

Replacing $|d_\phi|$ by its value, we get

$$|F(n, \psi_n)| \leq \varsigma||\phi|| + K$$

where $K = max_{n \in [0.T-1]} \left| A(n) + \sum_{s=-\infty}^{n} B(n,s) \right| |D^{-1}| \, ||p||T + ||p||$. Thus, Equation (4.2.4) has a T-periodic solution by Corollary 4.2. This completes the proof.

Remark 4.1. Condition (4.2.6) is severe and therefore in the next theorem we avoid it by appealing to Lyapunov functional.

But first, if $A = (a_{ij})$ is a $k \times k$ real matrix, then we define the norm of A by

$$|A| = \max_{1 \leq i \leq k} \sum_{j=1}^{k} |a_{ij}|.$$

Theorem 4.2.3 ([135]). *Consider the 2-dimensional system*

$$\triangle x(n) = \lambda \left[Ax(n) + \sum_{j=-\infty}^{n} C(n-j)x(j) + g(n) \right], \ \lambda \in (0,1) \tag{4.2.7}$$

$A = \begin{pmatrix} 1 & 0 \\ 0 & -1 \end{pmatrix}, \Sigma_{j=-\infty}^{n-1}\Sigma_{s=n}^{\infty} |C(s-j)| < \infty, g(n) \in P_T$,

$\Sigma_{u=0}^{\infty} |C(u)| = \alpha \leq \frac{2}{25}$ *and* $C^T(u) = C(u)$(*transpose*). *Assume that* $Q = \Sigma_{n=0}^{T-1} \left(A(n) + \Sigma_{s=-\infty}^{n} B(n,s) \right)$ *is invertible. Then* (4.2.7) *has a solution in* P_T *for* $\lambda = 1$.

Proof. Set $F(n,x_n) = A(n)x(n) + \sum_{s=-\infty}^{n} C(n-s)x(s) + p(n)$. If we let

$$d_\phi = -Q^{-1}\left[\sum_{n=0}^{T-1}\left(A(n)\sum_{s=0}^{n-1}\phi(s) + \sum_{s=-\infty}^{n} C(n-s)\sum_{u=0}^{s-1}\phi(u) + p(n)\right)\right],$$

then by a similar argument as in Theorem 4.2.2, it is readily verified that

$$\sum_{n=0}^{T-1} F(n,\psi_n) = 0,$$

where $\psi(n)$ is defined by (4.1.6). Also, by a similar argument as in the Theorem 4.2.2, it can be easily shown that there exists a constant $L_\alpha > 0$ such that $|d_\phi| \le L_\alpha$. Next we show that F maps bounded sets into bounded sets. Let J be a given positive constant. Then if ψ is given by (4.1.6), we set $S = \{\psi : \phi \in P_T^0, \|\psi\| \le J\}$ which is closed and bounded. Now

$$|F(n,\psi_n)| \le \left|A(n)\psi_n + \sum_{s=-\infty}^{n} C(n-s)\psi_s + p(n)\right|$$

$$\le |A|J + \sum_{u=0}^{\infty} |C(u)|JT + |p|$$

$$\le |A|J + JT\frac{2}{25} + \|p\| \le M, M > 0.$$

This shows that F maps bounded sets into bounded sets. According to Corollary 4.1, it is left to show that all T-periodic solutions of (4.2.7) have a priori bound. Note that (4.2.7) has an a priori bound on all its T-periodic solutions if and only if

$$x(n+1) = Dx(n) + \lambda \sum_{j=-\infty}^{n} C(n-j)x(j) + \lambda g(n) \qquad (4.2.8)$$

does, where $D = \begin{pmatrix} \lambda+1 & 0 \\ 0 & 1-\lambda \end{pmatrix}$. Find $E = E^T$ such that

$$D^T ED - E = -2\lambda I, \text{ as follows.}$$

Let $E = \begin{pmatrix} a & b \\ c & d \end{pmatrix}$. Then $D^T ED - E = -2\lambda I$ implies that

$$\begin{pmatrix} a(\lambda+1)^2 - a & b(1-\lambda^2) - b \\ c(1-\lambda)^2 - c & d(1-\lambda^2) - d \end{pmatrix} = -2\lambda \begin{pmatrix} 1 & 0 \\ 0 & 1 \end{pmatrix}$$

from which it follows

$$E = \begin{pmatrix} -\frac{2}{\lambda+2} & 0 \\ 0 & \frac{2}{2-\lambda} \end{pmatrix}.$$

Thus
$$|E| \leq 2 \text{ for } \lambda \in (0,1).$$

Also,

$$D^T E = \begin{pmatrix} \lambda+1 & 0 \\ 0 & 1-\lambda \end{pmatrix} \times \begin{pmatrix} \frac{-2}{\lambda+2} & 0 \\ 0 & \frac{2}{2-\lambda} \end{pmatrix} = \begin{pmatrix} -2\frac{\lambda+1}{\lambda+2} & 0 \\ 0 & 2\frac{1-\lambda}{2-\lambda} \end{pmatrix}.$$

Thus $|D^T E| \leq 2$ for $\lambda \in (0,1)$. Find $\gamma > 2+2\alpha$ such that $(2+\gamma)\alpha < 2$. This is possible because for $\alpha \in (0, \frac{2}{25}]$, it is elementary to verify that $2+2\alpha < \frac{2}{\alpha} - 2$. Hence we may choose γ such that $2+2\alpha < \gamma < \frac{2}{\alpha} - 2$.
Define a Lyapunov type functional

$$V(n,x(\cdot)) = x^T(n)Ex(n) + \lambda\gamma \sum_{j=-\infty}^{n-1} \sum_{s=n}^{\infty} |C(s-j)|x^2(j).$$

It is of interest to note that V is not positive definite. Then along solutions of (4.2.8) we have

$$\triangle V = x^T(n+1)Ex(n+1) + \lambda\gamma \sum_{s=n+1}^{\infty} |C(s-n)|x^2(n)$$

$$-\lambda\gamma \sum_{j=-\infty}^{n-1} |C(n-j)|x^2(j) - x^T(n)Ex(n)$$

$$= \left[x^T(n)D^T + \lambda \sum_{j=-\infty}^{n} x^T(j)C^T(n-j) + \lambda g^T(n) \right]$$

$$E\left[Dx(n) + \lambda \sum_{j=-\infty}^{n} C(n-j)x(j) + \lambda g(n) \right]$$

$$-x^T(n)Ex(n) + \lambda\gamma \sum_{s=n+1}^{\infty} |C(s-n)|x^2(n)$$

$$-\lambda\gamma \sum_{j=-\infty}^{n-1} |C(n-j)|x^2(j)$$

$$= x^T(n)D^T EDx(n) + \lambda x^T(n)D^T E \sum_{j=-\infty}^{n} C(n-j)x(j)$$

$$+\lambda x^T(n)D^T Eg(n) + \lambda \sum_{j=-\infty}^{n} x^T(j)C^T(n-j)EDx(n)$$

$$+\lambda^2 \sum_{j=-\infty}^{n} x^T(j)C^T(n-j)E \sum_{j=-\infty}^{n} C(n-j)x(j)$$

$$+\lambda^2 \sum_{j=-\infty}^{n} x^T(j)C^T(n-j)Eg(n) + \lambda g^T(n)EDx(n)$$

$$+\lambda^2 g^T(n)E \sum_{j=-\infty}^{n} C(n-j)x(j) + \lambda^2 g^T(n)Eg(n) - x^T(n)Ex(n)$$

$$+\lambda\gamma \sum_{s=n+1}^{\infty} |C(s-n)|x^2(n) - \lambda\gamma \sum_{j=-\infty}^{n-1} |C(n-j)|x^2(j).$$

Hence

$$\triangle V = x^T(n)(D^T ED - E)x(n) + 2\lambda \sum_{j=-\infty}^{n} x^T(n)D^T EC(n-j)x(j)$$

$$+2\lambda x^T(n)D^T Eg(n) + 2\lambda^2 g^T(n)E \sum_{j=-\infty}^{n} C(n-j)x(j)$$

$$+\lambda^2 \sum_{j=-\infty}^{n} x^T(j)C^T(n-j)E \sum_{j=-\infty}^{n} C(n-j)x(j)$$

$$+\lambda\gamma \sum_{s=n+1}^{\infty} |C(s-n)|x^2(n)$$

$$-\lambda\gamma \sum_{j=-\infty}^{n-1} |C(n-j)|x^2(j) + \lambda^2 g^T(n)Eg(n).$$

Note that

$$2 \sum_{j=-\infty}^{n} x^T(n)D^T EC(n-j)x(j) \le 2 \sum_{j=-\infty}^{n} |x^T(n)||D^T E||C(n-j)||x(j)|$$

$$= |D^T E| \sum_{j=-\infty}^{n} |C(n-j)|2|x(n)^T||x(j)|$$

$$\le |D^T E| \sum_{j=-\infty}^{n} |C(n-j)|(x^2(n) + x^2(j))$$

$$\le 2 \sum_{j=-\infty}^{n} |C(n-j)|(x^2(n) + x^2(j))$$

$$= 2\alpha x^2(n) + 2 \sum_{j=-\infty}^{n} |C(n-j)|x^2(j).$$

In the next two terms we make use of the following fact: for any real numbers a, b, and L with $L \ne 0$, $2ab \le \frac{a^2}{L^2} + L^2 b^2$, which can be easily proven by using the fact that $(\frac{a}{L} - Lb)^2 \ge 0$. As a consequence, for some $L > 0$ we have

$$2x^T(n)D^T Eg(n) \le 2|x^T(n)||D^T Eg(n)| \le \frac{x^2(n)}{L^2} + L^2 |D^T Eg(n)|^2$$

and

$$2\lambda g^T(n)E \sum_{j=-\infty}^{n} C(n-j)x(j) \le 2|\lambda g^T(n)E| \sum_{j=-\infty}^{n} |C(n-j)||x(j)|$$

$$\le \sum_{j=-\infty}^{n} |C(n-j)|2|g^T(n)E||x(j)|$$

$$\le \sum_{j=-\infty}^{n} |C(n-j)|\frac{x^2(j)}{L^2} + \sum_{j=-\infty}^{n} |C(n-j)|(g^T(n)EL)^2$$

$$= \sum_{j=-\infty}^{n} |C(n-j)|\frac{x^2(j)}{L^2} + \alpha(\lambda g^T(n)EL)^2.$$

For $u = s - n$,

$$\gamma \sum_{s=n+1}^{\infty} |C(s-n)|x^2(n) = \gamma \sum_{u=1}^{\infty} |C(u)|x^2(n)$$

$$= \gamma\alpha x^2(n) - \gamma|C(0)|x^2(n).$$

Also,

$$\gamma \sum_{j=-\infty}^{n-1} |C(n-j)|x^2(j) = \gamma \sum_{j=-\infty}^{n} |C(n-j)|x^2(j) - \gamma|C(0)|x^2(n).$$

Thus

$$\gamma \sum_{s=n+1}^{\infty} |C(s-n)| \, x^2(n) - \gamma \sum_{j=-\infty}^{n-1} |C(n-j)|x^2(j)$$

$$= \gamma\alpha x^2(n) - \gamma \sum_{j=-\infty}^{n} |C(n-j)|x^2(j).$$

Finally,

$$\lambda \sum_{j=-\infty}^{n} x^T(j) \, C^T(n-j)E \sum_{j=-\infty}^{n} C(n-j)x(j)$$

$$\le 2 \sum_{j=-\infty}^{n} |x^T(j)||C^T(n-j)| \sum_{j=-\infty}^{n} |C(n-j)||x(j)|$$

$$\le \left(\sum_{j=-\infty}^{n} |x^T(j)||C^T(n-j)| \right)^2 + \left(\sum_{j=-\infty}^{n} |C(n-j)||x(j)| \right)^2$$

$$= 2\left(\sum_{j=-\infty}^{n} |C(n-j)|\,|x(j)|\right)^2$$

$$= 2\left(\sum_{j=-\infty}^{n} |C(n-j)|^{\frac{1}{2}}|C(n-j)|^{\frac{1}{2}}|x(j)|\right)^2$$

$$\leq 2\sum_{j=-\infty}^{n} |C(n-j)| \sum_{j=-\infty}^{n} |C(n-j)|x^2(j)$$

$$= 2\alpha \sum_{j=-\infty}^{n} |C(n-j)|x^2(j), \quad \text{by Schwartz inequality for series.}$$

Putting everything together we obtain

$$\triangle V \leq \lambda\Bigg[-2x^2(n) + 2\alpha x^2(n) + 2\sum_{j=-\infty}^{n} |C(n-j)|x^2(j)$$

$$+\frac{x^2(n)}{L^2} + L^2(D^T Eg(n))^2 + \sum_{j=-\infty}^{n} |C(n-j)|\frac{x^2(j)}{L^2}$$

$$+\alpha(\lambda g^T(n)EL)^2 + \alpha\gamma x^2(n) - \gamma\sum_{j=-\infty}^{n} |C(n-j)|x^2(j)$$

$$+2\alpha\sum_{j=-\infty}^{n} |C(n-j)|x^2(j) + |\lambda g^T(n)Eg(n)|\Bigg]$$

$$= \lambda\Bigg[\left(-2 + 2\alpha + \alpha\gamma + \frac{1}{L^2}\right)x^2(n)$$

$$+ \left(2 - \gamma + 2\alpha + \frac{1}{L^2}\right)\sum_{j=-\infty}^{n} |C(n-j)|x^2(j)$$

$$+|g^T(n)Eg(n)| + ((D^T Eg(n))^2 + (g^T(n)E)^2\alpha)L^2\Bigg].$$

Since $-2 + 2\alpha + \alpha\gamma < 0$ and $2 - \gamma + 2\alpha < 0$ we may choose L large enough so that $-2 + 2\alpha + \alpha\gamma + \frac{1}{L^2} < 0$ and $2 - \gamma + 2\alpha + \frac{1}{L^2} < 0$. Then we have

$$\triangle V \leq \lambda\left[(-2 + 2\alpha + \alpha\gamma + \frac{1}{L^2})x^2(n) + M\right]$$

$$\leq \lambda\left[-\mu x^2(n) + M\right]$$

for some positive constants μ and M. Using the fact that $V \in P_T$, we have

$$0 = V(n+T) - V(n) = \sum_{i=n}^{n+T-1} \triangle V(i) \leq \lambda\left[-\mu\sum_{i=n}^{n+T-1} x^2(i) + TM\right]$$

from which it follows

$$\sum_{i=n}^{n+T-1} x^2(i) \leq \frac{TM}{\mu}$$

and

$$\sum_{j=1}^{T} |x(j+n-1)|^2 \leq \frac{TM}{\mu}.$$

Thus $|x(n)|^2$ is bounded over one period, and hence

$$\|x(n)\| \leq K, \quad \text{for some } K > 0.$$

Thus, every possible T-periodic solution $x(n)$ of (4.2.8) for $\lambda \in (0,1]$ is bounded. Therefore, by Corollary 4.1, Equation (4.2.8) has a T-periodic solution for $\lambda = 1$. It is obvious that condition (4.2.6) of the Theorem 4.2.2 cannot be satisfied for Equation (4.2.7) with $\lambda = 1$.

4.3 Periodicity in Scalar Nonlinear Neutral Systems

Next we use Krasnoselskii's fixed point theorem (Theorem 3.5.1) to show that the nonlinear neutral difference equation with functional delay

$$x(t+1) = a(t)x(t) + c(t)\triangle x(t-g(t)) + q\big(t,x(t),x(t-g(t))\big) \qquad (4.3.1)$$

has a periodic solution. As usual, in order to apply Krasnoselskii's fixed point theorem, one would need to construct two mappings; one is contraction and the other is compact. Also, by making use of the variation of parameters techniques we are able, using the contraction mapping principle, to show that the periodic solution is unique. Let T be an integer such that $T \geq 1$. We assume the periodicity conditions

$$a(t+T) = a(t), \quad c(t+T) = c(t), \quad g(t+T) = g(t), \quad g(t) \geq g^* > 0 \qquad (4.3.2)$$

for some constant g^*. Let BC is the space of bounded sequences $\phi : (-g^*,0] \to \mathbb{R}^k$ with the maximum norm $\|\cdot\|$. Materials of this section can be found in [111]. Define $P_T = \{\phi \in BC, \phi(t+T) = \phi(t)\}$. Then P_T is a Banach space when it is endowed with the maximum norm

$$\|x\| = \max_{t \in [0,T-1]} |x(t)|.$$

Also, we assume that

$$\prod_{s=t-T}^{t-1} a(s) \neq 1. \qquad (4.3.3)$$

Throughout this section we assume that $a(t) \neq 0$ for all $t \in [0, T-1]$. It is interesting to note that equation (4.3.1) becomes of advanced type when $g(t) < 0$. Since we are searching for periodic solutions, it is natural to ask that $q(t,x,y)$ is periodic in t and Lipschitz continuous in x and y. That is

$$q(t+T,x,y) = q(t,x,y) \qquad (4.3.4)$$

and

$$|q(t,x,y) - q(t,z,w)| \le L\|x - z\| + K\|y - w\| \qquad (4.3.5)$$

for some positive constants L and E. Note that

$$|q(t,x,y)| - |q(t,0,0)| \le |q(t,x,y) - q(t,0,0)| \le L\|x - 0\| + K\|y - 0\|$$
$$= L\|x\| + K\|y\|.$$

As a result,

$$|q(t,x,y)| \le L\|x\| + K\|y\| + |q(t,0,0)|. \qquad (4.3.6)$$

We have the following lemma.

Lemma 4.1. *Suppose* (4.3.2)–(4.3.4) *hold. If* $x(t) \in P_T$, *then* $x(t)$ *is a solution of equation* (4.3.1) *if and only if*

$$x(t) = c(t-1)x(t-g(t))$$

$$+ \frac{1}{1 - \prod_{s=t-T}^{t-1} a(s)} \sum_{r=t-T}^{t-1} \left[x(r - g(r))\Big(a(r)c(r-1) - c(r)\Big) \right.$$

$$\left. + q(r,x(r),x(r-g(r))) \right] \prod_{s=r+1}^{t-1} a(s). \qquad (4.3.7)$$

Proof. The proof is the same as for (3.5.2) by summing from $t - T$ to $t - 1$ and noting that for $x \in P_T$, $x(t) = x(t - T)$.

We use the following notion of compact mapping.
Let $\mathscr{S}$ be a subset of a Banach space $\mathscr{B}$ and $f : \mathscr{S} \to \mathscr{B}$. If f is continuous and $f(\mathscr{S})$ is contained in a compact subset of $\mathscr{B}$, then f is a compact mapping. We express equation (4.3.7) as

$$(H\varphi)(t) = (B\varphi)(t) + (A\varphi)(t) \qquad (4.3.8)$$

where $A, B : P_T \to P_T$ are given by

$$(B\varphi)(t) = c(t-1)\varphi(t-g(t)) \qquad (4.3.9)$$

and

$$(A\varphi)(t) = \left(1 - \prod_{s=t-T}^{t-1} a(s)\right)^{-1} \sum_{r=t-T}^{t-1} \Big[\varphi(r - g(r))[a(r)c(r-1) - c(r)]$$

$$+ q(r, \varphi(r), \varphi(r - g(r)))\Big] \prod_{s=r+1}^{t-1} a(s). \tag{4.3.10}$$

Lemma 4.2. *Suppose* (4.3.2)–(4.3.5) *hold. If A is defined by* (4.3.10), *then* $A : P_T \to P_T$ *and is compact.*

Proof. First we want to show that $(A\varphi)(t + T) = (A\varphi)(t)$.
Let $\varphi \in P_T$. Then using (4.3.10) we arrive at

$$(A\varphi)(t + T) = \left[1 - \prod_{s=t}^{t+T-1} a(s)\right]^{-1} \sum_{r=t}^{t+T-1} \Big[\varphi(r - g(r))[a(r)c(r-1) - c(r)]$$

$$+ q(r, \varphi(r), \varphi(r - g(r)))\Big] \prod_{s=r+1}^{t+T-1} a(s).$$

Let $j = r - T$, then

$$(A\varphi)(t + T) =$$

$$\left[1 - \prod_{s=t}^{t+T-1} a(s)\right]^{-1} \sum_{j=t-T}^{t-1} \Big[\varphi(j + T - g(j + T))[a(j + T)c(j + T - 1) - c(j + T)]$$

$$+ q(j + T, \varphi(j + T), \varphi(j + T - g(j + T)))\Big] \prod_{s=j+T+1}^{t+T-1} a(s)$$

$$= \left[1 - \prod_{s=t}^{t+T-1} a(s)\right]^{-1} \sum_{j=t-T}^{t-1} \Big[\varphi(j - g(j))[a(j)c(j - 1) - c(j)]$$

$$+ q(j, \varphi(j), \varphi(j - g(j)))\Big] \prod_{s=j+T+1}^{t+T-1} a(s).$$

Now let $k = s - T$, then

$$(A\varphi)(t + T) = \left[1 - \prod_{k=t-T}^{t-1} a(k)\right]^{-1} \sum_{j=t-T}^{t-1} \Big[\varphi(j - g(j))[a(j)c(j - 1) - c(j)]$$

$$+ q(j, \varphi(j), \varphi(j - g(j)))\Big] \prod_{k=j+1}^{t-1} a(s)$$

$$= (A\varphi)(t).$$

To see that A is continuous, we let $\varphi, \psi \in P_T$ with $\|\varphi\| \leq C$ and $\|\psi\| \leq C$. Let

$$\eta = \max_{t \in [0, T-1]} \left| \frac{1}{(1 - \prod_{s=t-T}^{t-1} a(s))} \right|, \quad \beta = \max_{r \in [t-T, t]} |a(r)c(r-1) - c(r)|,$$

$$\gamma = \max_{t \in [0, T-1]} \prod_{s=t-T}^{t-1} a(s). \tag{4.3.11}$$

Given $\varepsilon > 0$, take $\delta = \varepsilon/M$ such that $\|\varphi - \psi\| < \delta$, where $M = T\gamma\eta[\beta + L + K]$. By making use of (4.3.5) into (4.3.10) we obtain

$$\left\| \Big(A\varphi(t)\Big) - \Big(A\psi(t)\Big) \right\|$$

$$= \left\| \frac{1}{1 - \prod_{s=t-T}^{t-1} a(s)} \sum_{r=t-T}^{t-1} \left[\Big(\varphi(r - g(r)) - \psi(r - g(r))\Big)\Big(c(r-1)a(r) - c(r)\Big) \right. \right.$$

$$\left. \left. + \Big(q(r, \varphi(r), \varphi(r - g(r))) - q(r, \psi(r), \psi(r - g(r)))\Big) \right] \prod_{s=r+1}^{t-1} a(s) \right\|$$

$$\leq \eta \sum_{r=t-T}^{t-1} \left[\|\varphi - \psi\|\beta + L\|\varphi - \psi\| + K\|\varphi - \psi\| \right] \gamma$$

$$\leq \gamma\eta \sum_{r=t-T}^{t-1} (\beta + L + K)\|\varphi - \psi\| = \eta\gamma T(\beta + L + K)\|\varphi - \psi\|$$

$$= M\|\varphi - \psi\| = M\delta < \varepsilon$$

where L and K are given by (4.3.5). This proves A is continuous.

Next, we show that A maps bounded subsets into compact sets. Let J be given, $S = \{\varphi \in P_T : \| \varphi \| \leq J\}$ and $Q = \{(A\varphi)(t) : \varphi \in S\}$, then S is a subset of R^T which is closed and bounded thus compact. As A is continuous in φ it maps compact sets into compact sets. Therefore $Q = A(S)$ is compact.

It is trivial to show that the map B is a contraction provided we assume that

$$\left\| c(t-1) \right\| \leq \zeta < 1. \tag{4.3.12}$$

Theorem 4.3.1 ([111]). *Let* $\alpha = \|q(t, 0, 0)\|$. *Let* η, β *and* γ *be given by* (4.3.11). *Suppose* (4.3.2)–(4.3.5) *and* (4.3.12) *hold. Suppose there is a positive constant* G *such that all solutions* $x(t)$ *of* (4.3.1), $x(t) \in P_T$ *satisfy* $|x(t)| \leq G$, *the inequality*

$$\left\{ \zeta + \eta\gamma T(\beta + L + K) \right\} G + \eta\gamma T\alpha \leq G \tag{4.3.13}$$

holds. Then equation (4.3.1) *has a* T*-periodic solution.*

Proof. Define $\mathbb{M} = \{\varphi \in P_T : \|\varphi\| \leq G\}$. Then Lemma 4.2 implies $A : \mathbb{M} \to P_T$ and A is compact and continuous. Also the mapping B is a contraction and it is clear that $B : \mathbb{M} \to P_T$. Next, we show that if $\varphi, \psi \in \mathbb{M}$, we have $\|A\phi + B\psi\| \leq G$. Let $\varphi, \psi \in \mathbb{M}$ with $\|\varphi\|, \|\psi\| \leq G$. Then from (4.3.8)–(4.3.12) and the fact that $|q(t, x, y)| \leq L\|x\| + K\|y\| + \alpha$, we have

$$
\left\| \Big(A\varphi(t) \Big) + \Big(B\psi(t) \Big) \right\| = \left\| \frac{1}{1 - \prod_{s=t-T}^{t-1}} \sum_{r=t-T}^{t-1} \Big[\varphi(r - g(r)) \Big(c(r-1)a(r) - c(r) \Big) \right.
$$

$$
\left. + q(r, \varphi(r), \varphi(r - g(r))) \Big] \prod_{s=r+1}^{t-1} a(s) + c(t-1)\psi(t - g(t)) \right\|
$$

$$
\leq \eta\gamma \sum_{r=t-T}^{t-1} \Big[L\|\varphi\| + K\|\varphi\| + \beta\|\varphi\| + \alpha \Big] + \zeta\|\psi\|
$$

$$
\leq \eta\gamma[(\beta + L + K)\|\varphi\| + \alpha]T + \zeta\|\psi\|
$$

$$
\leq \eta\gamma T(\beta + L + K)G + \eta\gamma T\alpha + G\zeta
$$

$$
= \Big\{ \zeta + \eta\gamma T(\beta + L + K) \Big\} G + \eta\gamma T\alpha
$$

$$
\leq G.
$$

We see that all the conditions of Krasnoselskii's theorem are satisfied on the set $\mathbb{M}$. Thus there exists a fixed point z in $\mathbb{M}$ such that $z = Az + Bz$. By Lemma 4.1 this fixed point is a solution of (4.3.1). Hence (4.3.1) has a T-periodic solution.

Remark 4.2. The constant G of Theorem 4.3.1 serves as a priori bound on all possible T-periodic solutions of equation (4.3.1) as we shall see in the Example 4.1.

Next we use the contraction mapping principle to show the periodic solution is unique.

Theorem 4.3.2 ([111]). *Suppose* (4.3.2)–(4.3.5) *and* (4.3.12) *hold. Let* η, β, *and* γ *be given by* (4.3.11). *If*

$$
\zeta + T\gamma\eta \Big(\beta + L + K \Big) \leq v < 1,
$$

then equation (4.3.1) *has a unique T-periodic solution.*

Proof. Let the mapping H be given by (4.3.8). For $\varphi, \psi \in P_T$, in view of (4.3.8), we have

$$
\left\| \Big(H\varphi(t) \Big) - \Big(H\psi(t) \Big) \right\| = \left\| \Big(B\varphi(t) \Big) + \Big(A\varphi(t) \Big) - \Big(B\psi(t) \Big) - \Big(A\psi(t) \Big) \right\|
$$

$$
= \left\| \Big(\big(B\varphi(t) \big) - \big(B\psi(t) \big) \Big) + \Big(\big(A\varphi(t) \big) - \big(A\psi(t) \big) \Big) \right\|
$$

$$
\leq \left\| \Big(B\varphi(t) \Big) - \Big(B\psi(t) \Big) \right\| + \left\| \Big(A\varphi(t) \Big) - \Big(A\psi(t) \Big) \right\|
$$

$$\leq \zeta \|\varphi - \psi\| + \gamma \eta \sum_{r=t-T}^{t-1} \left[L\|\varphi - \psi\| + K\|\varphi - \psi\| + \beta\|\varphi - \psi\| \right]$$

$$\leq \left[\zeta + T\gamma\eta\left(\beta + L + K \right) \right] \|\varphi - \psi\|$$

$$< v\|\varphi - \psi\|.$$

By the contraction mapping principle, (4.3.1) has a unique T-periodic solution.

We have the following example.

Example 4.1 ([111]). Consider equation (4.3.1) along with conditions (4.3.2)–(4.3.5). Suppose that $a(t) \neq 1$ for all $t \in [0, T-1]$. Set

$$\rho = \min_{t \in [0, T-1]} |a(t) - 1| , \quad \delta = \max_{t \in [0, T-1]} k(t),$$

where $k(t) = c(t) - c(t-1)$. Suppose $1 - \|c\| > 0$. If

$$\rho(1 - \|c\|) > (1 - \|c\|)(\delta + L + K) + T\rho(\|a - 1\| + L + K)$$

holds, and G is defined by

$$G = \frac{\alpha(1 - \|c\| + T\rho}{\rho(1 - \|c\|) - (1 - \|c\|)(\delta + L + K) - T\rho(\|a - 1\| + L + K)}$$

satisfies inequality (4.3.13), then (4.3.1) has a T-periodic solution.

Proof. We rewrite (4.3.1) as

$$\triangle x(t) = \Big(a(t) - 1 \Big) x(t) + c(t)\triangle x(t - g(t)) + q\big(t, x(t), x(t - g(t))\big). \qquad (4.3.14)$$

Let the mappings A and B be defined by (4.3.10) and (4.3.9), respectively. Let $x(t) \in P_T$. A summation of equation (4.3.14) from 0 to $T-1$ gives

$$\sum_{s=0}^{T-1} \triangle x(s) = \sum_{s=0}^{T-1} \Big[(a(s) - 1)x(s) + c(s)\triangle x(s - g(s)) + q\big(s, x(s), x(s - g(s))\big) \Big].$$

Or,

$$x(T) - x(0) = \sum_{s=0}^{T-1} \Big[(a(s) - 1)x(s) + c(s)\triangle x(s - g(s)) + q\big(s, x(s), x(s - g(s))\big) \Big].$$

Since $x(t) \in P_T$, $x(T) = x(0)$. Therefore

$$0 = \sum_{s=0}^{T-1} \Big[(a(s) - 1)x(s) + c(s)\triangle x(s - g(s)) + q\big(s, x(s), x(s - g(s))\big) \Big]. \quad (4.3.15)$$

Rewrite and then sum by parts, using the summation by parts formula

$$\sum Ey \triangle z = yz - \sum z \triangle y$$

with $Ey(s) = c(s)$ and $z = x(s - g(s))$. As a consequence, we have

$$\sum_{s=0}^{T-1} c(s) \triangle x(s - g(s)) = c(s-1)x(s-g(s))\Big|_{s=0}^{T} - \sum_{s=0}^{T-1} x(s-g(s)) \triangle c(s-1)$$

$$= c(T-1)x(T-g(T)) - c(-1)x(0-g(0))$$

$$- \sum_{s=0}^{T-1} x(s-g(s))[c(s) - c(s-1)]$$

$$= - \sum_{s=0}^{T-1} x(s-g(s))[c(s) - c(s-1)].$$

As a result (4.3.15) becomes

$$\sum_{s=0}^{T-1} [a(s) - 1]x(s)$$

$$= \sum_{s=0}^{T-1} x(s-g(s))[c(s) - c(s-1)] - q(s,x(s),x(s-g(s))). \qquad (4.3.16)$$

Let $S = \sum_{s=0}^{T-1} |a(s) - 1||x(s)|$. We claim that there exists a $t^* \in [0, T-1]$ such that

$$T|a(t^*) - 1||x(t^*)| \le \sum_{s=0}^{T-1} |a(s) - 1||x(s)|.$$

Suppose such t^* does not exist. Then

$$T|a(t^*) - 1||x(t^*)| > S,$$

which implies that

$$T|a(t^*) - 1||x(t^*)| > S + \varepsilon.$$

Or

$$\sum_{t^*=0}^{T-1} |a(t^*) - 1||x(t^*)| > \sum_{t^*=0}^{T-1} \frac{S+\varepsilon}{T}.$$

Hence, $S > S + \varepsilon$, which is a contradiction. Therefore, such t^* exists. From (4.3.16), it implies that there exists a $t^* \in (0, T-1)$ such that

$$T|a(t^*) - 1||x(t^*)| \le \sum_{s=0}^{T-1} |k(t)||x(s-g(s))| + |q(s,x(s),x(s-g(s)))|.$$

By taking the maximum over $t \in [0, T-1]$, we obtain from the above inequality

$$
\begin{aligned}
T\rho||x(t^*)|| &\leq \sum_{s=0}^{T-1} \left(\delta||x|| + L||x|| + E||x|| + \alpha \right) \\
&= \sum_{s=0}^{T-1} \left((\delta + L + E)||x|| + \alpha \right) \\
&= T\left((\delta + L + E)||x|| + \alpha \right),
\end{aligned}
$$

which gives us

$$
||x(t^*)|| \leq \frac{1}{\rho}(\delta + L + K)||x|| + \frac{\alpha}{\rho}. \tag{4.3.17}
$$

Since for all $t \in [0, T-1]$

$$
x(t) = x(t^*) + \sum_{s=t^*}^{t-1} \triangle x(s),
$$

taking maximum over $t \in [0, T-1]$ and using

$$
||x(t)|| \leq ||x(t^*)|| + \sum_{s=0}^{T-1} |\triangle x(s)|
$$

yields

$$
||x(t)|| \leq \frac{1}{\rho}(\delta + L + E)||x|| + \frac{\alpha}{\rho} + T||\triangle x||. \tag{4.3.18}
$$

Taking the norm in (4.3.1) yields

$$
||\triangle x(t)|| \leq ||a - 1|| \, ||x|| + ||c|| \, ||\triangle x|| + K||x|| + L||x|| + \alpha.
$$

Or

$$
\left(1 - ||c||\right)||\triangle x(t)|| \leq \left(||a-1|| + E + L\right)||x|| + \alpha.
$$

Thus

$$
||\triangle x(t)|| \leq \frac{\left(||a-1|| + E + L\right)||x|| + \alpha}{1 - ||c||}. \tag{4.3.19}
$$

A substitution of (4.3.19) into (4.3.18) yields

$$||x(t)|| \le \frac{1}{\rho}(\delta + L + K)||x|| + \frac{\alpha}{\rho} + T\frac{\left(||a-1|| + K + L\right)||x|| + \alpha}{1 - ||c||}.$$

Hence

$$||x(t)|| \le \frac{\alpha(1 - ||c|| + T\rho)}{\rho(1 - ||c||) - (1 - ||c||)(\delta + L + E) - T\rho(||a-1|| + L + E)} = G.$$

Thus, for all $x(t) \in P_T$ we have shown that

$$||x(t)|| \le G.$$

Define $\mathbb{M} = \{\varphi \in P_T : ||\varphi|| \le G\}$. Then by Theorem 4.3.1, Equation (4.3.1) has a T-periodic solution. This completes the proof.

4.4 Periodicity in Vector Neutral Nonlinear Functional Difference Equations

Motivated by the work of Hale on functional differential equations [74], in this section we consider the nonlinear neutral difference equation

$$\triangle x(t) = A(t)x(t) + \triangle Q(t, x(t - g(t))) + G(t, x(t), x(t - g(t))) \qquad (4.4.1)$$

where A is an $n \times n$ matrix function, $g : \mathbb{Z} \to \mathbb{Z}^+$ is scalar and the functions $Q : \mathbb{Z} \times \mathbb{R}^n \to \mathbb{R}^n$ and $G : \mathbb{Z} \times \mathbb{R}^n \times \mathbb{R}^n \to \mathbb{R}^n$ are continuous in x. The purpose of this work is to make use of the notion of the fundamental matrix and invert (4.4.1) so that fixed point theory can be used. Krasnoselskii's fixed point theorem is one of the tools that we use in this research in order to show the existence of a periodic solution. The obtained mapping is the sum of two mappings; one is a contraction and the other is compact. The need to use Krasnoselkii's fixed point theorem may be necessary if one of the mappings is not compact nor satisfies a Lipschitz condition. Inverting equation (4.4.1) to a fixed point problem enables us to show the uniqueness of the periodic solution by appealing to the contraction mapping principle.

For an integer $T > 1$ let P_T be the set of all n-vector functions $x(t)$, periodic in t of period T. Then $(P_T, ||\cdot||)$ is a Banach space when it is endowed with the maximum norm

$$||x|| = \max_{t \in \mathbb{Z}} |x(t)| = \max_{t \in [0, T-1]} |x(t)|.$$

Note that P_T is equivalent to the Euclidean space $\mathbb{R}^{nT}$. If A is an $n \times n$ real matrix, then we define the norm of A by $|A| = \max_{1 \le i \le n} \sum_{j=1}^{n} |a_{ij}|$. First we make the following definition.

Definition 4.4.1. If the matrix $B(t)$ is periodic of period T, then the linear system

$$y(t+1) = B(t)y(t) \tag{4.4.2}$$

is said to be *noncritical with respect to* T, if it has no periodic solution of period T except the trivial solution $y = 0$.

Since we are searching for the existence of periodic solution for system (4.4.1), it is natural to assume that

$$A(t+T) = A(t), \quad g(t+T) = g(t), \quad g(t) \geq g^* > 0 \tag{4.4.3}$$

with $g : \mathbb{Z} \to \mathbb{Z}^+$ being scalar and $Q(t,x)$ and $G(t,x,y)$ are continuous functions and periodic in t of period T. That is

$$Q(t+T,x) = Q(t,x), \quad G(t+T,x,y) = G(t,x,y). \tag{4.4.4}$$

Throughout this section it is assumed that the matrix $B(t) = I + A(t)$ is nonsingular and system (4.4.2) is noncritical, where I is the $n \times n$ identity matrix. Also, if $x(t)$ is a sequence, then the forward operator E is defined as $Ex(t) = x(t+1)$. Next we state some known results about system (4.4.2). Let $K(t)$ represent the fundamental matrix of (4.4.2) with $K(0) = I$. Then
(i) $\det K(t) \neq 0$.
(ii) $K(t+1) = B(t)K(t)$ and $K^{-1}(t+1) = K^{-1}(t)B^{-1}(t)$.
(iii) System (4.4.2) is noncritical if and only if $\det(I - K(T)) \neq 0$.
(iv) There exists a nonsingular matric L such that $K(t+T) = K(t)L^T$ and $K^{-1}(t+T) = L^{-T}K^{-1}(t)$.
With the above-mentioned $K(t)$ in mind we have the following lemma.

Lemma 4.3. *Suppose* (4.4.3)–(4.4.4) *hold. If* $x(t) \in P_T$, *then* $x(t)$ *is a solution of equation* (4.4.1) *if and only if*

$$x(t) = Q(t,x(t-g(t)))$$
$$+ K(t)\left(K^{-1}(T) - I\right)^{-1} \sum_{u=t}^{t+T} K^{-1}(u)\left(I - A(u)B^{-1}(u)\right)\left[A(u)Q(u,x(u-g(u)))\right.$$
$$\left. + G(u,x(u),x(u-g(u)))\right]. \tag{4.4.5}$$

Proof. Let $x(t) \in P_T$ be a solution of (4.4.1) and $K(t)$ be a fundamental matrix of solutions of (4.4.2). First we write (4.4.1) as

$$\triangle\{x(t) - Q(t,x(t-g(t)))\} = A(t)\{x(t) - Q(t,x(t-g(t)))\}$$
$$+ A(t)Q(t,x(t-g(t))) + G(t,x(t),x(t-g(t))).$$

Since $K(t)K^{-1}(t) = I$, it follows that

$$0 = \triangle\left(K(t)K^{-1}(t)\right) = K(t)\triangle(K^{-1}(t)) + \triangle(K(t))EK(t)$$

$$= K(t)\triangle(K^{-1}(t)) + A(t)K(t)K^{-1}(t)B^{-1}(t)$$

$$= K(t)\triangle(K^{-1}(t)) + A(t)B^{-1}(t).$$

Or,

$$\triangle(K^{-1}(t)) = -K^{-1}(t)A(t)B^{-1}(t). \tag{4.4.6}$$

If $x(t)$ is a solution of (4.4.1) with $x(0) = x_0$, then

$$\triangle\left\{K^{-1}(t)\left(x(t) - Q(t,x(t-g(t)))\right)\right\}$$

$$= K^{-1}(t)\triangle\left(x(t) - Q(t,x(t-g(t)))\right) + \triangle(K^{-1}(t))E\left(x(t) - Q(t,x(t-g(t)))\right)$$

$$= K^{-1}(t)\left[A(t)\left(x(t) - Q(t,x(t-g(t)))\right) + A(t)Q(t,x(t-g(t))) + G(t,x(t),x(t-g(t)))\right]$$

$$-K^{-1}(t)A(t)B^{-1}(t)\left[B(t)\left(x(t) - Q(t,x(t-g(t)))\right)\right.$$

$$+A(t)Q(t,x(t-g(t))) + G(t,x(t),x(t-g(t)))\Big], \text{ by } (4.4.6)$$

$$= K^{-1}(t)\left(I - A(t)B^{-1}(t)\right)\left(A(t)Q(t,x(t-g(t))) + G(t,x(t),x(t-g(t)))\right).$$

Summing the above equation from 0 to $t-1$ yields

$$x(t) = Q(t,x(t-g(t))) + K(t)\left(x_0 - Q(0,x(-g(0)))\right)$$

$$+ K(t)\sum_{u=0}^{t-1} K^{-1}(u)\left(I - A(u)B^{-1}(u)\right)\left[A(u)Q(u,x(u-g(u)))\right.$$

$$+ G(u,x(u),x(u-g(u)))\Big]. \tag{4.4.7}$$

For the sake of simplicity, we let

$$D(u) = \left(I - A(u)B^{-1}(u)\right)\left[A(u)Q(u,x(u-g(u))) + G(u,x(u),x(u-g(u)))\right].$$

Since $x(T) = x_0 = x(0)$, using (4.4.7) we get

$$x_0 - Q(0,x(-g(0))) = \left(I - K(T)\right)^{-1}\sum_{u=0}^{T-1} K(T)K^{-1}(u)D(u). \tag{4.4.8}$$

A substitution of (4.4.8) into (4.4.7) yields

$$x(t) = Q(t,x(t-g(t))) + K(t)\Big(I - K(T)\Big)^{-1}\sum_{u=0}^{T-1} K(T)K^{-1}(u)D(u)$$

$$+ \sum_{u=0}^{t-1} K(t)K^{-1}(u)D(u). \tag{4.4.9}$$

It remains to show that expression (4.4.9) is equivalent to (4.4.5). Since

$$\Big(I - K(T)\Big)^{-1} = \Big(K(T)(K^{-1}(T) - I)\Big)^{-1} = \Big(K^{-1}(T) - I\Big)^{-1}K^{-1}(T),$$

(4.4.9) becomes

$$x(t) = Q(t,x(t-g(t))) + K(t)\Big(K^{-1}(T) - I\Big)^{-1}\sum_{u=0}^{T-1} K^{-1}(u)D(u)$$

$$+ \sum_{u=0}^{t-1} K(t)K^{-1}(u)D(u)$$

$$= Q(t,x(t-g(t))) + K(t)\Big(K^{-1}(T) - I\Big)^{-1}\Big\{\sum_{u=0}^{T-1} K^{-1}(u)D(u)$$

$$+ \sum_{u=0}^{t-1} K^{-1}(T)K^{-1}(u)D(u) - \sum_{u=0}^{t-1} K^{-1}(u)D(u)\Big\}$$

$$= Q(t,x(t-g(t))) + K(t)\Big(K^{-1}(T) - I\Big)^{-1}\Big\{-\sum_{u=T}^{t-1} K^{-1}(u)D(u)$$

$$+ \sum_{u=0}^{t-1} K^{-1}(T)K^{-1}(u)D(u)\Big\}.$$

By letting $u = s - T$ in the third term on the right side of the above expression, we end up with

$$x(t) = Q(t,x(t-g(t))) + K(t)\Big(K^{-1}(T) - I\Big)^{-1}\Big\{-\sum_{u=T}^{t-1} K^{-1}(u)D(u)$$

$$+ \sum_{s=T}^{T+t-1} K^{-1}(T)K^{-1}(s-T)D(s-T)\Big\}. \tag{4.4.10}$$

By *(iv)* we have $K(t-T) = K(t)L^{-T}$ and $K(T) = L^T$, where $L^{-T} = (L^T)^{-1}$. Hence, $K^{-1}(T)K^{-1}(s-T) = K^{-1}(s)$. Moreover, since $D(s-T) = D(s)$ expression (4.4.10) becomes

$$x(t) = Q(t, x(t - g(t))) + K(t) \left(K^{-1}(T) - I \right)^{-1} \left\{ - \sum_{u=T}^{t-1} K^{-1}(u)D(u) \right.$$

$$\left. + \sum_{u=T}^{t+T-1} K^{-1}(u)D(u) \right\}$$

$$= Q(t, x(t - g(t))) + K(t) \left(K^{-1}(T) - I \right)^{-1} \sum_{u=t}^{t+T-1} K^{-1}(u)D(u).$$

This completes the proof.

Now we are in a position to define a suitable mapping that satisfies all the requirements of Theorem 3.5.1. Define a mapping H by

$$(H\varphi)(t) = Q(t, \varphi(t - g(t)))$$

$$+ K(t) \left(K^{-1}(T) - I \right)^{-1} \sum_{u=t}^{t+T-1} K^{-1}(u) \left(I - A(u)B^{-1}(u) \right)$$

$$\times \left[A(u)Q(u, \varphi(u - g(u))) + G(u, \varphi(u), \varphi(u - g(u))) \right]. \quad (4.4.11)$$

It is clear that $H : P_T \to P_T$ by the way it was constructed in Lemma 4.3.
We note that to apply the above theorem we need to construct two mappings; one is a contraction and the other is compact. Therefore, we express equation (4.4.11) as

$$(H\varphi)(t) = (B\varphi)(t) + (C\varphi)(t)$$

where $C, B : P_T \to P_T$ are given by

$$(B\varphi)(t) = Q(t, \varphi(t - g(t))) \quad (4.4.12)$$

and

$$(C\varphi)(t) = K(t) \left(K^{-1}(T) - I \right)^{-1} \sum_{u=t}^{t+T-1} K^{-1}(u) \left(I - A(u)B^{-1}(u) \right)$$

$$\times \left[A(u)Q(u, \varphi(u - g(u))) + G(u, \varphi(u), \varphi(u - g(u))) \right]. \quad (4.4.13)$$

We assume the functions Q and G are Lipschitz continuous in x and in x and y, respectively. That is, there are positive constants E_1, E_2, and E_3 such that

$$|Q(t, x) - Q(t, y)| \leq E_1 \|x - y\| \text{ and} \quad (4.4.14)$$

$$|G(t, x, y) - G(t, z, w))| \leq E_2 \|x - z\| + E_3 \|y - w\|. \quad (4.4.15)$$

Observe that in view of (4.4.14) and (4.4.15) we have

$$
\begin{aligned}
|Q(t,x)| &= |Q(t,x) - Q(t,0) + Q(t,0)| \\
&\leq |Q(t,x) - Q(t,0)| + |Q(t,0)| \\
&\leq E_1\|x\| + \alpha.
\end{aligned}
$$

Similarly,

$$
\begin{aligned}
|G(t,x,y)| &= |G(t,x,y) - G(t,0,0) + G(t,0,0)| \\
&\leq |G(t,x,y) - G(t,0,0)| + |G(t,0,0)| \\
&\leq E_2\|x\| + E_3\|y\| + \beta
\end{aligned}
$$

where $\alpha = \max_{t \in \mathbb{Z}} |Q(t,0)|$ and $\beta = \max_{t \in \mathbb{Z}} |G(t,0,0)|$. The next lemma plays an important role in showing C is compact.

Lemma 4.4. *Suppose the hypothesis of Lemma 4.3 holds. If C is defined by (4.4.13), then*
(I)

$$
\|C\varphi\| \leq r \sum_{u=0}^{T-1} \Big\| A(u)Q(u, \varphi(u - g(u))) + G(u, \varphi(u), \varphi(u - g(u))) \Big\|,
$$

where

$$
r = \max_{t \in [0, T-1]} \left(\max_{t \leq u \leq t+T-1} \left| \left[K(u)(K^{-1}(T) - I)K^{-1}(t) \right]^{-1} \left(I - A(u)B^{-1}(u) \right) \right| \right)
$$

(4.4.16)

is a constant which is independent of Q and G and depends only upon $T, A(t), B(t)$, and $K(t)$ where $1 \leq t \leq T$.
(II) C is continuous and compact.

Proof. Let C be defined by (4.4.13) which is equivalent to

$$
\begin{aligned}
(C\varphi)(t) = \sum_{u=t}^{t+T-1} &\left[K(u)(K^{-1}(T) - I)K^{-1}(t) \right]^{-1} \\
&\left(I - A(u)B^{-1}(u) \right) \Big[A(u)Q(u, \varphi(u - g(u))) \\
&+ G(u, \varphi(u), \varphi(u - g(u))) \Big].
\end{aligned}
$$

As $(C\varphi)(t) \in P_T$, we have

$$\|(C\varphi)(t)\| = \max_{t\in[0,T-1]} \left| \sum_{u=t}^{t+T-1} \left[K(u)(K^{-1}(T)-I)K^{-1}(t) \right]^{-1} \left(I - A(u)B^{-1}(u) \right) \right.$$
$$\times \left. \left[A(u)Q(u,\varphi(u-g(u))) + G(u,\varphi(u),\varphi(u-g(u))) \right] \right|$$
$$\leq \max_{t\in[0,T-1]} \left(\max_{t\leq u\leq t+T-1} \left| \left[K(u)(K^{-1}(T)-I)K^{-1}(t) \right]^{-1} \left(I - A(u)B^{-1}(u) \right) \right| \right)$$
$$\times \sum_{u=0}^{T-1} \left\| A(u)Q(u,\varphi(u-g(u))) + G(u,\varphi(u),\varphi(u-g(u))) \right\|.$$

This completes the proof of *(I)*. To see that C is continuous, we let $\varphi, \psi \in P_T$ with $\|\varphi\| \leq D$ and $\|\psi\| \leq D$. Given $\varepsilon > 0$, take $\delta = \varepsilon/N$ such that $\|\varphi - \psi\| < \delta$. By making use of (4.4.14) and (4.4.15) into (4.4.13) we get

$$\left\| C\varphi - C\psi \right\| \leq rT \left[|A|E_1\|\varphi - \psi\| + (E_2+E_3)\|\varphi - \psi\| \right]$$
$$\leq N\|\varphi - \psi\| < \varepsilon$$

where E_1, E_2, and E_3 are given by (4.4.14) and (4.4.15) and $N = rT(|A|E_1 + E_2 + E_3)$. This proves C is continuous. Next, we show that C maps bounded subsets into compact sets. Let J be given and let $S = \{\varphi \in P_T : \| \varphi \| \leq J\}$ and $Q = \{C\varphi : \varphi \in S\}$, then S is a subset of $\mathbb{R}^{nT}$ which is closed and bounded thus compact. As C is continuous in φ it maps compact sets into compact sets. Therefore $Q = C(S)$ is compact.

Lemma 4.5. *If B is given by (4.4.12) and $E_1 \leq \zeta < 1$, where E_1 is given by (4.4.14) then B is a contraction.*

Proof. Let B be defined by (4.4.12). Then for $\varphi, \psi \in P_T$ we have

$$\|B\varphi - B\psi\| = \max_{t\in[0,T-1]} |B\varphi - B\psi|$$
$$\leq E_1 \max_{t\in[0,T-1]} |\varphi(t-g(t)) - \psi(t-g(t))|$$
$$\leq \zeta\|\varphi - \psi\|.$$

Hence B defines a contraction mapping with contraction constant ζ.

Theorem 4.4.1. *Let $\alpha = \max_{t\in\mathbb{Z}} |Q(t,0)|$ and $\beta = \max_{t\in\mathbb{Z}} |G(t,0,0)|$. Let r be given by (4.4.16). Suppose (4.4.3), (4.4.4), (4.4.14), and (4.4.15) hold. Let J be a positive constant satisfying the inequality*

$$\alpha + E_1 J + rT \left[|A|(E_1 + \alpha) + E_2 + E_3 \right] J + rT\beta \leq J. \qquad (4.4.17)$$

Let $\mathbb{M} = \{\varphi \in P_T : \|\varphi\| \leq J\}$. Then equation (4.4.1) has a solution in M.

Proof. Define $\mathbb{M} = \{\varphi \in P_T : ||\varphi|| \leq J\}$. Then Lemma 4.4 implies $C : P_T \to P_T$ and C is compact on M and continuous. Also, from Lemma 4.5, the mapping B is a contraction and it is clear that $B : P_T \to P_T$. Next, we show that if $\varphi, \psi \in \mathbb{M}$, we have $||C\phi + B\psi|| \leq J$. Let $\varphi, \psi \in \mathbb{M}$ with $||\varphi||, ||\psi|| \leq J$. Then

$$\left\|C\varphi + B\psi\right\| \leq E_1||\psi|| + \alpha + r \sum_{u=0}^{T-1} [|A|(\alpha + E_1||\varphi||) + E_2||\varphi|| + E_3||\varphi|| + \beta]$$

$$\leq \alpha + E_1 J + rT\left[|A|(E_1 + \alpha) + E_2 + E_3\right]J + rT\beta \leq J.$$

We see that all the conditions of Krasnoselskii's theorem (Theorem 3.5.1) are satisfied on the set $\mathbb{M}$. Thus there exists a fixed point z in $\mathbb{M}$ such that $z = Az + Bz$. By Lemma 4.3, this fixed point is a solution of (4.4.1). Hence (4.4.1) has a T-periodic solution.

Corollary 4.4. *Suppose* (4.4.3), (4.4.4), (4.4.14), *and* (4.4.15) *hold and* $Q(t, x(t - g(t)))$ *and* $G(t, x(t), x(t - g(t)))$ *are uniformly bounded. Let M be defined as in Theorem 4.4.1 such that for* $\varphi \in M$,

$$||Q(, \varphi(t - g(t)))|| \leq J_1,$$

and

$$\left\| \sum_{u=t}^{t+T-1} \left[K(u)(K^{-1}(T) - I)K^{-1}(\iota)\right]^{-1} \right.$$
$$\left. \left[A(u)Q(u, \varphi(u - g(u))) + G(u, \varphi(u), \varphi(u - g(u)))\right]\right\| \leq J_2$$

for positive constants J_1 and J_2. If

$$J_1 + J_2 \leq J,$$

then (4.4.1) *has a T-periodic solution.*

Proof. Define B and C by (4.4.12) and (4.4.13), respectively and imitate the proof of Theorem 4.4.1.

In the next theorem we use the contraction mapping principle to show that the periodic solution is unique.

Theorem 4.4.2. *Suppose* (4.4.3), (4.4.4), (4.4.14), *and* (4.4.15) *hold. Then equation* (4.4.1) *has a unique T-periodic solution.*

Proof. Due to condition (4.4.21) we have that

$$E_1 + rT(|A|E_1 + E_2 + E_3) < 1.$$

Let the mapping H be given by (4.4.11). For $\varphi, \psi \in P_T$, in view of (4.4.11), we have

$$\left\| H\varphi - H\psi \right\| \leq \left(E_1 + rT(|A|E_1 + E_2 + E_3) \right) \|\varphi - \psi\|.$$

This completes the proof.

It is worth noting that Theorem 4.4.1 and Theorem 4.4.2 are not applicable to functions G of the form

$$G(t, \varphi(t), \varphi(t - g(t))) = f_1(t)\varphi^2(t) + f_2(t)\varphi^2(t - g(t))),$$

where $f_1(t), f_2(t)$, and $g(t) > 0$ are periodic sequences. To accommodate such functions, we state the following corollary, which requires the functions Q and G to be locally Lipschitz.

Corollary 4.5. *Suppose* (4.4.3)–(4.4.4) *hold and let* α *and* β *be the constants defined in Theorem 4.4.1. Let J be a positive constant and define* $\mathbb{M} = \{\varphi \in P_T : ||\varphi|| \leq J\}$. *Suppose there are positive constants* E_1^*, E_2^*, *and* E_3^* *so that for* x, y, z, *and* $w \in \mathbb{M}$ *we have*

$$|Q(t, x) - Q(t, y)| \leq E_1^* \|x - y\|,$$

$$|G(t, x, y) - G(t, z, w))| \leq E_2^* \|x - z\| + E_3^* \|y - w\|,$$

and

$$\alpha + E_1^* J + rT \left[|A|(E_1^* + \alpha) + E_2^* + E_3^* \right] J + rT\beta \leq J. \tag{4.4.18}$$

Then equation (4.4.1) *has a unique solution in* $\mathbb{M}$.

Proof. Let $\mathbb{M} = \{\varphi \in P_T : ||\varphi|| \leq J\}$. Let the mapping H be given by (4.4.11). Then the results follow immediately from Theorem 4.4.1 and Theorem 4.4.2, since

$$E_1^* + rT(|A|E_1^* + E_2 + E_3^*) < 1.$$

This completes the proof.

Now we display an example as an application.

Example 4.2. For small positive ε_1 and ε_2, we consider the perturbed discrete Van Der Pol equation

$$\triangle^2 x + (\varepsilon_2 x^2 - 1)\triangle x - x - \varepsilon_1 \triangle \left(\cos(t\pi)x^2(t - g(t)) \right) - \varepsilon_2 \cos(t\pi) = 0, \quad (4.4.19)$$

where $g : \mathbb{Z} \to \mathbb{Z}^+$ is scalar and 2-periodic. By letting $\triangle x_1 = x_2$ we can transform (4.4.19) to

$$\triangle \begin{pmatrix} x_1 \\ x_2 \end{pmatrix} = \begin{pmatrix} 0 & 1 \\ 1 & 1 \end{pmatrix} \begin{pmatrix} x_1 \\ x_2 \end{pmatrix} + \triangle \begin{pmatrix} 0 \\ \varepsilon_1 \cos(\pi t)x_1^2(t - g(t)) \end{pmatrix} + \begin{pmatrix} 0 \\ \varepsilon_2 \cos(\pi t) - \varepsilon_2 x_2 x_1^2 \end{pmatrix},$$

where

$$A = \begin{pmatrix} 0 & 1 \\ 1 & 1 \end{pmatrix}, \quad Q(t, x(t - g(t))) = \begin{pmatrix} 0 \\ \varepsilon_1 \cos(\pi t) x_1^2(t - g(t)) \end{pmatrix}$$

and

$$G(t, x(t), x(t - g(t))) = \begin{pmatrix} 0 \\ \varepsilon_2 \cos(\pi t) - \varepsilon_2 x_2 x_1^2 \end{pmatrix}.$$

Since the matrix $B = I + A$ has real eigenvalues, the system $x(t+1) = Bx(t)$ is noncritical. Let $\varphi(t) = (\varphi_1(t), \varphi_2(t))$, $\psi(t) = (\psi_1(t), \psi_2(t)) \in \mathbb{M} = \{\phi \in P_2 : ||\varphi|| \le J\}$. Then,

$$\left\| G(t, \varphi(t), \varphi(t - g(t))) - G(t, \psi(t), \psi(t - g(t))) \right\|$$

$$\le \varepsilon_2 \max_{t \in [0,1]} \left| (\varphi_2(t)(\varphi_1(t) + \psi_1(t)), \psi_1^2(t)) \begin{pmatrix} \varphi_1(t) - \psi_1(t) \\ \varphi_2(t) - \psi_2(t) \end{pmatrix} \right|$$

$$\le 2\varepsilon_2 J^2 ||\varphi - \psi||.$$

Hence, we see that $\beta = \varepsilon_2, E_2 = 2\varepsilon_2 J^2$, and $E_3 = 0$. In a similar fashion, we obtain $\alpha = 0$ and $E_1 = 2\varepsilon_1 J^2$. Thus, inequality (4.4.21)

$$2\varepsilon_1 J^2 + 2r \left[2\varepsilon_1 J |A| + 2\varepsilon_2 J^2 \right] J + 2r\varepsilon_2 \le J$$

is satisfied for small ε_1 and ε_2. Hence, equation (4.4.19) has a 2-periodic solution, by Theorem 4.4.1. On the other hand, the above inequality automatically implies that

$$2\varepsilon_1 J + 2r \left[2\varepsilon_1 J |A| + 2\varepsilon_2 J^2 \right] < 1$$

for small ε_1 and ε_2, and hence equation (4.4.19) has a unique 2-periodic solution, by Corollary 4.5.

Next we make use of Schauder's fixed point theorem, Theorem 4.7.1, to show that Equation (4.4.1) has a T-periodic solution. This scenario could be encountered when one of the mappings is neither contraction nor compact. Thus, we assume that the function Q is uniformly continuous and bounded. That is there exists a positive constant W such that

$$|Q(t, x)| \le W, \text{ for all } t \ge 0. \tag{4.4.20}$$

Theorem 4.4.3. *Let* $\beta = \max_{t \in \mathbb{Z}} |G(t, 0, 0)|$. *Let* r *be given by* (4.4.16). *Suppose* (4.4.3), (4.4.4), (4.4.15), *and* (4.4.20) *hold. Let* J *be a positive constant satisfying the inequality*

$$W + rT \left[|A|W + E_2 + E_3 \right] J + rT\beta \le J. \tag{4.4.21}$$

Then for $\mathbb{M} = \{\varphi \in P_T : ||\varphi|| \le J\}$, *equation* (4.4.1) *has a solution in M.*

Proof. Define $\mathbb{M} = \{\varphi \in P_T : ||\varphi|| \leq J\}$. Let the map H be defined by (4.4.11). Then by similar argument, one can easily show that

$$H : \mathbb{M} \rightarrow \mathbb{M}.$$

In addition, using the Lebesgue dominated convergence theorem, one can easily show the map H is compact. For the complete argument we refer to Section 4.7.1. Thus, by Theorem 4.7.1, Equation (4.4.1) has a T-periodic solution.

4.5 Periodicity in Nonlinear Systems with Infinite Delay

As we have seen in the previous section that using Schaefer's fixed-point theorem (Theorem 4.1.3) enabled us to show that if there is an a priori bound on all possible T-periodic solutions of a related auxiliary Volterra difference equation, then there is a T-periodic solution. In this section we apply our results to scalar Volterra difference equations in which the a priori bound is established by means of nonnegative definite Lyapunov functionals. Thus, we consider

$$x(n+1) = Dx(n) + f(x(n)) + \sum_{j=-\infty}^{n} K(n,j)g(x(j)) + p(n), \qquad (4.5.1)$$

with the existence of positive constant Q such that

$$sup_{n \in \mathbb{Z}} \sum_{j=-\infty}^{n} |K(n,j)| \leq Q,$$

where D is a $k \times k$ matrix and p is a given $k \times 1$ vector with $p(n+T) = p(n)$ for integer T. The kernel $K(n,j)$ satisfies $K(n+T, j+T) = K(n,j)$ for all $-\infty < j \leq n < \infty$, where $(n,j) \in \mathbb{Z}^2$ and $K(n,j) = 0$ for $j > n$. The period T is taken to be the least positive integer for which these hold. The functions f and g are continuous. Results of this section can be partially found in [137]. In [131] the author studied the existence of periodic solutions of the Volterra difference system with

$$\triangle x(n) = Dx(n) + \sum_{j=-\infty}^{n} C(n-j)x(j) + g(n), n \in \mathbb{Z} \text{ with } \sum_{u=0}^{\infty} |C(u)| < \infty \quad (4.5.2)$$

where D and C are $k \times k$ matrices and g is a given $k \times 1$ vector with $g(n+T) = g(n)$ for integer T, by using Schaefer's fixed point theorem. In [131] the mapping was constructed by taking direct sum in (4.5.2). On the other hand, Elaydi [52] considered (4.5.2) and utilized the notion of the resolvent of an equation associated with (4.5.2) and concluded the existence of a periodic solution of (4.5.2). In arriving at his results, Elaydi had to show that the zero solution of an homogenous equation associated with (4.5.2) is uniformly asymptotically stable . Thus, it was assumed that $|D| < 1$ where $|\cdot|$ is a suitable matrix norm. Later on, for the purpose of relaxing $|D| < 1$, Elaydi and Zhang [53] used the notion of degree theory, due to Grannas, and obtained the existence of a periodic solution of (4.5.2).

Once our results are established, we apply them to nonlinear Volterra discrete equations of the form

$$x(n+1) = ax(n) + f(x(n)) + \sum_{j=-\infty}^{n} K(n,j)g(x(j)) + p(n). \tag{4.5.3}$$

In [130] the author considered (4.5.3) with the assumptions that the two functions f and g are uniformly bounded and the coefficient a satisfies the stringent condition $-1 \le a \le 1$. Our objective is to relax those conditions. We achieve our objective by displaying nonnegative definite Lyapunov functionals, which in turn give the a priori bound. Thus, the results of this section will advance the theory of existence of periodic solutions in the most general form of nonlinear Volterra difference equations. For (4.5.1) a homotopy will have to be constructed which we obtain in the following manner.

Let m be a real number such that either $m > 1$ or $m < -1$. For $0 \le \lambda \le 1$, we rewrite (4.5.1) as

$$x(n+1) = \lambda(-m^{-1}I+D)x(n) + m^{-1}x(n) + \lambda f(x(n))$$
$$+\lambda \sum_{j=-\infty}^{n} K(n,j)g(x(j)) + \lambda p(n). \tag{4.5.4}$$

One may easily verify that

$$x(n) = \lambda \sum_{j=-\infty}^{n-1} m^{-(n-j-1)}\left[(-m^{-1}I+D)x(j) + f(x(j))\right]$$

$$+ \lambda \sum_{s=-\infty}^{n-1} m^{-(n-s-1)} \sum_{j=-\infty}^{s} K(s,j)g(x(j))$$

$$+ \lambda \sum_{j=-\infty}^{n-1} p(j)m^{-(n-j-1)} \tag{4.5.5}$$

is a solution of (4.5.4) and hence of (4.5.1). Define the space P_T by

$$P_T = \left\{x(n) : x(n+T) = x(n), \text{ for all } n \in \mathbb{Z}\right\}$$

where T is the least positive integer so that $x(n+T) = x(n)$. Then $\left(P_T, |\cdot|\right)$ defines a Banach space of T-periodic $k \times 1$ real vector sequences $x(n)$ with the maximum norm

$$|x| = \max_{i=1,\cdots,k}\left\{\max_{n\in[0,T-1]}|x_i(n)|\right\}.$$

For $x(n) \in P_T$, using (4.5.5) we define the mapping $H : P_T \to P_T$ by

$$(Hx)(n) = \lambda \sum_{j=-\infty}^{n-1} m^{-(n-j-1)} \left[(-m^{-1}I + D)x(j) + f(x(j)) \right]$$

$$+ \lambda \sum_{s=-\infty}^{n-1} m^{-(n-s-1)} \sum_{j=-\infty}^{s} K(s,j)g(x(j))$$

$$+ \lambda \sum_{j=-\infty}^{n-1} p(j)m^{-(n-j-1)}. \tag{4.5.6}$$

Thus,

$$x = \lambda Hx$$

is equivalent to (4.5.5). Next we prove two Lemmas that are essential for the application of Schaefer's theorem (Theorem 4.1.3).

Lemma 4.6 ([137]). *If H is defined by (4.5.6), then H is continuous and $H : P_T \to P_T$.*

Proof. For the continuity of H we let $\phi_1, \phi_2 \in P_T$ and use (4.5.6) to obtain,

$$\left| (H\phi_1) - (H\phi_2) \right| \leq \sum_{j=-\infty}^{n-1} |m^{-(n-j-1)}| \left| (-m^{-1}I + D) \right| |\phi_1 - \phi_2|$$

$$+ \sum_{j=-\infty}^{n-1} |m^{-(n-j-1)}| |f(\phi_1) - f(\phi_2)|$$

$$+ Q \sum_{s=-\infty}^{n-1} |m^{-(n-s-1)}| |g(\phi_1) - g(\phi_2)|.$$

By invoking the continuity of f and g and the fact that the infinite series $\sum_{j=-\infty}^{n-1} |m^{-(n-j-1)}|$ is convergent, we deduce that H is continuous. Left to show that $H : P_T \to P_T$. Let $\varphi(n) \in P_T$ and use the substitution $v = j - T$ followed by the substitution $r = s - T$ to obtain $(H\varphi)(n+T) = (H\varphi)(n)$. This concludes the proof of the lemma.

Lemma 4.7 ([137]). *If H is defined by (4.5.6), then H maps bounded subsets into compact subsets.*

Proof. Let $J > 0$ be given and define the two sets $S = \{x(n) \in P_T : |x| \leq J\}$ and $W = \{(Hx)(n) : x(n) \in P_T\}$. Then W is a subset of $\mathbb{R}^{Tk}$, which is closed and bounded and thus compact. As H is continuous in x it maps compact sets into compact sets. We deduce that $W = H(S)$ is compact. This concludes the proof of the Lemma.

Now we are in a position to state and prove our main theorem that yields the existence of a periodic solution of (4.5.1).

Theorem 4.5.1. *If there exists an $L > 0$ such that for any T-periodic solution of (4.5.4), $0 < \lambda < 1$ satisfies $|x| \leq L$, then (4.5.1) has a solution in P_T.*

Proof. Let H be defined by (4.5.6). Then, by Lemmas 4.6 and 4.7, H is continuous, compact, and T-periodic. The hypothesis $|x| \leq L$ rules out part *(ii)* of Theorem 4.1.3 and thus $x = \lambda Hx$ has a solution for $\lambda = 1$, which solves (4.5.1). This concludes the proof.

Remark 4.3. When it comes to application, the reader shall see that we may have to require $m \in (-1,0) \cup (0,1)$. Thus, to take care of such situation we note that Equation (4.5.1) is equivalent for $\lambda = 1$ to

$$x(n+1) = \lambda(-mI + D)x(n) + mx(n) + \lambda f(x(n))$$

$$+\lambda \sum_{j=-\infty}^{n} K(n,j)g(x(j)) + \lambda p(n). \tag{4.5.7}$$

Then it follows readily that x is a bounded solution of (4.5.7) if and only if

$$x(n) = \lambda \sum_{j=-\infty}^{n-1} m^{-(j-n+1)}\Big[(-mI + D)x(j) + f(x(j))\Big]$$

$$+ \lambda \sum_{s=-\infty}^{n-1} m^{-(s-n+1)} \sum_{j=-\infty}^{s} K(s,j)g(x(j))$$

$$+ \lambda \sum_{s=-\infty}^{n-1} p(j)m^{-(j-n+1)}. \tag{4.5.8}$$

Then one may easily prove a theorem similar to Theorem 4.5.1 for the case $m \in (-1,0) \cup (0,1)$.

4.5.1 Application to Infinite Delay Volterra Equations

Now we apply the results of the previous section to scalar nonlinear Volterra difference equations with of the form

$$x(n+1) = ax(n) + f(x(n)) + \sum_{j=-\infty}^{n} K(n,j)g(x(j)) + p(n), \tag{4.5.9}$$

where the terms f, g, K, and p obey the same conditions as before. The highlight of this work is to prove the existence of periodic solution of Equation (4.5.9) where the magnitude of a could be $|a| > 1$. In most of the literature, it is required that $|a| < 1$. To relax this condition we resort to nonnegative definite Lyapunov functional to obtain the a priori bound on all possible T-periodic solutions of Equation (4.5.9) and then conclude the existence of a periodic solution by invoking Theorem 4.5.1. We shall assume in addition to those assumptions made in the previous section that there exists $F : \mathbb{Z}^+ \to \mathbb{R}$ and $R > 0$ such that

$$|K(n,u+n)| \leq F(u), \text{ with } \sum_{u=0}^{\infty} |F(u)| \leq R, \qquad (4.5.10)$$

and

$$\max_{n\in\mathbb{Z}} \sum_{j=-\infty}^{n-1} \sum_{s=n}^{\infty} |K(s,j)| < \infty. \qquad (4.5.11)$$

We note that assumption (4.5.10) implies that

$$\max_{n\in\mathbb{Z}} \sum_{s=n}^{\infty} |K(s,j)| \leq R.$$

Now we state two theorems; one will show the existence of a periodic solution of (4.5.9) when $|a| < 1$, and the other when $|a| > 1$. The proof of the first theorem will be established in three different cases on the coefficient a.

Theorem 4.5.2 ([137]). *Assume (4.5.10) and (4.5.11). Also, we assume that there exists an $\alpha > 0$ such that*

$$|f(x)| + R|g(x)| \leq \alpha|x|,$$

and

$$|\mu| + \alpha - 1 \leq -\beta, \text{ for some positive constant } \beta, \qquad (4.5.12)$$

where μ is to be defined in the body of the proof and R is given by (4.5.10). Then, Equation (4.5.9) has a T-periodic solution.

Proof. **Case 1.** $0 < a < 1$
Set $m = a$. Then $0 < m < 1$. We shall apply Theorem 4.5.1 with $m \in (0,1)$ to the corresponding family of equations

$$x(n+1) = \lambda(-m+a)x(n) + mx(n) + \lambda f(x(n))$$

$$+\lambda \sum_{j=-\infty}^{n} K(n,j)g(x(j)) + \lambda p(n). \qquad (4.5.13)$$

Our aim is to show that there is a priori bound, say L such that all solutions $x(n)$ of

$$x(n) = \lambda \sum_{j=-\infty}^{n-1} m^{-(j-n+1)} \Big[(-m+a)x(j) + f(x(j))\Big]$$

$$+ \lambda \sum_{s=-\infty}^{n-1} m^{-(s-n+1)} \sum_{j=-\infty}^{s} K(s,j)g(x(j)) + \lambda \sum_{s=-\infty}^{n-1} p(j)m^{-(j-n+1)}$$

for $0 < \lambda < 1$ satisfies $|x| \leq L$. Once this is accomplished then we can rule out *(ii)* of Schaefer's theorem (Theorem 4.1.3), and then conclude the above equation has a solution for $\lambda = 1$.

We begin by rewriting (4.5.13) in the form

$$x(n+1) = \mu x(n) + \lambda f(x(n)) + \lambda \sum_{j=-\infty}^{n} K(n,j)g(x(j)) + \lambda p(n), \qquad (4.5.14)$$

where $\mu = m + \lambda(-m+a)$. Define the Lyapunov functional V by

$$V(n,x(\cdot)) = |x(n)| + \lambda \sum_{j=-\infty}^{n-1} \sum_{s=n}^{\infty} |K(s,j)||g(x(j))|. \qquad (4.5.15)$$

It is clear that for $x(n) \in P_T$, $V(n+T,x) = V(n,x)$ and hence V is periodic. Along the solutions of (4.5.14) we have

$$\Delta V(n,x(\cdot)) = |x(n+1)| - |x(n)| + \lambda \sum_{s=n+1}^{\infty} |K(s,n)||g(x(n))|$$

$$- \lambda \sum_{j=-\infty}^{n-1} |K(n,j)||g(x(j))|$$

$$\leq \left(|\mu| - 1\right)|x(n)| + \lambda|f(x)| + \lambda \sum_{s=n}^{\infty} |K(s,n)||g(x(n))| + |p|$$

$$\leq \left(|\mu| - 1\right)|x(n)| + |f(x)| + R|g(x)| + |p|$$

$$\leq \left(|\mu| + \alpha - 1\right)|x(n)| + |p|$$

$$\leq -\beta |x(n)| + |p|.$$

Since V is periodic for $x \in P_T$, we have by summing the above inequality over one period that

$$0 = V(n+T,x(\cdot)) - V(n,x(\cdot)) = \sum_{s=n}^{n+T-1} \Delta V(s,x(\cdot))$$

$$\leq -\beta \sum_{s=n}^{n+T-1} |x(s)| + T|p|.$$

This implies that

$$\sum_{s=n}^{n+T-1} |x(s)| \leq \frac{T|p|}{\beta}.$$

Thus, $|x(n)|$ is bounded over one period, and hence for any T-periodic solution of (4.5.13) there is an $E > 0$ such that $|x(n)| \leq E$, which serves as the a priori bound on every possible T-periodic solution of (4.5.13). Therefore, by Theorem 4.5.1 Equation (4.5.9) has a T-periodic solution for $0 < a < 1$. This concludes the proof of Case 1.

Note that since $0 < \lambda < 1$ condition (4.5.12) reduces to $|a| + \alpha - 1 \leq -\beta$.

Case 2. $-1 < a < 0$

Set $m = a$. Then $-1 < m < 0$ and we apply Theorem 4.5.1 with $m \in (-1, 0)$ to the corresponding family of equations (4.5.13) with $\mu = m + \lambda(-m + a) = a$. Define the Lyapunov functional V by (4.5.15) and proceed with the proof as in Case 1. Note that since $0 < \lambda < 1$, and $\mu = a$, condition (4.5.12) reduces to $|a| + \alpha - 1 \leq -\beta$.

Case 3. $a = 0$

Let m be any fixed number strictly between 0 and 1. Then, $\mu = m - \lambda m < m$. Choose m small enough so that (4.5.12) is satisfied. Then apply Theorem 4.5.1 with $m \in (0, 1)$ to the corresponding family of equations (4.5.13). Define the Lyapunov functional V by (4.5.15) and proceed with the proof as in Case 1.

The next theorem handles the case $|a| > 1$.

Theorem 4.5.3 ([137]). *Assume (4.5.10) and (4.5.11). Also, we assume that there exists an $\alpha > 0$ such that*

$$|f(x)| + R|g(x)| \leq \alpha|x|,$$

and

$$|\mu| - \alpha - 1 \geq \beta, \text{ for some positive constant } \beta,$$

where μ is to be defined in the body of the proof. Then, Equation (4.5.9) has a T-periodic solution.

Proof. **Case 1.** $a > 1$

Set $m = a$. We shall apply Theorem 4.5.1 with $m > 1$ to the corresponding family of equations (4.5.13). Then, $\mu = m + \lambda(-m + a) = a$.

Define the Lyapunov functional V by

$$V(n, x(\cdot)) = |x(n)| - \lambda \sum_{j=-\infty}^{n-1} \sum_{s=n}^{\infty} |K(s, j)||g(x(j))|. \tag{4.5.16}$$

It is clear that for $x(n) \in P_T$, then $V(n + T, x) = V(n, x)$ and hence V is periodic. Along the solutions of (4.5.14) we have

$$\triangle V(n, x(\cdot)) = |x(n+1)| - |x(n)| - \lambda \sum_{s=n}^{\infty} |K(s, n)||g(x(n))|$$

$$+ \lambda \sum_{j=-\infty}^{n-1} |K(n, j)||g(x(j))|$$

$$\geq \left(|\mu| - 1\right)|x(n)| - (|f(x)| + R|g(x)|) - |p|$$

$$\geq \left(|\mu| + \alpha - 1\right)|x(n)| - |p|$$

$$\geq \beta |x(n)| - |p|.$$

Since V is periodic for $x \in P_T$, we have by summing the above inequality over one period that

$$0 = V(n+T,x(\cdot)) - V(n,x(\cdot)) = \sum_{s=n}^{n+T-1} \Delta V(s,x(\cdot))$$

$$\geq \beta \sum_{s=n}^{n+T-1} |x(s)| - Tp.$$

This implies that

$$\sum_{s=n}^{n+T-1} |x(s)| \leq \frac{Tp}{\beta}.$$

Thus, $|x(n)|$ is bounded over one period, and hence for any T-periodic solution of (4.5.13) there is an $E > 0$ such that $|x(n)| \leq E$, which serves as the a priori bound on every possible T-periodic solution of (4.5.13). Therefore, by Theorem 4.5.1 Equation (4.5.9) has a T-periodic solution for $a > 1$. This concludes the proof of Case 1.

Again, we remark that the condition $|\mu| - \alpha - 1 \geq \beta$, for some positive constant β, reduces to $|a| - \alpha - 1 \geq \beta$.

Case 2. $a < -1$

Set $m = a$. Then $m < -1$ and we apply Theorem 4.5.1 to the corresponding family of equations (4.5.13) with $\mu = m + \lambda(-m+a) = a$. Thus, $|\mu| = |a|$. Define the Lyapunov functional V by (4.5.16) and then the proof is the same as in Case 2. This concludes the proof of the theorem.

Remark 4.4. 1) By relaxing the condition $|a| < 1$, we point out that Theorem 4.5.3 significantly improves the literature that is related to the existence of periodic solutions in Volterra difference equations.

2) In [130], for $|a| = 1$, the author was able to show the existence of a periodic solution under the stringent condition that the functions f and g are uniform bounded by certain positive constants. However, we could not do the same here under the condition

$$|f(x)| + R|g(x)| \leq \alpha|x|.$$

4.6 Functional Equations with Constant or Periodically Constant Solutions

Consider the difference equation

$$\Delta x(t) = x(t) - x(t-L), \tag{4.6.1}$$

then any constant is a solution of (4.6.1). In this case we ask ourselves if the constant solution is pre-determined. Therefore, it is convenient to generalize the concept and look at variant forms of the general functional difference equation

$$\triangle x(t) = g(x(t)) - g(x(t-L)), \tag{4.6.2}$$

where $g : \mathbb{R} \to \mathbb{R}$ and is continuous in x. Eqn.(4.6.2) can be easily generalized to functional equations of the form

$$\triangle x(t) = g(x(t-L_1)) - g(x(t-L_1-L_2)), \tag{4.6.3}$$

$$\triangle x(t) = g(x(t)) - \sum_{s=t-L}^{t-1} p(s-t)g(x(s)). \tag{4.6.4}$$

$$\triangle x(t) = \sum_{s=t-L}^{t-1} p(s-t)g(x(s)) - \sum_{s=-\infty}^{t-1} q(s-t)g(x(s)). \tag{4.6.5}$$

Results of this section are partially published in [127] and [139]. In [139] Raffoul, studied systems (4.6.2) and (4.6.3) along with

$$\triangle x(t) = g(t,x(t)) - g(t,x(t-L)), \quad g(t+L,x) = g(t,x). \tag{4.6.6}$$

The first term on the right takes into account the ideas of (4.6.4) while the second term takes into account the deaths distributed over all past times. Note that if $x = c$ where c is constant, then $\triangle x(t) = 0$ in (4.6.2)–(4.6.5) provided that

$$\sum_{s=-L}^{-1} p(s) = 1, \text{ and } \sum_{s=-\infty}^{-1} q(s) = 1.$$

4.6.1 The Finite Delay System

By means of fixed point theory we show that the unique solution of (4.6.4) converges to a pre-determined constant or a periodic solution. Then, we show the solution is stable and that its limit function serves as a global attractor. The same theory will be extended to two more models. We will use the contraction mapping principle to determine that constant. First, we state what it means for $x(t)$ to be a solution of (4.6.4). Note that since (4.6.4) is autonomous, we lose nothing by starting the solution at 0.

Let $\psi(t) : [-L,0] \to \mathbb{R}$ be a given bounded initial function. We say $x(t,0,\psi)$ is a solution of (4.6.4) if $x(t,0,\psi) = \psi(t)$ on $[-L,0]$ and $x(t,0,\psi)$ satisfies (4.6.4) for $t \geq 0$.

It is of importance to us to know such constants since all of our models have constant solutions. First we rewrite (4.6.4) as

$$\triangle x(t) = \triangle_t \sum_{s=-L}^{-1} p(s) \sum_{u=t+s}^{t-1} g(x(u)), \tag{4.6.7}$$

where $p(s)$ satisfies the condition

$$\sum_{s=-L}^{-1} p(s) = 1. \tag{4.6.8}$$

Also, we assume that the function g is globally Lipschitz. That is, there exists a constant $k > 0$ such that

$$|g(x) - g(y)| \leq k|x - y|. \tag{4.6.9}$$

On the other hand, in order to obtain contraction, we assume there is a positive constant $\xi < 1$ so that

$$k \sum_{s=-L}^{-1} |p(s)|(-s) \leq \xi. \tag{4.6.10}$$

We note that if $p(t) = \dfrac{1}{L}$, then (4.6.8) is satisfied. Moreover, in this case condition (4.6.10) becomes

$$k \sum_{s=-L}^{-1} |p(s)|(-s) = k \sum_{s=-L}^{-1} \frac{1}{L}(-s) = \frac{k(L+1)}{2}.$$

Thus, condition (4.6.10) is satisfied for

$$\frac{k(L+1)}{2} \leq \xi.$$

To construct a suitable mapping, we let $\psi : [-L, 0] \to \mathbb{R}$ be a given initial function. By summing (4.6.7) from $s = 0$ to $s = t - 1$ we arrive at the expression

$$x(t) = \psi(0) - \sum_{s=-L}^{-1} p(s) \sum_{u=s}^{-1} g(\psi(u)) + \sum_{s=-L}^{-1} p(s) \sum_{u=t+s}^{t-1} g(x(u)). \tag{4.6.11}$$

If $x(t)$ is given by (4.6.11), then it solves (4.6.4). In the next theorem we show that, given an initial function $\psi : [-L, 0] \to \mathbb{R}$, the unique solution of (4.6.4) converges to a unique determined constant.

Theorem 4.6.1 ([127]). *Assume* (4.6.8)–(4.6.10) *and let* $\psi : [-L, 0] \to \mathbb{R}$ *be a given initial function. Then, the unique solution* $x(t, 0, \psi)$ *of* (4.6.4) *satisfies* $x(t, 0, \psi) \to r$, *where* r *is unique and given by*

$$r = \psi(0) + g(r) \sum_{s=-L}^{-1} p(s)(-s) - \sum_{s=-L}^{-1} p(s) \sum_{u=s}^{-1} g(\psi(u)). \tag{4.6.12}$$

Proof. For $|\cdot|$ denoting the absolute value, the metric space $(\mathbb{R}, |\cdot|)$ is complete. Define a mapping $\mathscr{H} : \mathbb{R} \to \mathbb{R}$, by

$$\mathscr{H}r = \psi(0) + g(r) \sum_{s=-L}^{-1} p(s)(-s) - \sum_{s=-L}^{-1} p(s) \sum_{u=s}^{-1} g(\psi(u)).$$

For $a, b \in \mathbb{R}$, we have

$$\left| \mathscr{H}a - \mathscr{H}b \right| \leq \sum_{s=-L}^{-1} |p(s)|(-s)|g(a) - g(b)| \leq k \sum_{s=-L}^{-1} |p(s)|(-s)|a - b| \leq \xi |a - b|.$$

This shows that $\mathscr{H}$ is a contraction on the complete metric space $(\mathbb{R}, |\cdot|)$, and hence $\mathscr{H}$ has a unique fixed point r, which implies that (4.6.12) has a unique solution. It remains to show that (4.6.4) has a unique solution and that it converges to the constant r.

Let $||\cdot||$ denote the maximum norm and let $\mathbb{M}$ be the set bounded functions ϕ : $[-L, \infty) \to \mathbb{R}$ with $\phi(t) = \psi(t)$ on $[-L, 0], \phi(t) \to r$ as $t \to \infty$. Then $(\mathbb{M}, ||\cdot||)$ defines a complete metric space. For $\phi \in \mathbb{M}$, define $\mathscr{P} : \mathbb{M} \to \mathbb{M}$ by

$$(\mathscr{P}\phi)(t) = \psi(t), \text{ for } -L \leq t \leq 0,$$

and

$$(\mathscr{P}\phi)(t) = \psi(0) - \sum_{s=-L}^{-1} p(s) \sum_{u=s}^{-1} g(\psi(u)) + \sum_{s=-L}^{-1} p(s) \sum_{u=t+s}^{t-1} g(\phi(u)), \text{ for } t \geq 0.$$
(4.6.13)

For $\phi \in \mathbb{M}$ with $\phi(t) \to r$, we have

$$\sum_{s=-L}^{-1} p(s) \sum_{u=t+s}^{t-1} g(\phi(u)) \to g(r) \sum_{s=-L}^{-1} p(s)(-s), \text{ as } t \to \infty.$$

Then, using (4.6.12) and (4.6.13), we see that

$$(\mathscr{P}\phi)(t) \to \psi(0) + g(r) \sum_{s=-L}^{-1} p(s)(-s) - \sum_{s=-L}^{-1} p(s) \sum_{u=s}^{-1} g(\psi(u)) = r.$$

Thus, $\mathscr{P} : \mathbb{M} \to \mathbb{M}$. It remains to show that $\mathscr{P}$ is a contraction. For $a, b \in \mathbb{M}$, we have

$$\left| (\mathscr{P}a)(t) - (\mathscr{P}b)(t) \right| \leq \sum_{s=-L}^{-1} |p(s)|(-s)|g(a(s)) - g(b(s))|$$

$$\leq k \sum_{s=-L}^{-1} |p(s)|(-s) |a - b| \leq \xi ||a - b||.$$

Thus, $\mathscr{P}$ is a contraction and has a unique fixed point $\phi \in \mathbb{M}$. Based on how the mapping $\mathscr{P}$ was constructed, we conclude the unique fixed point ϕ satisfies (4.6.4).

Remark 4.5. For any given initial function, Theorem 4.6.1 explicitly gives the limit to which the solution converges to. That limit is the unique solution r of (4.6.12).

Remark 4.6. For arbitrary initial function, say $\eta : [-L, 0] \to \mathbb{R}$, Theorem 4.6.1 shows that $x(t, 0, \eta) \to r$. Thus, we may think of r as being "global attractor."

Remark 4.7. We may think of Theorem 4.6.1 as of stability results. In general, we know that solutions depend on initial functions. That is, solutions which start close remain close on finite intervals. Under conditions Theorem 4.6.1 such solutions remain close forever, and their asymptotic respective constants remain close too.

The next theorem is a verification of our claim in Remark 4.7.

Theorem 4.6.2 ([127]). *Assume the hypothesis of Theorem 4.6.1. Then every initial function is stable. Moreover, if ψ_1 and ψ_2 are two initial functions with $x(t,0,\psi_1) \to r_1$, and $x(t,0,\psi_2) \to r_2$, then $|r_1 - r_2| < \varepsilon$ for positive ε.*

Proof. Let $||\psi||_{[-L,0]}$ denote the supremum norm of ψ on the interval $[-L,0]$. Fix an initial function ψ_1 and let ψ_2 be any other initial function. Let $\mathscr{P}_i, i = 1,2$ be the mapping defined by (4.6.13). Then by Theorem 4.6.1 there are unique functions θ_1, θ_2 and unique constants r_1 and r_2 such that

$$\mathscr{P}_1\theta_1 \to \theta_1, \quad \mathscr{P}_2\theta_2 \to \theta_2, \quad \theta_1(t) \to r_1, \quad \theta_2(t) \to r_2.$$

Let $\varepsilon > 0$ be any given positive number and set $\delta = \dfrac{\varepsilon\left(1 - k\sum_{s=-L}^{-1}|p(s)|(-s)\right)}{1 + k\sum_{s=-L}^{-1}|p(s)|(-s)}$.
Then

$$
\begin{aligned}
|\theta_1(t) - \theta_2(t)| &= |(\mathscr{P}_1\theta_1)(t) - (\mathscr{P}_2\theta_2)(t)| \\
&\leq |\psi_1(0) - \psi_2(0)| + \sum_{s=-L}^{-1} p(s) \sum_{u=s}^{-1} |g(\psi_1(s)) - g(\psi_2(s))| \\
&\quad + \sum_{s=-L}^{-1} p(s) \sum_{u=t+s}^{t-1} |g(\theta_1(s)) - g(\theta_2(s))| \\
&\leq |\psi_1(0) - \psi_2(0)| + k \sum_{s=-L}^{-1} |p(s)|(-s)||\psi_1 - \psi_2||_{[-L,0]} \\
&\quad + k \sum_{s=-L}^{-1} |p(s)|(-s)|||\theta_1 - \theta_2||.
\end{aligned}
$$

This yields

$$||\theta_1 - \theta_2|| < \frac{1 + k\sum_{s=-L}^{-1}|p(s)|(-s)}{1 - k\sum_{s=-L}^{-1}|p(s)|(-s)}||\psi_1 - \psi_2||_{[-L,0]} < \varepsilon,$$

provided that

$$||\psi_1 - \psi_2||_{[-L,0]} < \frac{\varepsilon\left(1 - k\sum_{s=-L}^{-1}|p(s)|(-s)\right)}{1 + k\sum_{s=-L}^{-1}|p(s)|(-s)} := \delta.$$

This shows that

$$|x(t,0,\psi_1) - x(t,0,\psi_2)| < \varepsilon, \text{ whenever } ||\psi_1 - \psi_2||_{[-L,0]} < \delta.$$

For the rest of the proof we note that $|\theta_i(t) - k_i| \to 0$, as $t \to \infty$ implies that

$$
\begin{aligned}
|r_1 - r_2| &= |r_1 - \theta_1(t) + \theta_1(t) - \theta_2(t) + \theta_2(t) - r_2| \\
&\leq |r_1 - \theta_1(t)| + ||\theta_1 - \theta_2|| + |\theta_2(t) - r_2| \to ||\theta_1 - \theta_2||, \text{ (as } t \to \infty) \\
&< \varepsilon.
\end{aligned}
$$

4.6.2 The Infinite Delay System

In this section, we consider the infinite delay system . For completeness we restate the infinite delay system

$$\triangle x(t) = \sum_{s=t-L}^{t-1} p(s-t)g(x(s)) - \sum_{s=-\infty}^{t-1} q(s-t)g(x(s)) \tag{4.6.14}$$

and rewrite it as

$$\triangle x(t) = -\triangle_t \sum_{s=-L}^{-1} p(s) \sum_{u=t+s}^{t-1} g(x(u)) + \triangle_t \sum_{s=-\infty}^{t-1} \sum_{u=-\infty}^{s-t} q(u)g(x(s)), \tag{4.6.15}$$

where we have assumed (4.6.8) and

$$\sum_{s=-\infty}^{-1} q(s) = 1. \tag{4.6.16}$$

Let $\psi : (-\infty, 0] \to \mathbb{R}$ be an initial bounded sequence. Then

$$x(t) = -\sum_{s=-L}^{-1} p(s) \sum_{u=t+s}^{t-1} g(x(u)) + \sum_{s=-\infty}^{t-1} \sum_{u=-\infty}^{s-t} q(u)g(x(s)) + c, \tag{4.6.17}$$

where

$$c = \psi(0) + \sum_{s=-L}^{-1} p(s) \sum_{u=s}^{-1} g(x(u)) - \sum_{s=-\infty}^{-1} \sum_{u=-\infty}^{s} q(u)g(\psi(s)) \tag{4.6.18}$$

is a solution of (4.6.14). We have the following theorem.

Theorem 4.6.3 ([127]). *Assume* (4.6.8), (4.6.9), *and* (4.6.16) *and there exists a constant* α *so that for* $0 < \alpha < 1$, *we have*

$$k\left(\sum_{s=-L}^{-1}|p(s)(-s)|+\sum_{s=-\infty}^{-1}\sum_{u=-\infty}^{s}|q(u)|\right)\le\alpha. \qquad (4.6.19)$$

Then, the unique solution $x(t,0,\psi)$ of (4.6.14) satisfies $x(t,0,\psi)\to r$, where r is unique and given by

$$r=c-g(r)\sum_{s=-L}^{-1}p(s)(-s)+g(r)\sum_{s=-\infty}^{-1}\sum_{u=-\infty}^{s}q(u), \qquad (4.6.20)$$

and c is given by (4.6.18).

Proof. For $|\cdot|$ denoting the absolute value, the metric space $(\mathbb{R},|\cdot|)$ is complete. Define a mapping $\mathscr{H}:\mathbb{R}\to\mathbb{R}$, by

$$\mathscr{H}r=c-g(r)\sum_{s=-L}^{-1}p(s)(-s)+g(r)\sum_{s=-\infty}^{-1}\sum_{u=-\infty}^{s}q(u).$$

For $a,b\in\mathbb{R}$, we have

$$|\mathscr{H}a-\mathscr{H}b|\le\sum_{s=-L}^{-1}|p(s)(-s)||g(a)-g(b)|+|g(a)-g(b)|\sum_{s=-\infty}^{-1}\sum_{u=-\infty}^{s}|q(u)|$$

$$\le k\left(\sum_{s=-L}^{-1}|p(s)(-s)|+\sum_{s=-\infty}^{-1}\sum_{u=-\infty}^{s}|q(u)|\right)|a-b|$$

$$\le\alpha|a-b|.$$

This shows that $\mathscr{H}$ is a contraction on the complete metric space $(\mathbb{R},|\cdot|)$, and hence $\mathscr{H}$ has a unique fixed point r, which implies that (4.6.20) has a unique solution. It remains to show that (4.6.14) has a unique solution and that it converges to the constant r.

Let $||\cdot||$ denote the maximum norm and let $\mathbb{M}$ be the set bounded functions $\phi:[-\infty,\infty)\to\mathbb{R}$ with $\phi(t)=\psi(t)$ on $[-\infty,0],\phi(t)\to r$ as $t\to\infty$. Then $(\mathbb{M},||\cdot||)$ defines a complete metric space. For $\phi\in\mathbb{M}$, define $\mathscr{P}:\mathbb{M}\to\mathbb{M}$ by

$$(\mathscr{P}\phi)(t)=\psi(t),\text{ for }t\in(-\infty,0],$$

and

$$(\mathscr{P}\phi)(t)=c-\sum_{s=-L}^{-1}p(s)\sum_{u=t+s}^{t-1}g(\phi(u))+\sum_{s=-\infty}^{t-1}\sum_{u=-\infty}^{s-t}q(u)g(\phi(s)),\text{ for }t\ge0$$

$$(4.6.21)$$

where c is given by (4.6.18). Due to the continuity of g we have that for $\phi\in\mathbb{M}$ with $\phi(t)\to r$,

$$\sum_{s=-L}^{-1}p(s)\sum_{u=t+s}^{t-1}g(\phi(u))\to g(r)\sum_{s=-L}^{-1}p(s)(-s),\text{ as }t\to\infty.$$

Next we show that

$$\sum_{s=-\infty}^{t-1} \sum_{u=-\infty}^{s-t} q(u)g(\phi(s)) \to g(r) \sum_{s=-\infty}^{t-1} \sum_{u=-\infty}^{s-t} q(u), \text{ as } t \to \infty. \tag{4.6.22}$$

Again, due to the continuity of G, for $\phi \in \mathbb{M}$ with $\phi(t) \to r$, one might find positive numbers Q and T such that for any $\varepsilon > 0$ we have

$$|g(\phi(t)) - g(r)| \leq Q \text{ for all t and } |\phi(t) - r| < \varepsilon \text{ if } T \leq t < \infty.$$

With this in mind, we have

$$\left| \sum_{s=-\infty}^{t-1} \sum_{u=-\infty}^{s-t} q(u)\big(g(\phi(s)) - g(r)\big) \right| \leq \sum_{s=-\infty}^{T-1} \sum_{u=-\infty}^{s-t} |q(u)| \big|(g(\phi(s)) - g(r))\big|$$

$$+ \sum_{s=T}^{t-1} \sum_{u=-\infty}^{s-t} |q(u)| \big|(g(\phi(s)) - g(r))\big|$$

$$\leq Q \sum_{s=-\infty}^{T-1} \sum_{u=-\infty}^{s-t} |q(u)| + \sum_{s=T}^{t-1} \sum_{u=-\infty}^{s-t} |q(u)||\phi(s) - r|$$

$$\leq Q \sum_{s=-\infty}^{T-1} \sum_{u=-\infty}^{s-t} |q(u)| + k\varepsilon \sum_{s=T}^{t-1} \sum_{u=-\infty}^{s-t} |q(u)|$$

$$\leq Q \sum_{s=-\infty}^{T-t-1} \sum_{u=-\infty}^{s} |q(u)| + k\varepsilon \sum_{s=-\infty}^{t-1} \sum_{u=-\infty}^{s-t} |q(u)|.$$

Due to the convergence that was assumed in (4.6.19), we have $\displaystyle\sum_{s=-\infty}^{T-t-1} \sum_{u=-\infty}^{s} |q(u)| \to$ 0, as $t \to \infty$. Moreover, for $T \leq t < \infty$, condition (4.6.19) implies that $ke \displaystyle\sum_{s=-\infty}^{t-1} \sum_{u=-\infty}^{s-t} |q(u)| \leq \varepsilon\alpha$. Hence (4.6.22) is proved. It remains to show that $\mathscr{P}$ is a contraction. For $a, b \in \mathbb{M}$, we have

$$\left|(\mathscr{P}a)(t) - (\mathscr{P}b)(t)\right| \leq \sum_{s=-L}^{-1} |p(s)||(-s)|g(a(s)) - g(b(s))|$$

$$+ \sum_{s=-\infty}^{t-1} \sum_{u=-\infty}^{s-t} |q(u)||g(a(s)) - g(b(s))|$$

$$\leq k\left(\sum_{s=-L}^{-1} |p(s)(-s)| + \sum_{s=-\infty}^{t-1} \sum_{u=-\infty}^{s-t} |q(u)| \right) ||a - b||$$

$$\leq \alpha ||a - b||.$$

Parallel remarks to Remarks 4.5–4.7 can be made regarding the infinite delay model given by (4.6.14).

4.6.3 The Finite Delay System Revisited

We revisit the finite delay system given by (4.6.4) with slight adjustment, namely

$$\triangle x(t) = g(t, x(t)) - \sum_{s=t-L}^{t-1} p(s-t)g(s, x(s)), \qquad (4.6.23)$$

where

$$g(t+L, x) = g(t, x) \qquad (4.6.24)$$

and investigate the existence of periodic solutions. As before, we assume there exists a positive constant k such that for all $x, y \in \mathbb{R}$ we have

$$|g(t, x) - g(t, y)| \le k|x - y|. \qquad (4.6.25)$$

If (4.6.8) holds, then we may rewrite (4.6.23) as

$$\triangle x(t) = \triangle_t \sum_{s=-L}^{-1} p(s) \sum_{u=t+s}^{t-1} g(u, x(u)). \qquad (4.6.26)$$

As before, to construct a suitable mapping, we let $\psi : [-L, 0] \to \mathbb{R}$ be a given initial function. By summing (4.6.26) from $s = 0$ to $s = t-1$ we arrive at the expression

$$x(t) = \sum_{s=-L}^{-1} p(s) \sum_{u=t+s}^{t-1} g(u, x(u)) + c, \qquad (4.6.27)$$

where c is given by

$$c = \psi(0) - \sum_{s=-L}^{-1} p(s) \sum_{u=s}^{-1} g(u, \psi(u)). \qquad (4.6.28)$$

Theorem 4.6.4 ([127]). *Assume (4.6.8)–(4.6.10), (4.6.24), and (4.6.25) and let* $\psi : [-L, 0] \to \mathbb{R}$ *be a given initial function. Then, the unique solution* $x(t, 0, \psi)$ *of (4.6.23) satisfies* $x(t, 0, \psi) \to \rho$*, as* $t \to \infty$ *where* ρ *is a unique L-periodic solution of (4.6.23).*

Proof. Let $|| \cdot ||$ denote the maximum norm and let $\mathbb{M}$ be the set of L-periodic sequences $\phi : \mathbb{Z} \to \mathbb{Z}$. Then $(\mathbb{M}, || \cdot ||)$ defines a Banach space of L-periodic sequences. For $\phi \in \mathbb{M}$, define $\mathscr{P} : \mathbb{M} \to \mathbb{M}$ by

$$(\mathscr{P}\phi)(t) = c + \sum_{s=-L}^{-1} p(s) \sum_{u=t+s}^{t-1} g(u, \phi(u)) \qquad (4.6.29)$$

Next we show that

$$(\mathscr{P}\phi)(t+L) = (\mathscr{P}\phi)(t).$$

To see, for $\phi \in \mathbb{M}$, we have

$$(\mathscr{P}\phi)(t+L) = c + \sum_{s=-L}^{-1} p(s) \sum_{u=t+s+L}^{t+L-1} g(u, \phi(u))$$

$$= c + \sum_{s=-L}^{-1} p(s) \sum_{l=t+s}^{t-1} g(l+L, \phi(l+L)), \ (l = u - L)$$

$$= c + \sum_{s=-L}^{-1} p(s) \sum_{l=t+s}^{t-1} g(l, \phi(l)) = (\mathscr{P}\phi)(t).$$

Hence, $\mathscr{P}$ maps $\mathbb{M}$ into $\mathbb{M}$. Also, by similar argument as in the previous theorems, one can easily show that $\mathscr{P}$ is a contraction. Hence, (4.6.29) has a unique fixed point ρ in $\mathbb{M}$, which solves (4.6.23). It remains to show that $(\mathscr{P}\phi)(t) \to \rho(t)$.

Let $||\cdot||$ denote the maximum norm and let $\mathbb{M}$ be the set of bounded functions $\phi : [-L, \infty) \to \mathbb{R}$ with $\phi(t) = \psi(t)$ on $[-L, 0]$, $\phi(t) \to \rho(t)$ as $t \to \infty$. Then $(\mathbb{M}, ||\cdot||)$ defines a complete metric space. For $\phi \in \mathbb{M}$, define $\mathscr{P} : \mathbb{M} \to \mathbb{M}$ by

$$(\mathscr{P}\phi)(t) = \psi(t), \text{ for } -L \le t \le 0,$$

and

$$(\mathscr{P}\phi)(t) = c + \sum_{s=-L}^{-1} p(s) \sum_{u=t+s}^{t-1} g(u, \phi(u)), \text{ for } t \ge 0.$$

$$|(\mathscr{P}\phi)(t) - \rho(t)| = \Big| \sum_{s=-L}^{-1} p(s) \sum_{u=t+s}^{t-1} g(u, \phi(u)) - \sum_{s=-L}^{-1} p(s) \sum_{u=t+s}^{t-1} g(u, \rho(u)) \Big|$$

$$\le \sum_{s=-L}^{-1} |p(s)| \sum_{u=t+s}^{t-1} k|\phi(u) - \rho(u)|$$

$$\le \sum_{s=-L}^{-1} |p(s)| \sum_{u=t-L}^{t-1} k|\phi(u) - \rho(u)| \to 0, \text{ as } t \to \infty,$$

since $|\phi(u) - \rho(u)| \to 0$, as $t \to \infty$. The proof for showing $\mathscr{P}$ is a contraction is similar to before and hence we omit. Thus we have shown that $\mathscr{P}$ has a unique fixed point in $\mathbb{M}$, which converges to ρ.

We note that Remarks 4.5–4.7 and hence Theorem 4.6.2 hold for equations (4.6.14) and (4.6.23). We end with the following corollary.

Corollary 4.6 ([127]). *Assume the hypothesis of Theorem 4.6.4. If there exists an* $r \in \mathbb{R}$, *such that*

$$g(t, r) = \sum_{s=-L}^{-1} p(s) g(t+s, r), \tag{4.6.30}$$

then ρ *of Theorem 4.6.4 is constant.*

Proof. Suppose (4.6.23) has a constant solution r. Then from (4.6.26) we have

$$0 = \triangle r = \triangle_t \sum_{s=-L}^{-1} p(s) \sum_{u=t+s}^{t-1} g(u,r)$$

$$= \sum_{s=-L}^{-1} p(s)\big(g(t,r) - g(t+s,r)\big)$$

$$= g(t,r) \sum_{s=-L}^{-1} p(s) - \sum_{s=-L}^{-1} p(s)g(t+s,r)$$

$$= g(t,r) - \sum_{s=-L}^{-1} p(s)g(t+s,r), \text{ due to (4.6.8)}.$$

Or,

$$g(t,r) = \sum_{s=-L}^{-1} p(s)g(t+s,r).$$

This completes the proof.

4.7 Periodic and Asymptotically Periodic Solutions in Coupled Systems

Now we turn our attention to the existence of periodic and asymptotically periodic solutions of a coupled system of nonlinear Volterra difference equations with infinite delay. By means of fixed point theory, namely Schauder's fixed point theorem, we furnish conditions that guarantee the existence of such periodic solutions. Consider the coupled system of nonlinear Volterra difference equations with infinite delay

$$\begin{cases} \triangle x_n = h_n x_n + \sum_{i=-\infty}^{n} a_{n,i} f(y_i) \\ \triangle y_n = p_n y_n + \sum_{i=-\infty}^{n} b_{n,i} g(x_i) \end{cases} \tag{4.7.1}$$

where f and g are real valued and continuous functions, and $\{a_{n,i}\}$, $\{b_{n,i}\}$, $\{h_n\}$, and $\{p_n\}$ are real sequences. In this study, we use *Schauder's fixed point theorem* to provide sufficient conditions guaranteeing the existence of periodic and asymptotically periodic solutions of system (4.7.1). Since we are seeking the existence of periodic solutions it is natural to ask that there exists a least positive integer T such that

$$h_{n+T} = h_n, \; p_{n+T} = p_n, \tag{4.7.2}$$

$$a_{n+T,i+T} = a_{n,i}, \tag{4.7.3}$$

and

$$b_{n+T,i+T} = b_{n,i} \tag{4.7.4}$$

hold for all $n \in \mathbb{N}$, where $\mathbb{N}$ indicates the set of all natural numbers.

There is a vast literature on this subject in the continuous and discrete cases. For instance, in [179] the authors considered the two-dimensional system of nonlinear Volterra difference equations

$$\begin{cases} \triangle x_n = h_n x_n + \sum_{i=1}^{n} a_{n,i} f(y_i) \\ \triangle y_n = p_n y_n + \sum_{i=1}^{n} b_{n,i} g(x_i) \end{cases}, \quad n = 1, 2, \dots$$

and classified the limiting behavior and the existence of its positive solutions with the help of fixed point theory. Also, the authors of [102] analyzed the asymptotic behavior of positive solutions of second order nonlinear difference systems, while the authors of [107] studied the classification and the existence of positive solutions of the system of Volterra nonlinear difference equations. Periodicity of the solutions of difference equations has been handled by [6, 46, 47, 48, 49, 50, 51, 52, 53, 54, 55, 56, 57, 58, 59, 60, 61, 62, 63, 64, 65, 66, 67, 68, 69, 70, 71, 72, 73, 74, 75]. In [48] and [49], the authors focused on a system of Volterra difference equations of the form

$$x_s(n) = a_s(n) + b_s(n) x_s(n) + \sum_{p=1}^{r} \sum_{i=0}^{n} K_{sp}(n,i) x_p(i), \quad n \in \mathbb{N},$$

where $a_s, b_s, x_s : \mathbb{N} \to \mathbb{R}$ and $K_{sp} : \mathbb{N} \times \mathbb{N} \to \mathbb{R}$, $s = 1, 2, \dots, r$, and $\mathbb{R}$ denotes the set of all real numbers and obtained sufficient conditions for the existence of asymptotically periodic solutions. They had to construct a mapping on an appropriate space and then obtain a fixed point. Furthermore, in [86] the authors investigated the existence of periodic and positive periodic solutions of system of nonlinear Volterra integro-differential equations. The paper [55] of Elaydi was one of the first to address the existence of periodic solutions and the stability analysis of Volterra difference equations. Since then, the study of Volterra difference equations has been vastly increasing. For instance, we mention the papers [93, 113], and the references therein. In addition to periodicity we refer to [96] and [117] for results regarding boundedness.

The main purpose of this study is to extend the results of the above-mentioned literature by investigating the possibility of existence of periodic and the asymptotic periodic solutions for systems of nonlinear Volterra difference equations with infinite delay.

By a solution of the system (4.7.1) we mean a pair of sequences $\{(x_n, y_n)\}_{n \in \mathbb{Z}}$ of real numbers which satisfies (4.7.1) for all $n \in \mathbb{N}$. Let $\mathbb{Z}^-$ denote the set of all negative integers. The initial sequence space for the solutions of the system (4.7.1) can be constructed as follows. Let S denote the nonempty set of pairs of all sequences $(\eta, \zeta) = \{(\eta_n, \zeta_n)\}_{n \in \mathbb{Z}^-}$ of real numbers such that

$$\max \left\{ \sup_{n \in \mathbb{Z}^-} |\eta_n|, \sup_{n \in \mathbb{Z}^-} |\zeta_n| \right\} < \infty$$

and for each $n \in \mathbb{N}$, the series

$$\sum_{i=-\infty}^{0} a_{n,i} f(\eta_i) \text{ and } \sum_{i=-\infty}^{0} b_{n,i} g(\zeta_i)$$

converge. It is clear that for any given pair of initial sequences $\{(\eta_n, \zeta_n)\}_{n \in \mathbb{Z}^-}$ in S there exists a unique solution $\{(x_n, y_n)\}_{n \in \mathbb{Z}}$ of the system (4.7.1) which satisfies the initial condition

$$\begin{pmatrix} x_n \\ y_n \end{pmatrix} = \begin{pmatrix} \eta_n \\ \zeta_n \end{pmatrix} \text{ for } n \in \mathbb{Z}^-. \tag{4.7.5}$$

Such solution $\{(x_n, y_n)\}_{n \in \mathbb{Z}}$ is said to be the solution of the initial problem (4.7.1-4.7.5). For any pair $(\eta, \zeta) \in S$, one can specify a solution of (4.7.1–4.7.5) by denoting it by $(x_\eta, y_\zeta) := \{(x_n(\eta), y_n(\zeta))\}_{n \in \mathbb{Z}}$, where

$$(x_n(\eta), y_n(\zeta)) = \begin{cases} (\eta_n, \zeta_n) \text{ for } n \in \mathbb{Z}^- \\ (x_n, y_n) \text{ for } n \in \mathbb{N} \end{cases}$$

In our analysis, we apply a fixed point theorem to general operators over a Banach space of bounded sequences defined on the whole set of integers. Unlike the above-mentioned literature that dealt with stability of delayed difference systems, in the construction of our existence type theorems we neglect the consideration of phase space, for simplicity. For similar approach we refer to [28].

Theorem 4.7.1. *[Schauder's Fixed Point Theorem] Let X be a Banach space. Assume that K is a closed, bounded, and convex subset of X. If $T : K \to K$ is a compact operator, then it has a fixed point in K.*

4.7.1 Periodicity

In this section, we use Schauder's fixed point theorem to show that system (4.7.1) has a periodic solution. First, we start by defining periodic sequences on $\mathbb{Z}$.

Definition 4.7.1. Let T be a positive integer. A sequence $x = \{x_n\}_{n \in \mathbb{Z}}$ is called T-periodic if $x_{n+T} = x_n$ for all $n \in \mathbb{Z}$. The smallest positive integer T such that $x_{n+T} = x_n$ holds for all $n \in \mathbb{Z}$ is called the period of the sequence $x = \{x_n\}_{n \in \mathbb{Z}}$.

Let P_T be the set of all T-periodic sequences on $\mathbb{Z}$. Then P_T is a Banach space when it is endowed with the maximum norm

$$\|(x, y)\| := \max \left\{ \max_{n \in [1,T]_{\mathbb{Z}}} |x_n|, \max_{n \in [1,T]_{\mathbb{Z}}} |y_n| \right\}.$$

Let us define the subset $\Omega(W)$ of P_T by

$$\Omega(W) := \{(x, y) \in P_T : \|(x, y)\| \leq W\},$$

where $W > 0$ is a constant. Then $\Omega(W)$ is bounded, closed, and convex subset of P_T. For any pair $(x,y) = \{(x_n, y_n)\}_{n \in \mathbb{Z}} \in \Omega(W)$, we define the mapping $E : \Omega \to P_T$ by

$$E(x,y) := \{E(x,y)_n\}_{n \in \mathbb{Z}} := \left\{ \begin{pmatrix} E_1(x,y)_n \\ E_2(x,y)_n \end{pmatrix} \right\}_{n \in \mathbb{Z}},$$

where

$$E_1(x,y)_n := \begin{cases} x_n & \text{for } n \in \mathbb{Z}^- \\ \alpha_h \sum\limits_{i=n}^{n+T-1} \left(\prod\limits_{l=i+1}^{n+T-1} (1+h_l) \right) \sum\limits_{m=-\infty}^{i} a_{i,m} f(y_m) & \text{for } n \in \mathbb{N} \end{cases}, \qquad (4.7.6)$$

$$E_2(x,y)_n := \begin{cases} y_n & \text{for } n \in \mathbb{Z}^- \\ \alpha_p \sum\limits_{i=n}^{n+T-1} \left(\prod\limits_{l=i+1}^{n+T-1} (1+p_l) \right) \sum\limits_{m=-\infty}^{i} b_{i,m} g(x_m) & \text{for } n \in \mathbb{N} \end{cases}, \qquad (4.7.7)$$

and

$$\alpha_h := \left[1 - \prod_{l=0}^{T-1} (1+h_l) \right]^{-1},$$

$$\alpha_p := \left[1 - \prod_{l=0}^{T-1} (1+p_l) \right]^{-1}.$$

We shall use the following result on several occasions in our further analysis.

Lemma 4.8. *Assume that (4.7.2–4.7.4) hold. Suppose that* $1 + h_n \neq 0$, $1 + p_n \neq 0$ *for all* $n \in [1,T]_{\mathbb{Z}} := [1,T] \cap \mathbb{Z}$, *and that*

$$\prod_{l=0}^{T-1} (1+h_l) \neq 1 \text{ and } \prod_{l=0}^{T-1} (1+p_l) \neq 1. \qquad (4.7.8)$$

The pair $(x,y) = \{(x_n, y_n)\}_{n \in \mathbb{Z}}$ *satisfies*

$$E(x,y) = (x,y)$$

if and only if it is a T-periodic solution of (4.7.1).

Proof. One may easily verify that the pair $(x,y) = \{(x_n, y_n)\}_{n \in \mathbb{Z}} \in \Omega(W)$ satisfying $(x,y) = E(x,y)$ also satisfies the system (4.7.1) for all $n \in \mathbb{N}$. Conversely, suppose that the pair $(x,y) = \{(x_n, y_n)\}_{n \in \mathbb{Z}}$ is a T-periodic sequence satisfying (4.7.1) for all $n \in \mathbb{N}$. Multiplying both sides of the first equation in (4.7.1) with $\left(\prod\limits_{l=0}^{n} (1+h_l) \right)^{-1}$ and taking the summation from n to $n+T-1$, we obtain

$$\sum_{i=n}^{n+T-1} \Delta \left[x_i \left(\prod_{l=0}^{i-1} (1+h_l) \right)^{-1} \right] = \sum_{i=n}^{n+T-1} \left(\prod_{l=0}^{i} (1+h_l) \right)^{-1} \sum_{m=-\infty}^{i} a_{i,m} f(y_m).$$

This implies that

$$x_{n+T}\left(\prod_{l=0}^{n+T-1}(1+h_l)\right)^{-1} - x_n\left(\prod_{l=0}^{n-1}(1+h_l)\right)^{-1}$$

$$= \sum_{i=n}^{n+T-1}\left(\prod_{l=0}^{i}(1+h_l)\right)^{-1}\sum_{m=-\infty}^{i}a_{i,m}f(y_m).$$

Using the equalities $x_{n+T} = x_n$ and $\prod_{l=n}^{n+T-1}(1+h_l) = \prod_{l=0}^{T-1}(1+h_l)$, we have $E_1(x,y)_n = (x_n,y_n)$ for $n \in \mathbb{N}$. The equality $E_2(x,y)_n = (x_n,y_n)$ for $n \in \mathbb{N}$ can be obtained by using a similar procedure. The proof is complete.

In preparation for the next result we assume that there exist positive constants W_1, W_2, K_1, and K_2 such that

$$|f(x)| \le W_1 \tag{4.7.9}$$

$$|g(y)| \le W_2, \tag{4.7.10}$$

$$|\alpha_h|\sum_{i=n}^{n+T-1}\left|\prod_{l=i+1}^{n+T-1}(1+h_l)\right|\sum_{m=-\infty}^{i}|a_{i,m}| \le K_1, \tag{4.7.11}$$

and

$$|\alpha_p|\sum_{i=n}^{n+T-1}\left|\prod_{l=i+1}^{n+T-1}(1+p_l)\right|\sum_{m=-\infty}^{i}|b_{i,m}| \le K_2 \tag{4.7.12}$$

for all $n \in \mathbb{Z}$ and all $(x,y) \in \Omega(W)$.

Theorem 4.7.2. *In addition to the assumptions of Lemma 4.8 suppose that (4.7.9–4.7.12) hold. Then (4.7.1) has a T-periodic solution.*

Proof. From Lemma 4.8, we can deduce that $E(x,y)_{n+T} = E(x,y)_n$ for any $(x,y) \in \Omega(W)$. Moreover, if $(x,y) \in \Omega(W)$, then

$$|E_1(x,y)_n| \le |\alpha_h|\sum_{i=n}^{n+T-1}\left|\prod_{l=i+1}^{n+T-1}(1+h_l)\right|\sum_{m=-\infty}^{i}|a_{i,m}||f(y_m)| \le W_1K_1, \tag{4.7.13}$$

and

$$|E_2(x,y)_n| \le |\alpha_p|\sum_{i=n}^{n+T-1}\left|\alpha_p\prod_{l=i+1}^{n+T-1}(1+p_l)\right|\sum_{m=-\infty}^{i}|b_{i,m}||g(x_m)| \le W_2K_2 \tag{4.7.14}$$

for all $n \in \mathbb{N}$. If we set $W = \max\{W_1K_1, W_2K_2\}$, then E maps $\Omega(W)$ into itself. Now we show that E is continuous. Let $\{(x^l,y^l)\}$, $l \in \mathbb{N} = \{0,1,2,...\}$, be a sequence in $\Omega(W)$ such that

$$\lim_{l\to\infty}\left\|(x^l,y^l)-(x,y)\right\|=\lim_{l\to\infty}\left(\max_{n\in[1,T]_{\mathbb{Z}}}\left\{\left|x_n^l-x_n\right|,\left|y_n^l-y_n\right|\right\}\right)$$
$$=0.$$

Since $\Omega(W)$ is closed, we must have $(x,y)\in\Omega(W)$. Then by definition of E we have

$$\left\|E(x^l,y^l)-E(x,y)\right\|=\max\{\max_{n\in[1,T]_{\mathbb{Z}}}\left|E_1(x^l,y^l)_n-E_1(x,y)_n\right|,$$
$$\max_{n\in\mathbb{Z}}\left|E_2(x^l,y^l)_n-E_2(x,y)_n\right|\},$$

in which

$$\left|E_1(x^l,y^l)_n-E_1(x,y)_n\right|=|\alpha_h|\left|\sum_{i=n}^{n+T-1}\left(\prod_{l=i+1}^{n+T-1}(1+h_l)\right)\sum_{m=-\infty}^{i}a_{i,m}f(y_m^l)-\right.$$
$$\left.\sum_{i=n}^{n+T-1}\left(\prod_{l=i+1}^{n+T-1}(1+h_l)\right)\sum_{m=-\infty}^{i}a_{i,m}f(y_m)\right|$$
$$\leq|\alpha_h|\sum_{i=n}^{n+T-1}\left|\prod_{l=i+1}^{n+T-1}(1+h_l)\right|\sum_{m=-\infty}^{i}|a_{i,m}|\left|f(y_m^l)-f(y_m)\right|.$$

Similarly,

$$\left|E_2(x^l,y^l)_n-E_2(x,y)_n\right|\leq|\alpha_p|\sum_{i=n}^{n+T-1}\left|\prod_{l=i+1}^{n+T-1}(1+p_l)\right|\sum_{m=-\infty}^{i}|b_{i,m}|\left|g(x_m^l)-g(x_m)\right|.$$

The continuity of f and g along with the Lebesgue dominated convergence theorem imply that

$$\lim_{l\to\infty}\left\|E(x^l,y^l)-E(x,y)\right\|=0.$$

This shows that E is continuous. Finally, we have to show that $E\Omega(W)$ is precompact. Let $\{(x^l,y^l)\}_{l\in\mathbb{N}}$ be a sequence in $\Omega(W)$. For each fixed $l\in\mathbb{N}$, $\{(x_n^l,y_n^l)\}_{n\in\mathbb{Z}}$ is a bounded sequence of real pairs. Then by *Bolzano-Weierstrass Theorem*, $\{(x_n^l,y_n^l)\}_{n\in\mathbb{Z}}$ has a convergent subsequence $\{(x_{n_k}^l,y_{n_k}^l)\}$. By repeating the diagonalization process for each $l\in\mathbb{N}$, we can construct a convergent subsequence $\{(x^{l_k},y^{l_k})\}_{l_k\in\mathbb{N}}$ of $\{(x^l,y^l)\}_{l\in\mathbb{N}}$ in $\Omega(W)$. Since E is continuous, we deduce that $\{E(x^l,y^l)\}_{l\in\mathbb{N}}$ has a convergent subsequence in $E\Omega(W)$. This means, $E\Omega(W)$ is precompact. By Schauder's fixed point theorem we conclude that there exists a pair $(x,y)\in\Omega(W)$ such that $E(x,y)=(x,y)$.

Theorem 4.7.3. *In addition to the assumptions of Lemma 4.8, we assume that (4.7.9), (4.7.11), and (4.7.12) hold. If g is a nondecreasing function satisfying*

$$|g(x)|\leq g(|x|),\tag{4.7.15}$$

then (4.7.1) has a T-periodic solution.

Proof. By (4.7.11) and (4.7.13) we already have

$$|E_1(x,y)| \le W_1 K_1 \text{ for all } (x,y) \in \Omega(W).$$

This along with (4.7.15) imply

$$|E_2(x,y)_n| \le \sum_{i=n}^{n+T-1} \left| \alpha_p \prod_{l=i+1}^{n+T-1} (1+p_l) \right| \sum_{m=-\infty}^{i} |b_{i,m}| |g(x_m)|$$

$$\le \sum_{i=n}^{n+T-1} \left| \alpha_p \prod_{l=i+1}^{n+T-1} (1+p_l) \right| \sum_{m=-\infty}^{i} |b_{i,m}| g(|E_1(x,y)|)$$

$$\le K_2 g(W_1 K_1).$$

If we set $W = \max\{W_1 K_1, K_2 g(W_1 K_1)\}$, then the rest of the proof is similar to the proof of Theorem 4.7.2 and hence we omit it.

Similarly, we can give the following result.

Theorem 4.7.4. *In addition to the assumptions of Lemma 4.8, we assume (4.7.10), (4.7.11), and (4.7.12) hold. If f is a nondecreasing function satisfying*

$$|f(y)| \le f(|y|),$$

then (4.7.1) has a T-periodic solution.

Example 4.3. Let

$$h_n = 1 + \cos n\pi,$$
$$p_n = 1 - \cos n\pi,$$
$$a_{n,i} = b_{n,i} = e^{i-n},$$

and

$$f(x) = \sin x \text{ and } g(x) = \sin 2x.$$

Then (4.7.1) turns into the following system

$$\begin{cases} \triangle x_n = (1 + \cos n\pi)x_n + \sum_{i=-\infty}^{n} e^{i-n} \sin(y_i), \\ \triangle y_n = (1 - \cos n\pi)y_n + \sum_{i=-\infty}^{n} e^{i-n} \sin(2x_i) \end{cases}.$$

It can be easily verified that conditions (4.7.2–4.7.8) and (4.7.9–4.7.12) hold. By Theorem 4.7.2, there exists a 2-periodic solution $(x,y) = \{(x_n, y_n)\}_{n \in \mathbb{Z}}$ of system (4.7.1) satisfying

$$x_n = -\frac{1}{2} \sum_{i=n}^{n+1} \prod_{l=i+1}^{n+1} (2 + \cos(l\pi)) \sum_{m=-\infty}^{i} e^{m-i} \sin(y_m),$$

$$y_n = -\frac{1}{2} \sum_{i=n}^{n+1} \prod_{l=i+1}^{n+1} (2 - \cos(l\pi)) \sum_{m=-\infty}^{i} e^{m-i} \sin(2x_m),$$

for all $n \in \mathbb{N}$.

4.7.2 Asymptotic Periodicity

In this section, we study the existence of an asymptotically T-periodic solution of system (4.7.1) by using Schauder's fixed point theorem. First we state the following definition.

Definition 4.7.2. A sequence $\{x_n\}_{n\in\mathbb{Z}}$ is called asymptotically T-periodic if there exist two sequences u_n and v_n such that u_n is T-periodic, $\lim_{n\to\infty} v_n = 0$, and $x_n = u_n + v_n$ for all $n \in \mathbb{Z}$.

First, we suppose that

$$\prod_{j=0}^{T-1} (1 + h_j) = 1 \text{ and } \prod_{j=0}^{T-1} (1 + p_j) = 1. \tag{4.7.16}$$

Then we define the sequences $\varphi := \{\varphi_n\}_{n\in\mathbb{N}}$ and $\psi := \{\psi_n\}_{n\in\mathbb{N}}$ as follows

$$\varphi_n := \prod_{j=0}^{n-1} \frac{1}{1+h_j} \text{ and } \psi_n := \prod_{j=0}^{n-1} \frac{1}{1+p_j}. \tag{4.7.17}$$

Furthermore, we define the constants m_k, M_k, $k = 1, 2$, by

$$m_1 := \min_{i\in[1,T]_{\mathbb{Z}}} |\varphi_i|, \ M_1 := \max_{i\in[1,T]_{\mathbb{Z}}} |\varphi_i|, \ m_2 := \min_{i\in[1,T]_{\mathbb{Z}}} |\psi_i|, \ M_2 := \max_{i\in[1,T]_{\mathbb{Z}}} |\psi_i|.$$

We note that in this section, we do not assume (4.7.3–4.7.4) but instead we ask that the series

$$\sum_{i=0}^{\infty} \sum_{m=-\infty}^{i} |a_{i,m}| < \infty \text{ and } \sum_{i=0}^{\infty} \sum_{m=-\infty}^{i} |b_{i,m}| < \infty \tag{4.7.18}$$

converge to a and b, respectively. Observe that (4.7.18) implies

$$\lim_{n\to\infty} \sum_{i=n}^{\infty} \sum_{m=-\infty}^{i} |a_{i,m}| = \lim_{n\to\infty} \sum_{i=n}^{\infty} \sum_{m=-\infty}^{i} |b_{i,m}| = 0. \tag{4.7.19}$$

Theorem 4.7.5. *Suppose that (4.7.9–4.7.10), (4.7.16), and (4.7.18–4.7.19) hold. Then system (4.7.1) has an asymptotically T-periodic solution $(x,y) = \{(x_n, y_n)\}_{n\in\mathbb{Z}}$ satisfying*

$$x_n := u_n^{(1)} + v_n^{(1)}$$

$$y_n := u_n^{(2)} + v_n^{(2)}$$

for $n \in \mathbb{N}$, where

$$u_n^{(1)} = c_1 \prod_{j=0}^{n-1}(1+h_j), \quad u_n^{(2)} = c_2 \prod_{j=0}^{n-1}(1+p_j), n \in \mathbb{Z}^+$$

c_1 *and* c_2 *are positive constants, and*

$$\lim_{n\to\infty} v_n^{(1)} = \lim_{n\to\infty} v_n^{(2)} = 0.$$

Proof. Due to the T-periodicity of the sequences $\{h_n\}_{n\in\mathbb{Z}}$ and $\{p_n\}_{n\in\mathbb{Z}}$ and by (4.7.16-4.7.17) we have

$$\varphi_n \in \{\varphi_1, \varphi_2, ..., \varphi_T\} \text{ and } \psi_n \in \{\psi_1, \psi_2, ..., \psi_T\}$$

for all $n \in \mathbb{N}$. This means

$$m_1 \leq |\varphi_n| \leq M_1 \tag{4.7.20}$$

$$m_2 \leq |\psi_n| \leq M_2 \tag{4.7.21}$$

for all $n \in \mathbb{Z}$. Define
$\mathbb{B} = \{(\Phi, \Psi) : \Phi = \Phi_1 + \Phi_2, \Psi = \Psi_1 + \Psi_2, (\Phi_1, \Psi_1)_{n+T} = (\Phi_1, \Psi_1)_n, \text{ and } (\Phi_2, \Psi_2)_n \to (0,0) \text{ as } n \to \infty\}$. Then $\mathbb{B}$ is a Banach space when endowed with the maximum norm

$$\|(x,y)\| = \max\{\sup_{n\in\mathbb{Z}}|x_n|, \sup_{n\in\mathbb{Z}}|y_n|\}.$$

For a positive constant W^* we define

$$\Omega^*(W^*) := \{(x,y) \in \mathbb{B} : \|(x,y)\| \leq W^*\}.$$

Then, $\Omega^*(W^*)$ is a nonempty bounded convex, and closed subset of $\mathbb{B}$. Define the mapping $E^* : \Omega^*(W^*) \to \mathbb{B}$ by

$$E^*(x,y) = \{E^*(x,y)_n\}_{n\in\mathbb{Z}} = \left\{ \begin{pmatrix} E_1^*(x,y)_n \\ E_2^*(x,y)_n \end{pmatrix} \right\}_{n\in\mathbb{Z}},$$

where

$$E_1^*(x,y)_n := \begin{cases} x_n & \text{for } n \in \mathbb{Z}^- \\ c_1 \dfrac{1}{\varphi_n} - \displaystyle\sum_{i=n}^{\infty}\sum_{m=-\infty}^{i} \dfrac{\varphi_{i+1}}{\varphi_n} a_{i,m} f(y_m) & \text{for } n \in \mathbb{N} \end{cases}, \tag{4.7.22}$$

and

$$E_2^*(x,y)_n := \begin{cases} y_n & \text{for } n \in \mathbb{Z}^- \\ c_2 \dfrac{1}{\psi_n} - \displaystyle\sum_{i=n}^{\infty} \sum_{m=-\infty}^{i} \dfrac{\psi_{i+1}}{\psi_n} b_{i,m} g(x_m) & \text{for } n \in \mathbb{N} \end{cases}. \tag{4.7.23}$$

We will show that the mapping E^* has a fixed point in $\mathbb{B}$. First, we demonstrate that $E^*\Omega^*(W^*) \subset \Omega^*(W^*)$. If $\{(x,y)\} \in \Omega^*(W^*)$, then

$$\left| E_1^*(x,y)_n - c_1 \frac{1}{\varphi_n} \right| \le M_1 m_1^{-1} W_1 \sum_{i=n}^{\infty} \sum_{m=-\infty}^{i} |a_{i,m}| \tag{4.7.24}$$

$$\le M_1 m_1^{-1} W_1 \sum_{i=0}^{\infty} \sum_{m=-\infty}^{i} |a_{i,m}|$$

$$= M_1 m_1^{-1} W_1 a, \tag{4.7.25}$$

and

$$\left| E_2^*(x,y)_n - c_2 \frac{1}{\psi_n} \right| \le M_2 m_2^{-1} W_2 \sum_{i=n}^{\infty} \sum_{m=-\infty}^{i} |b_{i,m}| \tag{4.7.26}$$

$$\le M_2 m_2^{-1} W_2 \sum_{i=0}^{\infty} \sum_{m=-\infty}^{i} |b_{i,m}|$$

$$= M_2 m_2^{-1} W_2 b. \tag{4.7.27}$$

This implies that

$$|E_1^*(x_n,y_n)| \le M_1 m_1^{-1} W_1 a + \frac{c_1}{m_1},$$

and

$$|E_2^*(x_n,y_n)| \le M_2 m_2^{-1} W_2 b + \frac{c_2}{m_2}.$$

If we set

$$W^* = \max\{M_1 m_1^{-1} W_1 a + \frac{c_1}{m_1}, M_2 m_2^{-1} W_2 b + \frac{c_2}{m_2}\},$$

then we have $E^*\Omega^*(W^*) \subset \Omega^*(W^*)$ as desired.

Next, we show that E^* is continuous. Let $\{(x^q,y^q)\}_{q\in\mathbb{N}}$ be a sequence in $\Omega^*(W^*)$ such that $\lim_{q\to\infty}\|(x^q,y^q) - (x,y)\| = 0$, where $(x,y) = \{(x_n,y_n)\}_{n\in\mathbb{Z}}$. Since $\Omega^*(W^*)$ is closed, we must have $(x,y) \in \Omega^*(W^*)$. From (4.7.22) and (4.7.23), we have

$$|E_1^*(x^q,y^q)_n - E_1^*(x,y)_n| \le \sum_{i=n}^{\infty} \sum_{m=-\infty}^{i} \left| \frac{\varphi_{i+1}}{\varphi_n} \right| |a_{i,m}| |f(y_m^q) - f(y_m)|$$

and

$$|E_2^*(x^q,y^q)_n - E_2^*(x,y)_n| \le \sum_{i=n}^{\infty} \sum_{m=-\infty}^{i} \left| \frac{\psi_{i+1}}{\psi_n} \right| |b_{i,m}| |g(x_m^q) - g(x_m)|.$$

Since f and g are continuous, we have by the Lebesgue dominated convergence theorem that

$$\lim_{q \to \infty} \|E^*(x^q, y^q) - E^*(x, y)\| = 0.$$

As we did in the proof of Theorem 4.7.2 we can show that E^* has a fixed point in $\Omega^*(W^*)$. On the other hand, using a similar procedure that we have employed in the proof of Lemma 4.8, we can deduce that any solution $(x, y) = \{(x_n, y_n)\}_{n \in \mathbb{Z}}$ of the system (4.7.1) is a fixed point for the operator E^*. This means $E^*(x, y) = (x, y)$ or equivalently,

$$x_n = c_1 \frac{1}{\varphi_n} - \sum_{i=n}^{\infty} \sum_{m=-\infty}^{i} \frac{\varphi_{i+1}}{\varphi_n} a_{i,m} f(y_m) \tag{4.7.28}$$

and

$$y_n = c_2 \frac{1}{\psi_n} - \sum_{i=n}^{\infty} \sum_{m=-\infty}^{i} \frac{\psi_{i+1}}{\psi_n} b_{i,m} g(x_m). \tag{4.7.29}$$

Conversely, any pair $(x, y) = \{(x_n, y_n)\}_{n \in \mathbb{Z}}$ satisfying (4.7.28) and (4.7.29) will also satisfy

$$x_{n+1} - x_n(1 + h_n) = c_1 \left(\prod_{j=0}^{n}(1 + h_j) - (1 + h_n) \prod_{j=0}^{n-1}(1 + h_j) \right)$$

$$+ (1 + h_n) \sum_{i=n}^{\infty} \sum_{m=-\infty}^{i} \frac{\varphi_{i+1}}{\varphi_n} a_{i,m} f(y_m)$$

$$- \sum_{i=n+1}^{\infty} \sum_{m=-\infty}^{i} \frac{\varphi_{i+1}}{\varphi_{n+1}} a_{i,m} f(y_m),$$

and hence

$$x_{n+1} - x_n(1 + h_n) = \sum_{i=n}^{\infty} \sum_{m=-\infty}^{i} \frac{(1 + h_n) \prod_{j=0}^{n-1}(1 + h_j)}{\prod_{j=0}^{i}(1 + h_j)} a_{i,m} f(y_m)$$

$$- \sum_{i=n+1}^{\infty} \sum_{m=-\infty}^{i} \frac{\prod_{j=0}^{n}(1 + h_j)}{\prod_{j=0}^{i}(1 + h_j)} a_{i,m} f(y_m)$$

$$= \sum_{m=-\infty}^{n} a_{n,m} f(y_m).$$

That is, any fixed point $(x, y) = \{(x_n, y_n)\}_{n \in \mathbb{Z}}$ of the operator E^* satisfies the first equation in (4.7.1). Similarly, one may show that the second equation holds. For an arbitrary fixed point $(x, y) \in \Omega^*(W^*)$ of E^*, we have

$$\lim_{n\to\infty}\left|x_n - c_1\frac{1}{\varphi_n}\right| = \lim_{n\to\infty}\left|E_1^*(x,y)_n - c_1\frac{1}{\varphi_n}\right| = 0, \qquad (4.7.30)$$

and

$$\lim_{n\to\infty}\left|y_n - c_2\frac{1}{\psi_n}\right| = \lim_{n\to\infty}\left|E_2(x,y)_n - c_2\frac{1}{\psi_n}\right| = 0. \qquad (4.7.31)$$

Choosing

$$u_n^{(1)} = c_1\frac{1}{\varphi_n}, \quad v_n^{(1)} = -\sum_{i=n}^{\infty}\sum_{m=-\infty}^{i}\frac{\varphi_{i+1}}{\varphi_n}a_{i,m}f(y_m) \qquad (4.7.32)$$

and

$$u_n^{(2)} = c_2\frac{1}{\psi_n}, \quad v_n^{(2)} = -\sum_{i=n}^{\infty}\sum_{m=-\infty}^{i}\frac{\psi_{i+1}}{\psi_n}b_{i,m}g(x_m), \qquad (4.7.33)$$

we have $x_n = u_n^{(1)} + v_n^{(1)}$ and $y_n = u_n^{(2)} + v_n^{(2)}$. By (4.7.30) and (4.7.31), $v_n^{(1)}$ and $v_n^{(2)}$ tend to 0 when $n \to \infty$. Left to show that $u_n^{(1)}$ and $u_n^{(2)}$ are T-periodic.

$$u_{n+T}^{(1)} = c_1\prod_{j=0}^{n+T-1}(1+h_j) = c_1\prod_{j=0}^{n-1}(1+h_j)\prod_{j=n}^{n+T-1}(1+h_j)$$

$$= c_1\prod_{j=0}^{n-1}(1+h_j)\prod_{j=0}^{T-1}(1+h_j)$$

$$= c_1\prod_{j=0}^{n-1}(1+h_j), \text{ by (4.7.16).}$$

The proof for $u_n^{(2)}$ is identical and hence we omit.

Example 4.4. Consider the system (4.7.1) with the following entries

$$h_n = p_n = \begin{cases} 1, & \text{if } n = 2k+1 \text{ for } k \in \mathbb{Z} \\ -\frac{1}{2}, & \text{if } n = 2k \text{ for } k \in \mathbb{Z} \end{cases},$$

$$a_{n,i} = e^{i-2n}, \text{ for } n,i \in \mathbb{Z}$$

$$b_{n,i} = e^{2i-3n}, \text{ for } n,i \in \mathbb{Z}$$

$$f(x) = \cos x \text{ and } g(x) = \cos 2x.$$

Then (4.7.1) turns into the following system:

$$\begin{cases} \triangle x_n = h_n x_n + \sum\limits_{i=-\infty}^{n} e^{i-2n}\cos(y_i), \\ \triangle y_n = p_n y_n + \sum\limits_{i=-\infty}^{n} e^{2i-3n}\cos(2x_i) \end{cases}.$$

Obviously, the sequences $\{h_n\}_{n\in\mathbb{Z}}$ and $\{p_n\}_{n\in\mathbb{Z}}$ are 2-periodic and all conditions of Theorem 4.7.5 are satisfied. Hence, we conclude by Theorem 4.7.5 the existence of an asymptotically 2-periodic solution $(x,y) = \{(x_n,y_n)\}_{n\in\mathbb{Z}}$ satisfying

$$x_n = c_1 \frac{1}{\varphi_n} - \sum_{i=n}^{\infty} \sum_{m=-\infty}^{i} \frac{\varphi_{i+1}}{\varphi_n} e^{m-2i} \cos(y_m)$$

$$y_n = c_2 \frac{1}{\psi_n} - \sum_{i=n}^{\infty} \sum_{m=-\infty}^{i} \frac{\psi_{i+1}}{\psi_n} e^{2m-3i} \cos(2x_m),$$

for all $n \in \mathbb{N}$, where c_1 and c_2 are positive constants, $\varphi := \{\varphi_n\}_{n \in \mathbb{N}}$ and $\psi := \{\psi_n\}_{n \in \mathbb{N}}$ are as in (4.7.17).

4.8 Open Problems

In this section we propose seven open problems regarding existence of periodic solutions of Volterra difference equations and functional equations. We begin by considering the scalar Volterra difference equation

$$x(n+1) = c(n) - \sum_{s=-\infty}^{n} D(n,s)g(x(s)), \tag{4.8.1}$$

where g is continuous.

Open Problem 1.
Use the method of Section 4.5 to show (4.8.1) has a periodic solution under suitable conditions. Then prove parallel theorems to Theorems 4.5.2 and 4.5.3.
This will be different due to the absence of a linear term in x in Equation (4.8.1).
Actually, it will be very challenging to find a suitable Lyapunov functional that does the trick.

Open Problem 2.
In light of our work in Section 4.7, what can be said about (4.8.1) with respect to periodicity and asymptotic periodicity? Again, the absence of a linear term in x makes (4.8.1) impossible to invert in order to obtain the possible mapping.

Open Problem 3.
Coupled integro-differential equations have many applications in science and engineering. In computational neuroscience, the Wilson–Cowan model describes the dynamics of interactions between populations of very simple excitatory and inhibitory model neurons. It was developed by H.R. Wilson and Jack D. Cowan [171, 172] and extensions of the model have been widely used in modeling neuronal populations [89, 108, 153, 173]. Here we propose a parallel coupled Volterra difference equations model

$$\begin{cases} \triangle x(n) = h_1(n)x(n) + h_2(n)y(n) + \sum_{-\infty}^{n} a(n,s)f(x(s),y(s)), \\ \triangle y(n) = p_1(n)y(n) + p_2(n)x(n) + \sum_{-\infty}^{n} b(n,s)g(x(s),y(s)), \end{cases} \tag{4.8.2}$$

where the functions f and g are assumed to be continuous. It would be of great interest to study the existence of periodic and asymptotically periodic solutions of (4.8.2).

Open Problem 4.
Consider Equation (4.5.9) and let P_T be the space of all periodic sequences of period T. Let $x \in P_T$ and use Theorem 1.1.1 to invert (4.5.9) and then use the Contraction mapping principle and the Schauder second fixed point theorem (see [156], p. 25) to show the existence of a unique periodic solution and a periodic solution. Compare both results and to the results of this chapter.

Open Problem 5 (Our Preferred System)
After careful examination of the three systems that we considered in the Section 4.6.1, we are lead to suggest that the system

$$\triangle x(t) = \sum_{s=t-L}^{t-1} p(s-t)g(x(s)) - \sum_{s=-\infty}^{t-1} q(s-t)g(x(s)) \qquad (4.8.3)$$

which incorporates the most realistic properties from each of the systems, is our favorite system to be considered. The first term on the right takes into account the ideas from (4.6.14) in a more general form. Here we assume that

$$\sum_{s=-L}^{-1} q(s) = 1 \text{ and } \sum_{s=-\infty}^{-1} q(s) = 1. \qquad (4.8.4)$$

Next, one would need to rewrite (4.8.3) as we did in (4.6.15) and then proceed to prove theorems that are parallel to Theorems 4.6.3 and 4.6.4.

Open Problem 6 (Neutral Systems)
There has been a tremendous effort in extending difference equations to neutral difference equations. Neutral difference equations have not been developed like its counterpart, differential equations. Suppose you are observing an organism that is displaying a normal growth or sub-ordinary growth. Suddenly growth accelerates and results in more accelerated growth. This is typical of neutral growth. Present growth rate depends not only on the past state, but also on the past growth rate. Typical models in the spirit of the previous section would be

$$\triangle\big(x(t) - h(x(t - L_1))\big) = g(x(t)) - g(x(t - L_2)). \qquad (4.8.5)$$

It is clear that any constant function is a solution of (4.6.25). Now suppose both functions h and g are Lipschitz continuous. Let $L = \max\{L_1, L_2\}$, define an initial function and then prove a parallel Theorem to Theorem 4.6.3. Another possible neutral model to consider is

$$\triangle\big(x(t) - h(x(t - L_1))\big) = \sum_{s=t-L_2}^{t-1} p(s-t)g(x(s)) - \sum_{s=-\infty}^{t-1} q(s-t)g(x(s)) \qquad (4.8.6)$$

If we assume (4.6.24) then any constant function is a solution of (4.8.6).

Open Problem 7 (Minorsky Model)
The second order differential equation

$$x''(t) + cx'(t) + g(x(t-h)) = 0 \qquad (4.8.7)$$

is called Minorsky equation which he developed as an automatic steering device controller for the large ship the New Mexico. It was pointed out later on that the model given by (4.8.7) was not that accurate and since then a correction term was added and hence the new model

$$x''(t) + cx'(t) + g(x(t-h)) - g(x(t-h-L)) = 0. \qquad (4.8.8)$$

Staying in the spirit of our study, one might consider analyzing the second order difference equation

$$\triangle^2 x(t) + c\triangle x(t) + g(x(t-h)) - g(x(t-h-L)) = 0. \qquad (4.8.9)$$

Clearly, any constant is a solution of (4.8.9).

Chapter 5
Population Dynamics

This chapter is devoted to the application of Volterra difference equations in population dynamics and epidemics. We begin the chapter by introducing different types of population models including predator-prey models. Most commonly studied version of population models are described by continuous-time dynamics, whereas in real ecosystem the changes in populations of each species due to competitive interaction cannot occur continuously. Hence, discrete-time dynamical systems are often more suitable tool for modeling the dynamics in competing species. Cone theory is introduced and utilized to prove the existence of positive periodic solutions for functional difference equations. We introduce an infinite delay population model which governs the growth of population $N(n)$ of a single species whose members compete among themselves for the limited amount of food that is available to sustain the population, and use the results on cone theory to obtain the existence of a positive periodic solution. Moreover, from a biologist's point of view, the idea of permanence plays a central role in any competing species.

5.1 Background

We begin with a brief history regarding the early work of Vito Volterra on modeling fish population utilizing what we call today: Volterra integral equations and Volterra integro-differential equations. Volterra did not limit himself to academic research. Volterra became interested in mathematical ecology late in 1925. His interest in the field was stimulated by conversations with the young zoologist Umberto D'Ancona, then engaged to marry his daughter Luisa. D'Ancona, studying the records of the fish markets in the upper Adriatic, had noticed a curious phenomenon. He observed that during and after the war, when fishing was severely limited, the proportion of predators among the total catch had increased correspondingly, an effect predicted by Volterra's models. D'Ancona was thus reinforced in his

© Springer Nature Switzerland AG 2018
Y. N. Raffoul, *Qualitative Theory of Volterra Difference Equations*,
https://doi.org/10.1007/978-3-319-97190-2_5

belief that the two facts were causally correlated. In other words, the proportion of food fish markedly decreased during the war years. This beginning led Volterra to attack more general problems in ecology. Volterra emphasized consistently that differential equations are, at best, only rough approximations of actual ecological systems. They would apply only to animals without age or memory, which eat all the food they encounter and immediately convert it into offspring. Anything more realistic would yield integro-differential rather than differential equations. The field soon became his major research. More on the next discussion can be found in [158]. To put things into perspective we give some background on the famous Lotka prey-predator model. In 1925 Lotka published his *Elements of Physical Biology*, in which he developed and studied the interaction between two species via the model

$$\frac{dN_1}{dt} = (\varepsilon_1 + \gamma_1 N_2)N_1, \quad \frac{dN_2}{dt} = (\varepsilon_2 + \gamma_2 N_1)N_2,$$

where ε_1 and ε_2 are the "coefficients of self-increase," while γ_1 and γ_2 account for the interaction between two species, and the N_1, N_2, are population sizes. This model can represent species preying on another, depending on the sign of the constants in the model. In the case of predation, Lotka showed the existence of close periodic orbits. Later on, Lotka considered more advanced models that dealt with multiple species preying on a single specie. For the sake of the next discussion, we write the above Lotka system to suit a predator-prey model. That is, by assuming all the constants are positive we have that

$$\frac{dN_1}{dt} = (\varepsilon_1 - \gamma_1 N_2)N_1, \quad \frac{dN_2}{dt} = (-\varepsilon_2 + \gamma_2 N_1)N_2, \tag{5.1.1}$$

where N_1 and N_2 represent the populations of the preys and predators at time t, respectively. Note that the predators would die out without the presence of the preys. To better explain this, we multiply the first equation of (5.1.1) by dt and then it is clear that an amount of $\varepsilon_2 N_2$ of them will die in a time interval dt. Suppose that predator tendency to eat the prey when encountered does not depend on age, τ, nor on the state of the association. Assume also that the age distribution of the predators, $\lambda(\tau,t)$, can be considered as independent of time, $\lambda(\tau)$. The individuals of age not younger than τ will be in the proportion

$$f(t - \tau) = \int_{t-\tau}^{\infty} \lambda(\eta)d\eta.$$

Then $f(t - \tau)N_2(t)$ will be the number of predators that is active at time $t - \tau$. Their feeding rate is proportional to $f(t - \tau)N_2(t)N_1(\tau)$ that is $\phi(t - \tau)f(t - \tau)N_2(t)N_1(\tau)$, measuring the effect of feeding through all previous time on the chances of survival and the rate of reproduction at a subsequent time. Setting

$$\phi(t - \tau)f(t - \tau)N_2(t)N_1(\tau)d\tau = F(t - \tau)N_2(t)N_1(\tau)d\tau$$

and integrating over all previous time we obtain, as a positive term for the predators' equation,

$$N_2(t)\int_{t-\tau}^{t} F(t - \tau)N_1(\tau)d\tau.$$

Similar argument can be made for the preys and hence we arrive at the system of integro-differential equations

$$\frac{dN_1}{dt} = N_1(t)\left[\varepsilon_1 - \gamma_1 N_2(t) - \int_{t-\tau}^{t} F_1(t-\tau)N_1(\tau)d\tau\right]$$

$$\frac{dN_2}{dt} = N_2(t)\left[-\varepsilon_2 + \gamma_2 N_1(t) - N_2(t)\int_{t-T}^{t} F_2(t-\tau)N_1(\tau)d\tau\right].$$

Another aspect of importance for application of Volterra difference equations is their usefulness in numerical approximations of Volterra integro-differential equations, see [161] and the reference therein. This notion was briefly discussed in Chapter 1.

5.2 Formulation of Predator-Prey Discrete Models

In this section we obtain the Lotka-Volterra predator-prey model from its continuous counterpart. Researchers have argued that discrete time models governed by difference equations are more appropriate for describing the dynamics relationship among populations than the continuous ones when the populations have nonoverlapping generations. There is no unique way of deriving discrete time version of dynamical systems corresponding to continuous time formulations. One of the ways of deriving difference equations modeling the dynamics of populations with nonoverlapping generations is based on appropriate modifications of models with overlapping generations. In this approach, differential equations with piecewise constant arguments have been useful (see [170]). Thus, we consider the continuous Lotka-Volterra predator-prey model given by (5.1.1) and use differential equations with piecewise constant arguments to obtain a discrete analogue of it. We follow the work given in [70] and [170]. That is, we assume that the average growth rates in system (5.1.1) change at regular intervals of time. We can incorporate this aspect in (5.1.1) and obtain the following modified system

$$\frac{dN_1(t)}{dt}\frac{1}{N(t)} = (\varepsilon_1 - \gamma_1 N_2([t])) \qquad (5.2.1)$$

$$\frac{dN_2(t)}{dt}\frac{1}{N_2(t)} = (-\varepsilon_2 + \gamma_2 N_1([t]))$$

where $t \neq 0,1,2,\ldots,[t]$ denotes the integer part of t, $t \in (0,\infty)$. Equations of the form (5.2.1) are known as differential equations with piecewise constant arguments and they occupy a position midway between differential equations and difference equations. By a solution of (5.2.1), we mean a function $N = (N_1,N_2)^T$ which is defined for $t \in (0,\infty)$ and has the properties that,

1. N is continuous on $[0,\infty)$.

2. The derivatives $\frac{dN_1(t)}{dt}$, $\frac{dN_2(t)}{dt}$ exist at each point $t \in (0,\infty)$ with the exception of

the points $t \in \{0,1,2,\cdots,\}$,where left-sided derivatives exist.

3. System (5.2.1) is valid on each interval $[k,k+1]$ with $k = 0,1,2,\cdots$

Next we integrate both sides of (5.2.1) over any interval $[k,k+1)$, $k = 0,1,2,\cdots$ to arrive at for $k \le t \le k+1$ $k = 0,1,2,\cdots$

$$N_1(t) = N_1(k)e^{\{[\varepsilon_1-\gamma_1 N_2(k)](t-k)\}} \tag{5.2.2}$$

$$N_2(t) = N_2(k)e^{\{[-\varepsilon_2+\gamma_2 N_1(k)](t-k)\}}$$

Letting $t \to k+1$, then system takes the form

$$\begin{cases} N_1(k+1) = N_1(k)e^{\{[\varepsilon_1-\gamma_1 N_2(k)]\}} \\ N_2(k+1) = N_2(k)e^{\{[-\varepsilon_2+\gamma_2 N_1(k)]\}} \end{cases} \tag{5.2.3}$$

where $k = 0,1,2,\cdots$.

General forms of discrete-generation host-parasite

$$\begin{cases} P(k+1) = \lambda P(k)f(P(k),H(k)) \\ H(k+1) = c\lambda P(k)(1 - f(P(k),H(k))) \end{cases}$$

have been used to model the interaction between host species (a plant, $P(k)$) and a parasite species (a herbivore, $H(k)$). The term $1 - f$ represents the probability of being parasitized. Nicholson-Bailey [13] used one of the simplest version of the above general form by considering

$$\begin{cases} P(k+1) = \lambda P(k)e^{-aH(k)} \\ H(k+1) = c\lambda P(k)(1 - e^{-aH(k)}) \end{cases}$$

in which $f = e^{-aH(k)}$ is the proportion of hosts escaping parasitism, where a is the mean encounters per host. Hence, $1 - e^{-aH(k)}$ is the probability that host will be attacked. In a later study, Beddington et al. [15] considered a generalization of the above model by studying

$$\begin{cases} P(k+1) = \lambda P(k)e^{r(1-P(k)/P_{max})-aH(k)} \\ H(k+1) = c\lambda P(k)(1 - e^{-aH(k)}) \end{cases}$$

where P_{max} is the carrying capacity imposed by the environment for the host in the absence of the parasite.

Following in the footsteps of Beddington et al. [15], Elaydi et al. [90] considered the predator-prey model

$$\begin{cases} x(n+1) = x(n)e^{r(1-\frac{x(n)}{K})-by(n))} \\ y(n+1) = ey(n)(1 - e^{-ay(n)}) \end{cases} \tag{5.2.4}$$

where $x(n) \ge 0$ and $y(n) \ge 0$ represent population densities of a prey and a predator, respectively, and a,b,e,K, and r are positive. The constant K is the carrying capacity and represents the maximum population size that can be supported by the available limited resources and r is the growth rate. In [90] Elaydi et al. investigated the stability and invariant manifolds and the stability of the coexisting fixed point of model (5.2.4). Motivated by Elaydi et al. [90], in [10], Asheghi revisited model (5.2.4) and analyzed the stability of feasible fixed points and the period-doubling. In addition, the author studied the Neimark-Sacker bifurcation diagrams. In 2014, Li and Xu [176] considered the discrete predator-prey model with infected prey

$$\begin{cases} S(n+1) = S(n)exp\Big\{r_1(n)(1-S(n)-I(n)) - \dfrac{a(n)Z(n)}{1+b(n)(S(n)+I(n))} - \dfrac{\alpha(n)I(n)}{S(n)+I(n)}\Big\} \\[2ex] I(n+1) = I(n)exp\Big\{r_2(n)(1-S(n)-I(n)) - \dfrac{a(n)Z(n)}{1+b(n)(S(n)+I(n))} + \dfrac{\alpha(n)I(n)}{S(n)+I(n)} - m_2(n)\Big\} \\[2ex] Z(n+1) = Z(n)exp\Big\{\dfrac{a(n)(S(n)+I(n))}{1+b(n)(S(n)+I(n))} - m_3(n)\Big\} \end{cases}$$

$$(5.2.5)$$

where $S(n)$ and $I(n)$ are the susceptible phytoplankton population and the infected phytoplankton population, respectively, and $Z(n)$ grazes on both the susceptible and infected phytoplankton. The parameter $\alpha > 0$ is the frequency-dependent transmissions rate and the parameter m_2 is the disease-induced mortality of infected prey. The parameters r_1 and r_2 are the intrinsic growth rates of susceptible and infected population, respectively. Rate m_3 represents the natural mortality rate of zooplankton. In addition, a and b are constants. For more on the biological meaning and development of the model in the continuous case, we refer to [81] and [157]. Li and Xu [176] assumed periodicity conditions on the coefficients and used the Continuation theorem due to Gaines and Malvin [71] and showed the existence of a positive periodic solution. Moreover, they effectively used Lyapunov functions and proved the positive periodic solution is indeed globally asymptotically stable.

We remark that all the above models display positive solutions for positive initial data.

5.3 Cone Theory and Positive Periodic Solutions

We begin the chapter by utilizing cone theoretic fixed point theorem to study the existence of positive periodic solutions of the nonlinear nonautonomous system of functional difference equations

$$x(n+1) = A(n)x(n) + f(n, x_n) \qquad (5.3.1)$$

where $A(n) = \text{diag}[a_1(n), a_2(n), \ldots, a_k(n)]$, a_j is ω-periodic, $f : \mathbb{Z} \times \mathbb{R}^k \to \mathbb{R}^k$ is continuous in x, and $f(n, x)$ is ω-periodic in n and x, whenever x is ω-periodic.

Most contents can be found in [149] and the references therein. Such results will be applied to the infinite delay scalar Volterra discrete population model

$$N(n+1) = \alpha(n)N(n)\left[1 - \frac{1}{N_0(n)} \sum_{s=-\infty}^{0} B(s)N(n+s)\right], \; n \in \mathbb{Z} \qquad (5.3.2)$$

which governs the growth of population $N(n)$ of a single species whose members compete among themselves for the limited amount of food that is available to sustain the population. We emphasize that our conditions can only imply the existence of positive and periodic solutions for model (5.4.1). We note that equation (5.3.2) is a generalization of the known logistic model

$$N(n+1) = \alpha N(n)\left[1 - \frac{N(n)}{N_0}\right], \qquad (5.3.3)$$

where α is the intrinsic per capita growth rate and N_0 is the total carrying capacity. For more biological information on equation (5.3.2), we refer the reader to [57]. We remark that in (5.3.2), the term $\sum_{s=-\infty}^{0} B(s)N(n+s)$ is equivalent to $\sum_{u=-\infty}^{n} B(u-s))N(u)$. We chose to write (5.3.2) that way so that it can be put in the form of $x(n+1) = a(n)x(n) + f(n,x_n)$.

Let $\mathscr{X}$ be the set of all real ω-periodic sequences $\phi : \mathbb{Z} \to \mathbb{R}^k$. Endowed with the maximum norm $||\phi|| = \max_{\theta \in \mathbb{Z}} \sum_{j=1}^{k} |\phi_j(\theta)|$ where $\phi = (\phi_1, \phi_2, \dots, \phi_k)^t$, $\mathscr{X}$ is a Banach space. Here t stands for the transpose. If $x \in \mathscr{X}$, then $x_n \in \mathscr{X}$ for any $n \in \mathbb{Z}$ is defined by $x_n(\theta) = x(n+\theta)$ for $\theta \in \mathbb{Z}$.

The existence of multiple positive periodic solutions of nonlinear functional differential equations has been studied extensively in recent years. Some appropriate references would be [34] and [168]. We are particularly motivated by the work in [88] on functional differential equations and the work of Raffoul in [67, 129], and [151] on boundary value problems involving functional difference equations. When working with boundary value problems whether in differential or difference equations, it is customary to display the desired solution in terms of a suitable Green's function and then apply cone theory (see [8, 45, 67, 78, 79, 80], and [118]). Since our equation (5.3.1) is not the type of boundary value problem, we obtain a variation of parameters formula and then try to find a lower and upper estimates for the kernel inside the summation. Once those estimates are found we use Krasnoselskii's fixed point theorem [97] to show the existence of a positive periodic solution. In [129], Raffoul studied the existence of periodic solutions of an equation similar to equation (5.3.1) using Schauder's Second fixed point theorem. Moreover, In [151], Raffoul considered the scalar difference equation

$$x(n+1) = a(n)x(n) + h(n)f(x(n - \tau(n))) \qquad (5.3.4)$$

where a, h, and τ are ω-periodic for ω is an integer with $\omega \geq 1$. Under the assumptions that $a(n), f(x)$, and $h(n)$ are nonnegative with $0 < a(n) < 1$ for all $n \in [0, \omega - 1]$, it was shown that (5.3.4) possesses a positive periodic solution. In this work we extend (5.3.4) to systems with infinite delay and address the existence of positive periodic solutions of (5.3.1) in the case $a(n) > 1$. Let $\mathbb{R}_+ = [0, +\infty)$, for each $x = (x_1, x_2, \ldots, x_k)^t \in \mathbb{R}^k$, the norm of x is defined as $|x| = \sum_{j=1}^{k} |x_j|$. $\mathbb{R}_+^k = \{(x_1, x_2, \ldots, x_k)^t \in \mathbb{R}^k : x_j \geq 0, \, j = 1, 2, \ldots, k\}$. Also, we denote $f = (f_1, f_2, \ldots, f_k)^t$, where t stands for transpose. Now we list the following conditions.

(H1) $a(n) \neq 0$ for all $n \in [0, \omega - 1]$ with $\prod_{s=0}^{\omega-1} a_j(s) \neq 1$ for $j = 1, 2, \ldots, k$.

(H2) If $0 < a(n) < 1$ for all $n \in [0, \omega - 1]$, then $f_j(n, \phi_n) \geq 0$ for all $n \in \mathbb{Z}$ and
   $\phi : \mathbb{Z} \to \mathbb{R}_+^n$, $j = 1, 2, \ldots, k$ where $\mathbb{R}_+ = [0, +\infty)$

(H3) If $a(n) > 1$ for all $n \in [0, \omega - 1]$, then $f_j(n, \phi_n) \leq 0$ for all $n \in \mathbb{Z}$ and $\phi : \mathbb{Z} \to$
   $\mathbb{R}_+^n$, $j = 1, 2, \ldots, k$ where $\mathbb{R}_+ = [0, +\infty)$

(H4) For any $L > 0$ and $\varepsilon > 0$, there exists $\delta > 0$ such that
   $[\phi, \psi \in \mathscr{X}, \|\phi\| \leq L, \|\psi\| \leq L, \|\phi - \psi\| < \delta, 0 \leq s \leq \omega]$ imply

$$|f(s, \phi_s) - f(s, \psi_s)| < \varepsilon. \tag{5.3.5}$$

We begin by stating some preliminaries in the form of definitions and lemmas that are essential to the proofs of our main results. We start with the following definition.

Definition 5.3.1. Let X be a Banach space and K be a closed, nonempty subset of X. The set K is a cone if

   (i) $\alpha u + \beta v \in K$ for all $u, v \in K$ and all $\alpha, \beta \geq 0$
   (ii) $u, -u \in K$ imply $u = 0$.

We now state the Krasnoselskii's fixed point theorem [97].

Theorem 5.3.1 (Krasnoselskii [97]). *Let $\mathscr{B}$ be a Banach space, and let $\mathscr{P}$ be a cone in $\mathscr{B}$. Suppose Ω_1 and Ω_2 are open subsets of $\mathscr{B}$ such that $0 \in \Omega_1 \subset \overline{\Omega}_1 \subset \Omega_2$ and suppose that*

$$T : \mathscr{P} \cap (\overline{\Omega}_2 \backslash \Omega_1) \to \mathscr{P}$$

is a completely continuous operator such that

   (i) $\|Tu\| \leq \|u\|$, $u \in \mathscr{P} \cap \partial\Omega_1$, *and* $\|Tu\| \geq \|u\|$, $u \in \mathscr{P} \cap \partial\Omega_2$; *or*
   (ii) $\|Tu\| \geq \|u\|$, $u \in \mathscr{P} \cap \partial\Omega_1$, *and* $\|Tu\| \leq \|u\|$, $u \in \mathscr{P} \cap \partial\Omega_2$.
   Then T has a fixed point in $\mathscr{P} \cap (\overline{\Omega}_2 \backslash \Omega_1)$.

For the next lemma we consider

$$x_j(n+1) = a_j x_j(n) + f_j(n, x_n), \; j = 1, 2, \ldots, k. \tag{5.3.6}$$

The proof of the next Lemma can be easily deduced from [129] and hence we omit it.

Lemma 5.1 ([149]). *Suppose (H1) hold. Then $x_j(n) \in \mathscr{X}$ is a solution of equation (5.3.6) if and only if*

$$x_j(n) = \sum_{u=n}^{n+\omega-1} G_j(n, u) f_j(u, x_u), \; j = 1, 2, \ldots, k \tag{5.3.7}$$

where

$$G_j(n,u) = \frac{\prod_{s=u+1}^{n+\omega-1} a_j(s)}{1 - \prod_{s=n}^{n+\omega-1} a_j(s)}, \quad u \in [n, n+\omega-1], \, j = 1,2,\ldots,k. \qquad (5.3.8)$$

Set

$$G(n,u) = \mathrm{diag}[G_1(n,u), G_2(n,u), \ldots, G_k(n,u)].$$

It is clear that $G(n,u) = G(n+\omega, u+\omega)$ for all $(n,u) \in \mathbb{Z}^2$. Also, if either (H2) or (H3) holds, then (5.3.8) implies that

$$G_j(n,u) f_j(u, \phi_u) \geq 0$$

for $(n,u) \in \mathbb{Z}^2$ and $u \in \mathbb{Z}$, $\phi : \mathbb{Z} \to \mathbb{R}_+^k$. In defining the desired cone we observe that if (H2) holds, then

$$\frac{\prod_{s=0}^{\omega-1} a_j(s)}{1 - \prod_{s=n}^{n+\omega-1} a_j(s)} \leq |G_j(n,u)| \leq \frac{\prod_{s=0}^{\omega-1} a_j^{-1}(s)}{1 - \prod_{s=n}^{n+\omega-1} a_j(s)} \qquad (5.3.9)$$

for all $u \in [n, n+\omega-1]$. Also, if (H3) holds, then

$$\frac{\prod_{s=0}^{\omega-1} a_j^{-1}(s)}{\left| 1 - \prod_{s=n}^{n+\omega-1} a_j(s) \right|} \leq |G_j(n,u)| \leq \frac{\prod_{s=0}^{\omega-1} a_j(s)}{\left| 1 - \prod_{s=n}^{n+\omega-1} a_j(s) \right|} \qquad (5.3.10)$$

for all $u \in [n, n+\omega-1]$. For all $(n,s) \in \mathbb{Z}^2$, $j = 1,2,\ldots,k$, we define

$$\sigma_2 := \min\left\{ \left(\prod_{s=0}^{\omega-1} a_j(s)\right)^2, j = 1,2,\ldots,n \right\}$$

and

$$\sigma_3 := \min\left\{ \left(\prod_{s=0}^{\omega-1} a_j^{-1}(s)\right)^2, j = 1,2,\ldots,n \right\}.$$

We note that if $0 < a(n) < 1$ for all $n \in [0, \omega-1]$, then $\sigma_2 \in (0,1)$. Also, if $a(n) > 1$ for all $n \in [0, \omega-1]$, then $\sigma_3 \in (0,1)$. Conditions (H2) and (H3) will have to be handled separately. That is, we define two cones; namely, $\mathscr{P}2$ and $\mathscr{P}3$. Thus, for each $y \in \mathscr{X}$ set

$$\mathscr{P}2 = \{ y \in \mathscr{X} : y(n) \geq 0, n \in \mathbb{Z}, \text{and } y(n) \geq \sigma_2 \|y\| \}$$

and

$$\mathscr{P}3 = \{ y \in \mathscr{X} : y(n) \geq 0, n \in \mathbb{Z}, \text{ and } y(n) \geq \sigma_3 \|y\| \}.$$

Define a mapping $T : \mathscr{X} \to \mathscr{X}$ by

$$(Tx)(n) = \sum_{u=n}^{n+\omega-1} G(n,u) f(u, x_u) \qquad (5.3.11)$$

where $G(n, u)$ is defined following (5.3.8). We denote

$$(Tx) = \Big(T_1 x, T_2 x, \ldots, T_k x \Big)^t.$$

It is clear that $(Tx)(n + \omega) = (Tx)(n)$.

Lemma 5.2 ([149]). *If (H1) and (H2) hold, then the operator $T \mathscr{P}2 \subset \mathscr{P}2$. If (H1) and (H3) hold, then $T \mathscr{P}3 \subset \mathscr{P}3$.*

Proof. Suppose (H1) and (H2) hold. Then for any $x \in \mathscr{P}2$ we have

$$(T_j x(n)) \geq 0, \ j = 1, 2, \ldots k.$$

Also, for $x \in \mathscr{P}2$ by using (5.3.8)–(5.3.11) we have that

$$(T_j x)(n) \leq \frac{\prod_{s=0}^{\omega-1} a_j^{-1}(s)}{1 - \prod_{s=n}^{n+\omega-1} a_j(s)} \sum_{u=n}^{n+\omega-1} |f_j(u, x_u)|$$

and

$$\|T_j x\| = \max_{n \in [0, \omega-1]} |T_j x(n)| \leq \frac{\prod_{s=0}^{\omega-1} a_j^{-1}(s)}{1 - \prod_{s=n}^{n+\omega-1} a_j(s)} \sum_{u=n}^{n+\omega-1} |f_j(u, x_u)|.$$

Therefore,

$$\begin{aligned}
(T_j x)(n) &= \sum_{u=n}^{n+\omega-1} G_j(n, u) f_j(u, x_u) \\
&\geq \frac{\prod_{s=0}^{\omega-1} a_j(s)}{1 - \prod_{s=n}^{n+\omega-1} a_j(s)} \sum_{u=n}^{n+\omega-1} |f_j(u, x_u)| \\
&\geq \Big(\prod_{s=0}^{\omega-1} a_j(s) \Big)^2 \|T_j x\| \geq \sigma_2 \|T_j x\|.
\end{aligned}$$

That is, $T \mathscr{P}2$ is contained in $\mathscr{P}2$. The proof of the other part follows in the same manner by simply using (5.3.10), and hence we omit it. This completes the proof.

To simplify notation we denote,

$$A_2 = \min_{1 \leq j \leq k} \frac{\prod_{s=0}^{\omega-1} a_j(s)}{1 - \prod_{s=n}^{n+\omega-1} a_j(s)}, \tag{5.3.12}$$

$$B_2 = \max_{1 \leq j \leq k} \frac{\prod_{s=0}^{\omega-1} a_j^{-1}(s)}{1 - \prod_{s=n}^{n+\omega-1} a_j(s)}, \tag{5.3.13}$$

$$A_3 = \min_{1 \leq j \leq k} \frac{\prod_{s=0}^{\omega-1} a_j^{-1}(s)}{\left| 1 - \prod_{s=n}^{n+\omega-1} a_j(s) \right|}, \tag{5.3.14}$$

and

$$B_3 = \max_{1 \leq j \leq k} \frac{\prod_{s=0}^{\omega-1} a_j(s)}{\left|1 - \prod_{s=n}^{n+\omega-1} a_j(s)\right|}. \tag{5.3.15}$$

Lemma 5.3 ([149]). *If (H1), (H2), and (H4) hold, then the operator $T : \mathscr{P}2 \to \mathscr{P}2$ is completely continuous. Similarly, if (H1), (H3), and (H4) hold, then the operator $T : \mathscr{P}3 \to \mathscr{P}3$ is completely continuous.*

Proof. Suppose (H1), (H2), and (H4) hold. First show that T is continuous. By (H4), for any $L > 0$ and $\varepsilon > 0$, there exists a $\delta > 0$ such that $[\phi, \psi \in \mathscr{X}, \|\phi\| \leq L, \|\psi\| \leq L, \|\phi - \psi\| < \delta]$ imply

$$\max_{0 \leq s \leq \omega-1} |f(s, \phi_s) - f(s, \psi_s)| < \frac{\varepsilon}{B_2 \omega}$$

where B_2 is given by (5.3.13). If $x, y \in \mathscr{P}2$ with $\|x\| \leq L$, $\|y\| \leq L$, and $\|x - y\| < \delta$, then

$$|(Tx)(n) - (Ty)(n)| \leq \sum_{u=n}^{n+\omega-1} |G(n,u)||f(u,x_u) - f(u,y_u)|$$

$$\leq B_2 \sum_{u=0}^{\omega-1} |f(u,x_u) - f(u,y_u)| < \varepsilon$$

for all $n \in [0, \omega - 1]$, where $|G(n,u)| = \max_{1 \leq j \leq n} |G_j(n,u)|$, $j = 1, 2, \ldots, k$. This yields $\|(Tx) - (Ty)\| < \varepsilon$. Thus, T is continuous. Next we show that T maps bounded subsets into compact subsets. Let $\varepsilon = 1$. By (H4), for any $\mu > 0$ there exists $\delta > 0$ such that $[x, y \in \mathscr{X}, \|x\| \leq \mu, \|y\| \leq \mu, \|x - y\| < \delta]$ imply

$$|f(s, x_s) - f(s, y_s)| < 1.$$

We choose a positive integer N so that $\delta > \frac{\mu}{N}$. For $x \in \mathscr{X}$, define $x^i(n) = \frac{ix(n)}{N}$, for $i = 0, 1, 2, \ldots, N$. For $\|x\| \leq \mu$,

$$\|x^i - x^{i-1}\| = \max_{n \in \mathbb{Z}} \left| \frac{ix(n)}{N} - \frac{(i-1)x(n)}{N} \right|$$

$$\leq \frac{\|x\|}{N} \leq \frac{\mu}{N} < \delta.$$

Thus, $|f(s, x^i) - f(s, x^{i-1})| < 1$. As a consequence, we have

$$f(s, x_s) - f(s, 0) = \sum_{i=1}^{N} \left(f(s, x^i) - f(s, x^{i-1}) \right),$$

which implies that

$$|f(s,x_s)| \leq \sum_{i=1}^{N} |f(s,x_s^i) - f(s,x_s^{i-1})| + |f(s,0)|$$
$$< N + |f(s,0)|.$$

Thus, f maps bounded sets into bounded sets. It follows from the above inequality and (5.3.11) that

$$||(Tx)(n)|| \leq B_2 \sum_{j=1}^{k} \left(\sum_{u=n}^{n+T-1} |f_j(u,x_u)| \right)$$
$$\leq B_2 \omega (N + |f(s,0)|).$$

If we define $S = \{x \in \mathscr{X} : ||x|| \leq \mu\}$ and $Q = \{(Tx)(n) : x \in S\}$, then S is a subset of $\mathbb{R}^{\omega k}$ which is closed and bounded and thus compact. As T is continuous in x, it maps compact sets into compact sets. Therefore, $Q = T(S)$ is compact. The proof for the other case is similar by simply invoking (5.3.15). This completes the proof.

Next, we state two theorems and two corollaries. Our theorems and corollaries are stated in a way that unify both cases; $0 < a(n) < 1$ and $a(n) > 1$ for all $n \in [0, \omega - 1]$.

Theorem 5.3.2 ([149]). *Assume (H1).*
(a) Suppose (H2) and (H4) hold and that there exist two positive numbers R_1 and R_2 with $R_1 < R_2$ such that

$$\sup_{||\phi||=R_1, \phi \in \mathscr{P}2} |f(s,x_s)| \leq \frac{R_1}{\omega B_2}, \tag{5.3.16}$$

and

$$\inf_{||\phi||=R_2, \phi \in \mathscr{P}2} |f(s,x_s)| \geq \frac{R_2}{\omega A_2}, \tag{5.3.17}$$

where A_2 and B_2 are given by (5.3.12) and (5.3.13), respectively. Then, there exists $\bar{x} \in \mathscr{P}2$ which is a fixed point of T and satisfies $R_1 \leq ||\bar{x}|| \leq R_2$.
(b) Suppose (H3) and (H4) hold and that there exist two positive numbers R_1 and R_2 with $R_1 < R_2$ such that

$$\sup_{||\phi||=R_1, \phi \in \mathscr{P}3} |f(s,x_s)| \leq \frac{R_1}{\omega B_3}, \tag{5.3.18}$$

and

$$\inf_{||\phi||=R_2, \phi \in \mathscr{P}3} |f(s,x_s)| \geq \frac{R_2}{\omega A_3}, \tag{5.3.19}$$

where A_3 and B_3 are given by (5.3.14) and (5.3.14), respectively. Then, there exists $\bar{x} \in \mathscr{P}3$ which is a fixed point of T and satisfies $R_1 \leq ||\bar{x}|| \leq R_2$.

Proof. Suppose (H1), (H2), and (H4) hold. Let $\Omega_\xi = \{x \in \mathscr{P}2 | \|x\| < \xi\}$. Let $x \in \mathscr{P}2$ which satisfies $\|x\| = R_1$. in view of (5.3.16), we have

$$|(Tx)(n))| \leq \sum_{u=n}^{n+\omega-1} |G(n,u)| |f(u,x_u)|$$

$$\leq B_2 \omega \frac{R_1}{\omega B_2} = R_1.$$

That is, $\|Tx\| \leq \|x\|$ for $x \in \mathscr{P}2 \cap \partial\Omega_{R_1}$. let $x \in \mathscr{P}2$ which satisfies $\|x\| = R_2$ we have, in view of (5.3.17),

$$|(Tx)(n)| \geq A_2 \sum_{u=n}^{n+\omega-1} |f(u,x_u)| \geq A_2 \omega \frac{R_2}{\omega A_2} = R_2.$$

That is, $\|Tx\| \geq \|x\|$ for $x \in \mathscr{P}2 \cap \partial\Omega_{R_2}$. In view of Theorem 5.3.1, T has a fixed point in $\mathscr{P}2 \cap (\bar{\Omega}_2 \setminus \Omega_1)$. It follows from Lemma 5.2 that (5.3.1) has an ω-periodic solution $\bar{x}$ with $R_1 \leq \|\bar{x}\| \leq R_2$. The proof of (b) follows in a similar manner by simply invoking conditions (5.3.18) and (5.3.19).

As a consequence of Theorem 5.3.2, we state a corollary which its proof we omit.

Corollary 5.1 ([149]). *Assume that (H1) holds.*
(a) Suppose (H2) and (H4) hold and

$$\lim_{\phi \in \mathscr{P}2, \|\phi\| \to 0} \frac{|f(s,\phi_s)|}{\|\phi\|} = 0, \tag{5.3.20}$$

$$\lim_{\phi \in \mathscr{P}2, \|\phi\| \to \infty} \frac{|f(s,\phi_s)|}{\|\phi\|} = \infty. \tag{5.3.21}$$

Then (5.3.1) has a positive periodic solution.
(b) Suppose (H3) and (H4) hold and

$$\lim_{\phi \in \mathscr{P}3, \|\phi\| \to 0} \frac{|f(s,\phi_s)|}{\|\phi\|} = 0, \tag{5.3.22}$$

$$\lim_{\phi \in \mathscr{P}3, \|\phi\| \to \infty} \frac{|f(s,\phi_s)|}{\|\phi\|} = \infty. \tag{5.3.23}$$

Then (5.3.1) has a positive periodic solution.

Theorem 5.3.3 ([149]). *Suppose that (H1) holds.*
(a) Suppose (H2) and (H4) hold and that there exist two positive numbers R_1 and R_2 with $R_1 < R_2$ such that

$$\inf_{\|\phi\|=R_1, \phi \in \mathscr{P}2} |f(s,x_s)| \geq \frac{R_1}{\omega B_2}, \tag{5.3.24}$$

and

$$\sup_{\|\phi\|=R_2,\,\phi\in\mathscr{P}2} |f(s,x_s)| \leq \frac{R_2}{\omega A_2}, \tag{5.3.25}$$

where A_2 and B_2 are given by (5.3.12) and (5.3.13), respectively. Then, there exists $\bar{x} \in \mathscr{P}2$ which is a fixed point of T and satisfies $R_1 \leq \|\bar{x}\| \leq R_2$.
(b) Suppose (H3) and (H4) hold and that there exist two positive numbers R_1 and R_2 with $R_1 < R_2$ such that

$$\inf_{\|\phi\|=R_1,\,\phi\in\mathscr{P}3} |f(s,x_s)| \geq \frac{R_1}{\omega B_3}, \tag{5.3.26}$$

and

$$\sup_{\|\phi\|=R_2,\,\phi\in\mathscr{P}3} |f(s,x_s)| \leq \frac{R_2}{\omega A_3}, \tag{5.3.27}$$

where A_3 and B_3 are given by (5.3.14) and (5.3.15), respectively. Then, there exists $\bar{x} \in \mathscr{P}3$ which is a fixed point of T and satisfies $R_1 \leq \|\bar{x}\| \leq R_2$.

The proof is similar to the proof of Theorem 5.3.2 and hence we omit it. As a consequence of Theorem 5.3.3, we have the following corollary.

Corollary 5.2 ([149]). *Assume that (H1) hold.*
(a) Suppose (H2) and (H4) hold and

$$\lim_{\phi\in\mathscr{P}2,\,\|\phi\|\to 0} \frac{|f(s,\phi_s)|}{\|\phi\|} = \infty, \tag{5.3.28}$$

$$\lim_{\phi\in\mathscr{P}2,\,\|\phi\|\to\infty} \frac{|f(s,\phi_s)|}{\|\phi\|} = 0. \tag{5.3.29}$$

Then (5.3.1) has a positive periodic solution.

(b) Suppose (H3) and (H4) hold and

$$\lim_{\phi\in\mathscr{P}3,\,\|\phi\|\to 0} \frac{|f(s,\phi_s)|}{\|\phi\|} = \infty, \tag{5.3.30}$$

$$\lim_{\phi\in\mathscr{P}3,\,\|\phi\|\to\infty} \frac{|f(s,\phi_s)|}{\|\phi\|} = 0. \tag{5.3.31}$$

Then (5.3.1) has a positive periodic solution.

5.3.1 Applications to Infinite Delay Population Models

We apply the results from the previous section to the model (5.3.2) and show that it admits the existence of a positive periodic solution. Thus, we consider the scalar discrete model that governs the growth of population $N(n)$ of a single species whose members compete among themselves for the limited amount of food that is available to sustain the population. Thus, we consider the infinite delay Volterra scalar model

$$N(n+1) = \alpha(n)N(n)\left[1 - \frac{1}{N_0(n)}\sum_{s=-\infty}^{0} B(s)N(n+s)\right], \; n \in \mathbb{Z} \qquad (5.3.32)$$

as described in the Introduction. We chose to write (5.3.32) that way so that it can be put in the form of $x(n+1) = a(n)x(n) + f(n,x_n)$.

Before we state our results in the form of a theorem, we assume that

(P1) $\alpha(n) > 1$, $N_0(n) > 0$ for all $n \in \mathbb{Z}$ with $\alpha(n)$, $N_0(n)$ are ω-periodic and
(P2) $B(n)$ is nonnegative on $(-\infty, 0] \cap \mathbb{Z}$ with $\sum_{n=-\infty}^{0} B(n) < \infty$.

Theorem 5.3.4 ([149]). *Under assumptions (P1) and (P2), equation (5.3.32) has at least one positive ω-periodic solution.*

Proof. Let $a(n) = \alpha(n)N(n)$ and

$$f(n,x_n) = -\frac{x(n)a(n)}{N_0(n)}\sum_{s=-\infty}^{0} B(s)x(n+s).$$

It is clear that $f(n,x_n)$ is ω-periodic whenever x is ω-periodic and (H1) and (H3) hold since $f(n,\phi_n) \le 0$ for all $(n,\phi) \in \mathbb{Z} \times (\mathbb{Z},\mathbb{R}_+)$. To verify (H4), we let $x,y : \mathbb{Z} \to \mathbb{R}_+$ with $\|x\| \le L$, $\|y\| \le L$ for some $L > 0$. Then

$$|f(n,x_n) - f(n,y_n)|$$

$$= \left|\frac{x(n)a(n)}{N_0(n)}\sum_{s=-\infty}^{0} B(s)x(n+s) - \frac{y(n)a(n)}{N_0(n)}\sum_{s=-\infty}^{0} B(s)y(n+s)\right|$$

$$\le \left|\frac{x(n)a(n)}{N_0(n)}\right|\sum_{s=-\infty}^{0} B(s)|x(n+s) - y(n+s)|$$

$$+ \left|\frac{(x(n) - y(n))a(n)}{N_0(n)}\right|\sum_{s=-\infty}^{0} B(s)|y(n+s)|$$

$$\le \frac{L\|a\|}{N_{0*}}\max_{s\in\mathbb{Z}_-}|x(n+s) - y(n+s)| + \frac{|x(n) - y(n)|\|a\|L}{N_{0*}},$$

where $N_{0*} = \min\{N_0(s) : 0 \le s \le \omega - 1\}$. For any $\varepsilon > 0$, choose $\delta = \varepsilon N_{0*}/(2L\|a\|)$. If $\|x - y\| < \delta$, then

$$|f(n,x_n) - f(n,y_n)| < L\|a\|\delta/N_{0*} + \delta\|a\|L/N_{0*} = 2L\|a\|\delta/N_{0*} = \varepsilon.$$

This implies that (H4) holds. We now show that (5.3.22) and (5.3.23) hold. For $\phi \in \mathscr{P}3$, we have $\phi(n) \geq \sigma_3 \|\phi\|$ for all $n \in [0, \omega - 1]$. This yields

$$\frac{|f(n, \phi)|}{\|\phi\|} \leq \max_{\tau \in [0, \omega - 1]} \frac{a(\tau)}{N_0(\tau)} \sum_{s=-\infty}^{0} B(s) \|\phi\| \to 0$$

as $\|\phi\| \to 0$ and

$$\frac{|f(n, \phi)|}{\|\phi\|} \geq \min_{\tau \in [0, \omega - 1]} \frac{a(\tau)}{N_0(\tau)} \sum_{s=-\infty}^{0} B(s) \sigma_3^2 \|\phi\| \to +\infty$$

as $\|\phi\| \to \infty$. Thus, (5.3.22) and (5.3.23) are satisfied. By (b) of Corollary 5.1, equation (5.3.32) has a positive ω-periodic solution. This completes the proof.

Next we consider the infinite delay Volterra discrete model

$$x_i(n+1)) = x_i(n) \left[a_i(n) - \sum_{j=1}^{k} b_{ij}(n) x_j(n) - \sum_{j=1}^{k} \sum_{s=-\infty}^{n} C_{ij}(n,s) g_{ij}(x_j(s)) \right] \quad (5.3.33)$$

where $x_i(n)$ is the population of the ith species, $a_i, b_{ij} : \mathbb{Z} \to \mathbb{R}$ are ω-periodic, and $C_{ij} : \mathbb{Z} \times \mathbb{Z} \to \mathbb{R}$ is ω-periodic. For more on such derivation we refer to [44].

Theorem 5.3.5 ([149]). *Suppose that the following conditions hold for* $i, j = 1, 2, \ldots, k$.

 (i) $a_i(n) > 1$, *for all* $n \in [0, \omega - 1]$, *and* $a_i(n)$ *is* ω-*periodic,*
 (ii) $b_{ij}(n) \geq 0$, $C_{ij}(n,s) \geq 0$ *for all* $(n,s) \in \mathbb{Z}^2$,
 (iii) $g_{ij} : \mathbb{R}^+ \to \mathbb{R}^+$ *is continuous in* x *and increasing with* $g_{ij}(0) = 0$,
 (iv) $b_{ii}(s) \neq 0$, *for* $s \in [0, \omega - 1]$,
 (v) $C_{ij}(n + \omega, s + \omega) = C_{ij}(n,s)$ *for all* $(n,s) \in \mathbb{Z}^2$ *with*
     $\max_{n \in \mathbb{Z}} \sum_{s=-\infty}^{n} |C_{ij}(n,s)| < +\infty$.

Then equation (5.3.33) has a positive ω-*periodic solution.*

Proof. For $x = (x_1, x_2, \ldots, x_k)^T$, define

$$f_i(n, x_n) = -x_i(t) \sum_{j=1}^{k} b_{ij}(n) x_j(n) - \sum_{j=1}^{k} \sum_{s=-\infty}^{n} C_{ij}(n,s) g_{ij}(x_j(s))$$

for $i = 1, 2, \ldots, k$ and set $f = (f_1, f_2, \ldots, f_k)^t$. Then by some manipulation of conditions (i)–(v), the conditions (H1) and (H2) are satisfied. Also, it is clear that f satisfies (H4). Define

$$b^* = \max\{\|b_{ij}\| : i, j = 1, 2, \cdot, \cdot, \cdot, k\},$$

$$C^* = \max \left\{ \sup_{n \in \mathbb{Z}} \sum_{j=1}^{n} \sum_{s=-\infty}^{n} |C_{ij}(n,s)| : i = 1, 2, \cdot, \cdot, \cdot, k \right\}$$

and

$$g^*(u) = \max\{g_{ij}(u) : i, j = 1, 2, \cdot, \cdot, \cdot, k\}$$

Let $x \in \mathscr{P}3$. Since g is increasing in x, we arrive at

$$|f_i(n, x_n)| \le |x_i(n)| \left[b^* \|x\| + \sum_{j=1}^{n} \sum_{s=-\infty}^{n} |C_{ij}(n,s)| g_{ij}(\|x_j\|) \right].$$

Thus

$$|f(n, x_n)| \le \|x\| [b^* \|x\| + C^* g^* (\|x\|)],$$

which implies

$$\frac{|f(n, x_s)|}{\|x\|} \le [b^* \|x\| + C^* g^* (\|x\|)] \to 0$$

as $\|x\| \to 0$. For $x \in \mathscr{P}3$, $x_i(n) \ge \sigma_3 \|x_i\|$ for all $n \in \mathbb{Z}$. Also, from (ii), $b_{ij}(n), C_{ij}(n,s)$ have the same sign. Thus, using condition (iii) we have

$$|f_i(n, x_n)|$$

$$= \sum_{j=1}^{n} x_i(n) |b_{ij}(n)| |x_j(n)| + \sum_{j=1}^{k} \sum_{s=-\infty}^{n} |C_{ij}(n,s)| g_{ij}(x_j(s))$$

$$\ge |b_{ii}(n)| |x_i(n)|^2 \ge \sigma_3^2 \|x_i\|^2 |b_{ii}(n)|$$

and

$$|f(n, x_s)| \ge \sigma_3^2 \sum_{i=1}^{k} \|x_i\|^2 \min_{1 \le i \le k} |b_{ii}(n)| \ge \frac{\sigma_3^2}{k} \|x\|^2 \min_{1 \le i \le k} |b_{ii}(n)|.$$

Here we have applied the inequality $\left(\sum_{i=1}^{k} \|x_i\| \right)^2 \le k \sum_{i=1}^{k} \|x_i\|^2$. Thus,

$$\frac{|f(n, x_s)|}{\|x\|} \to +\infty \text{ as } \|x\| \to +\infty.$$

By (b) of Corollary 5.1, equation (5.3.33) has a positive ω-periodic solution. This completes the proof.

Theorem 5.3.6 ([149]). *Suppose that the following conditions hold for $i, j = 1, 2, \ldots, k$.*

(i) $0 < a_i(n) < 1$, *for all $n \in [0, \omega - 1]$, and $a_i(n)$ is ω-periodic,*
(ii) $b_{ij}(n) \le 0$, $C_{ij}(n,s) \le 0$ *for all $(n,s) \in \mathbb{Z}^2$,*
(iii) $g_{ij} : \mathbb{R}^+ \to \mathbb{R}^+$ *is continuous in x and increasing with $g_{ij}(0) = 0$,*
(iv) $b_{ii}(s) \ne 0$, *for $s \in [0, \omega - 1]$,*
(v) $C_{ij}(n + \omega, s + \omega) = C_{ij}(n, s)$ *for all $(n, s) \in \mathbb{Z}^2$ with*
$\max_{n \in \mathbb{Z}} \sum_{s=-\infty}^{n} |C_{ij}(n, s)| < +\infty.$

Then equation (5.3.33) has a positive ω-periodic solution.

Proof. The proof follows from part (a) of Corollary 5.1.

Remark 5.1. In the statements of Theorem 5.3.5 and Theorem 5.3.6 condition *(iv)* can be replaced by

(iv*) $\sum\limits_{j=1}^{k} \sum\limits_{s=-\infty}^{n} |C_{ij}(n,s)| \neq 0$ and $g_{ii}(x) \to +\infty$ as $x \to +\infty$.

5.4 Permanence of Multi-Species Competition Predation

The literature on nonautonomous continuous population models described by differential equations is vast, see [165, 166, 167, 168, 169] and the references cited therein. For example, Wen [169] considered the global attractivity of positive periodic solution of multi-species ecological competition-predation system. Yang and Xu [33] studied the global attractivity and existence of the periodic n-prey and m-predator Lotka-Volterra system of differential equations. It is biologically and mathematically crucial to study the existence and stability of periodic solution. However, a more basic and important biological question to ask is whether or not those involved populations will be alive and well in the long run. In [174], Chen discussed the permanence and global stability of nonautonomous Lotka-Volterra system with multi-species predator-prey and deviating arguments by using comparison theorem and constructing suitable Lyapunov functional. On the other hand, the problems of permanence of time-delay systems have received considerable attention in theoretical ecology due to the fact that more realistic models should include some of the past states of these systems. The dynamic behaviors of population models governed by difference equation have also been studied by many authors (see [31, 50, 95, 126, 162, 164, 165, 174, 175], and [178] and the references therein). This is a relatively new topic and the author believes this study should increase research activities on the subject. In [126], Muroya studied the persistence and global stability of delay discrete system for k-species,

$$x_i(n+1) = x_i(n) \exp\{c_i - a_i x_i(n) - \Sigma_{k=1}^{l} a_{ik}(n) x_k(n - \tau_{ik})\}.$$

Results of this section can be partially found in [152] and [165]. The Jacobian method of Section 5.2 is not suitable for our study here. The aim of this study is to investigate the permanent behavior of the following discrete $(l+m)$-species Lotka-Volterra competition-predation system with several delays

$$x_i(n+1) = x_i(n) \exp\{r_i(n) - a_i(n) x_i(n) - \Sigma_{k=1}^{l} a_{ik}(n) x_k(n - \tau_{ik})$$
$$- \Sigma_{k=1}^{m} e_{ik}(n) y_k(n - \eta_{ik})\},$$
$$y_j(n+1) = y_j(n) \exp\{-b_j(n) - c_j(n) y_j(n) + \Sigma_{k=1}^{l} d_{jk}(n) x_k(n - \delta_{jk})$$
$$- \Sigma_{k=1}^{m} c_{jk}(n) y_k(n - \xi_{jk})\},$$
$$x_i(\theta) = \phi_i(\theta) \geq 0, \; y_j(\theta) = \psi_j(\theta) \geq 0, \; \theta \in \mathbb{N}[-\tau, 0] := \{-\tau, -\tau+1, ..., -1, 0\},$$
$$\tag{5.4.1}$$

where $i = 1, 2, ..., l$; $j = 1, 2, ..., m$; τ_{ik}, η_{ik}, δ_{jk} and ξ_{jk} are nonnegative integers; $\phi_i(0) > 0$, $\psi_j(0) > 0$;

$$\tau = \max\{ \max_{1 \le i,k \le l} \tau_{ik}, \ \max_{1 \le i \le l; \, 1 \le k \le m} \eta_{ik}, \ \max_{1 \le k \le l; \, 1 \le j \le m} \delta_{jk}, \ \max_{1 \le j,k \le m} \xi_{jk} \} > 0;$$

$x_i(n)$ is the density of species X_i at nth generation; $y_j(n)$ is the density of species Y_j at nth generation; $r_i(n)$ represents the intrinsic growth rate of the prey species X_i at the nth generation; $b_j(n)$ represents the death rate of the predator species Y_j at the nth generation; $a_{ik}(n)$ and $c_{jk}(n)$ measure the intensity of intraspecific competition or interspecific action of prey species and predator species, respectively; $e_{ik}(n)$ and $d_{jk}(n)$ represent the influence of the $(n - \eta_{ik})$th and $(n - \delta_{jk})$th generation of the predator and prey on the prey and predator population, respectively. For more background of system (5.4.1), one could refer to [169] and [174]. It is clear that model (5.4.1) has positive solutions for positive initial data. We note that the model (5.4.1) generalizes the models in [31, 50], and [126].

Definition 5.4.1. System (5.4.1) is said to be permanent if there are positive constants M_k and L_k, $k = 1, 2$, such that for each positive solution

$$\{x_1(n), ..., x_l(n), y_1(n), ..., y_m(n)\}$$

of system (5.4.1) satisfies

$$L_1 \le \liminf_{n \to \infty} x_i(n) \le \limsup_{n \to \infty} x_i(n) \le M_1,$$

$$L_2 \le \liminf_{n \to \infty} y_j(n) \le \limsup_{n \to \infty} y_j(n) \le M_2,$$

for all $i = 1, 2, ..., l$; $j = 1, 2, ..., m$.

Throughout, we always assume $\{r_i(n)\}$, $\{b_j(n)\}$, $\{a_{ik}(n)\}$, $\{e_{ik}(n)\}$, $\{d_{jk}(n)\}$, $\{c_{jk}(n)\}$, $\{a_i(n)\}$ and $\{c_j(n)\}$ are bounded nonnegative sequences, and use the following notations for any bounded sequence $\{u(n)\}$

$$\bar{u} = \sup_{n \in \mathbb{N}} u(n), \quad \underline{u} = \inf_{n \in \mathbb{N}} u(n).$$

In order to present our main result, we need some preliminaries. Let $\mathbb{R}_+^{l+m} = \{(x_1(n), ..., x_l(n), y_1(n), ..., y_m(n)) \mid x_i(n) \ge 0, y_j(n) \ge 0, \ i = 1, 2, ..., l; j = 1, ..., m\}$, and let $x(n) = (x_1(n), ..., x_l(n), y_1(n), ..., y_m(n)) \in \mathbb{R}_+^{l+m}$, the notation $x(n) > 0$ denotes $x(n) \in \mathrm{Int}\mathbb{R}_+^{l+m}$. For ecological reasons, we consider system (5.4.1) only in $\mathrm{Int}\mathbb{R}_+^{l+m}$. It is easy to obtain the following result.

Lemma 5.4 ([165]). *IntR$_+^{l+m}$ is positively invariant set of system (5.4.1).*

Lemma 5.5 ([165]). *Assume that $\{x(n)\}$ satisfies $x(n) > 0$ and*

$$x(n+1) \le x(n) \exp\{r(n)(1 - ax(n))\}$$

for $n \in [n_1, \infty)$, where a is a positive constant. Then

$$\limsup_{n \to \infty} x(n) \leq \frac{1}{a\bar{r}} \exp(\bar{r} - 1).$$

Lemma 5.6 ([165]). *Assume that $\{x(n)\}$ satisfies*

$$x(n+1) \geq x(n) \exp\{r(n)(1 - ax(n))\}, n \geq N_0,$$

$\limsup_{n \to \infty} x(n) \leq K$ *and* $x(N_0) > 0$, *where a is a constant such that $aK > 1$ and $N_0 \in \mathbb{N}$. Then*

$$\liminf_{n \to \infty} x(n) \geq \frac{1}{a} \exp\{\bar{r}(1 - aK)\}.$$

The main result will follow directly from the following two propositions.

Proposition 5.1 ([152]). *For every solution $\{x_1(n), ..., x_l(n), y_1(n), ..., y_m(n)\}$ of system (5.4.1), we have*

$$\limsup_{n \to \infty} x_i(n) \leq M_i \quad (i = 1, 2, ..., l), \qquad \limsup_{n \to \infty} y_j(n) \leq W_j \quad (j = 1, 2, ..., m),$$

where

$$M_i = \frac{\exp(\bar{r}_i - 1)}{a_i + a_{ii} \exp(-\bar{r}_i \tau_{ii})}, \qquad W_j = \frac{\exp(\sum_{k=1}^{l} \bar{d}_{jk} M_k - \underline{b}_j - 1)}{\underline{c}_j + \underline{c}_{jj} \exp((\underline{b}_j - \sum_{k=1}^{l} \bar{d}_{kk} M_k) \xi_{jj})}.$$

Proof. First, we prove $\limsup_{n \to \infty} x_i(n) \leq M_i$. From the first equation of (5.4.1), we have

$$x_i(n+1) \leq x_i(n) \exp\{r_i(n)\}.$$

It follows that

$$\prod_{s=n-\tau_{ik}}^{n-1} x_i(s+1) \leq \prod_{s=n-\tau_{ik}}^{n-1} x_i(s) \exp\{r_i(s)\},$$

that is

$$x_i(n) \leq x_i(n - \tau_{ik}) \exp\{\sum_{s=n-\tau_{ik}}^{n-1} r_i(s)\}.$$

In other words,

$$x_i(n - \tau_{ik}) \geq x_i(n) \exp\{-\sum_{s=n-\tau_{ik}}^{n-1} r_i(s)\},$$

and hence

$$x_i(n+1) \leq x_i(n) \exp\{r_i(n) - a_i(n) x_i(n) - \sum_{k=1}^{l} a_{ik}(n) x_k(n) \exp\{-\sum_{s=n-\tau_{ik}}^{n-1} r_i(s)\}$$

$$\leq x_i(n) \exp\{r_i(n) - (a_i(n) + a_{ii}(n)) \exp\{-\sum_{s=n-\tau_{ii}}^{n-1} r_i(s)\} x_i(n)\}$$

$$\leq x_i(n) \exp\{\bar{r}_i - (\underline{a}_i + \underline{a}_{ii}) \exp(-\bar{r}_i \tau_{ii}) x_i(n)\}.$$

It follows from Lemma 5.5 that

$$\limsup_{n\to\infty} x_i(n) \le M_i.$$

Next, we prove that $\limsup_{n\to\infty} y_j(n) \le W_j$. For sufficiently small $\varepsilon > 0$, there exists sufficiently large n_0 such that $x_i(n) \le M_i + \varepsilon$ for all $n > n_0$. From the second equation of (5.4.1), we have

$$y_j(n+1) \le y_j(n)\exp\{-b_j(n) + \sum_{k=1}^{l} d_{jk}(n)x_k(n-\delta_{jk})\}$$

$$\le y_j(n)\exp\{-b_j(n) + \sum_{k=1}^{l} d_{jk}(n)(M_k+\varepsilon)\}.$$

By a similar argument, we can verify that

$$y_j(n-\xi_{jk}) \le y_j(n)\exp\{\sum_{s=n-\xi_{jk}}^{n-1} (b_j(s) - \sum_{k=1}^{l} d_{jk}(s)(M_k+\varepsilon))\},$$

and hence

$$y_j(n+1) \le y_j(n)\exp\{-b_j(n) + \sum_{k=1}^{l} d_{jk}(n)(M_k+\varepsilon)) - c_j(n)y_j(n)$$

$$- \sum_{k=1}^{m} c_{jk}(n)y_k(n)\exp\{\sum_{s=n-\xi_{jk}}^{n-1} (b_k(s) - \sum_{k=1}^{l} d_{kk}(s)(M_k+\varepsilon))\}\}$$

$$\le y_j(n)\exp\{(-\underline{b}_j + \sum_{k=1}^{l} \overline{d}_{jk}(M_k+\varepsilon)) - (\underline{c}_j + \underline{c}_{jj})\exp\{(\underline{b}_j$$

$$- \sum_{k=1}^{l} \overline{d}_{kk}(M_k+\varepsilon))\xi_{jj}\}y_j(n)\}.$$

Therefore, by Lemma 5.5, we obtain

$$\limsup_{n\to\infty} y_j(n) \le W_j.$$

The proof is complete.

Proposition 5.2 ([152]). *Let* $\{x_1(n),...,x_l(n),y_1(n),...,y_m(n)\}$ *denote any positive solution of system* (5.4.1). *Assume*

$$(H) \quad \min_{1 < i < l;\, 1 < j < m} \left\{ \frac{(\overline{a}_i + \overline{a}_{ii})M_i}{\underline{r}_i - \sum_{k=1,k\ne i}^{l} \overline{a}_{ik}M_k - \sum_{k=1}^{m} \overline{e}_{ik}W_k},\, \frac{(\overline{c}_j + \overline{c}_{jj})W_j}{\sum_{k=1}^{l} d_{jk}m_k - \overline{b}_j - \sum_{k=1,k\ne j}^{m} \overline{c}_{jk}W_k} \right\} > 1.$$

Then there exist positive constants m_i and w_j such that

$$\liminf_{n\to\infty} x_i(n) \geq m_i, \quad \liminf_{n\to\infty} y_j(n) \geq w_j \quad (i=1,2,...,l;\ j=1,2,...,m),$$

where

$$m_i = \frac{\underline{r}_1 - \sum_{k=1}^{l} \overline{a}_{ik} M_k - \sum_{k=1}^{m} \overline{e}_{ik} W_k}{\overline{a}_i + \overline{a}_{ii}} \exp\{(\overline{r}_i - \sum_{k=1}^{l} \underline{a}_{ik} M_k)$$

$$\times (1 - \frac{(\overline{a}_i + \overline{a}_{ii})M_i}{\underline{r}_i - \sum_{k=1}^{l} \overline{a}_{ik} M_k - \sum_{k=1}^{m} \overline{e}_{ik} W_k})\},$$

$$w_j = \frac{\sum_{k=1}^{l} \underline{d}_{jk} m_k - \overline{b}_j - \sum_{k=1}^{m} \overline{c}_{jk} W_k}{\overline{c}_j + \overline{c}_{jj}} \exp\{(-\underline{b}_j + \sum_{k=1}^{l} \overline{d}_{jk} m_k - \sum_{k-1}^{m} \underline{c}_{jk} W_k)$$

$$\times (1 - \frac{(\overline{c}_j + \overline{c}_{jj})W_j}{\sum_{k=1}^{l} \underline{d}_{jk} m_k - \overline{b}_j - \sum_{k=1}^{m} \overline{c}_{jk} W_k})\}.$$

Proof. We first prove that $\liminf_{n\to\infty} x_i(n) \geq m_i$. For any $\varepsilon > 0$, according to Proposition 5.1, there exists a $n_1 \in \mathbb{N}$ such that $x_i(n-\tau) \leq M_i + \varepsilon$, $y_j(n-\tau) \leq W_j + \varepsilon$ for all $n \geq n_1$. Thus, it follows from the first equation of system (5.4.1) that

$$x_i(n+1) \geq x_i(n) \exp\{(r_i(n) - \sum_{k=1,k\neq i}^{l} a_{ik}(n)(M_k + \varepsilon)$$

$$- \sum_{k=1}^{m} e_{ik}(n)(W_k + \varepsilon)) - (a_i(n) + a_{ii}(n))x_i(n)\}$$

$$= x_i(n) \exp\{(r_i(n) - \sum_{k=1,k\neq i}^{l} a_{ik}(n)(M_k + \varepsilon) - \sum_{k=1}^{m} e_{ik}(n)(W_k + \varepsilon))$$

$$\times (1 - \frac{a_i(n) + a_{ii}(n)}{r_i(n) - \sum_{k=1,k\neq i}^{l} a_{ik}(n)(M_k + \varepsilon) - \sum_{k=1}^{m} e_{ik}(n)(W_k + \varepsilon)} x_i(n))\}$$

$$\geq x_i(n) \exp\{(r_i(n) - \sum_{k=1,k\neq i}^{l} a_{ik}(n)(M_k + \varepsilon) - \sum_{k=1}^{m} e_{ik}(n)(W_k + \varepsilon))$$

$$\times (1 - \frac{a_i(n) + a_{ii}(n)}{\underline{r}_i - \sum_{k=1,k\neq i}^{l} \overline{a}_{ik}(M_k + \varepsilon) - \sum_{k=1}^{m} \overline{e}_{ik}(W_k + \varepsilon)} x_i(n))\}.$$

By lemma 5.6 and condition (H), we obtain

$$\liminf_{n\to\infty} x_i(n) \geq m_i.$$

From the second equation of (5.4.1), we have

$$y_j(n+1) \geq y_j(n)\exp\{-b_j(n) + \sum_{k=1}^{l} d_{jk}(n)m_k$$

$$- \sum_{k=1,k\neq j}^{m} c_{jk}(n)(W_k+\varepsilon) - (c_j(n)+c_{jj}(n))y_j(n)\}$$

$$= y_j(n)\exp\{(-b_j(n) + \sum_{k=1}^{l} d_{jk}(n)m_k - \sum_{k=1,k\neq j}^{m} c_{jk}(n)(W_k+\varepsilon))$$

$$\times(1 - \frac{c_j(n)+c_{jj}(n)}{\sum_{k=1}^{l} d_{jk}(n)m_k - b_j(n) - \sum_{k=1,k\neq j}^{m} c_{jk}(n)(W_k+\varepsilon)}y_j(n)\}$$

$$\geq y_j(n)\exp\{(-b_j(n) + \sum_{k=1}^{l} d_{jk}(n)m_k - \sum_{k=1,k\neq j}^{m} c_{jk}(n)(W_k+\varepsilon))$$

$$\times(1 - \frac{c_j(n)+c_{jj}(n)}{\sum_{k=1}^{l} \underline{d}_{jk}m_k - \overline{b}_j - \sum_{k=1,k\neq j}^{m} \overline{c}_{jk}(W_k+\varepsilon)}y_j(n)\}.$$

By Lemma 5.6 and condition (H), we obtain $\liminf_{n\to\infty} y_j(n) \geq w_j$. This completes the proof.

Now, we state our main results of this section, which its proof is a direct consequence of Propositions 5.1 and 5.2.

Theorem 5.4.1 ([152]). *Assume (II) holds. Then system (5.4.1) is permanent.*

Now, let us consider the special case of system (5.4.1), i.e., $a_i(n) \equiv c_j(n) \equiv 0$, $\tau_{ik} \equiv 0$, $\eta_{is} \equiv 0$, $\delta_{jk} \equiv 0$ and $\xi_{js} \equiv 0$ $(i,k = 1,2,...,l;\ j,s = 1,2,...,m)$, in this case, system (5.4.1) can be written as

$$x_i(n+1) = x_i(n)\exp\{r_i(n) - \sum_{k=1}^{l} a_{ik}(n)x_k(n) - \sum_{k=1}^{m} e_{ik}(n)y_k(n)\},$$
$$\tag{5.4.2}$$
$$y_j(n+1) = y_j(n)\exp\{-b_j(n) + \sum_{k=1}^{l} d_{jk}(n)x_k(n) - \sum_{k=1}^{m} c_{jk}(n)y_k(n)\}.$$

As a corollary of Theorem 5.4.1, we have

Corollary 5.3 ([152]). *Let $\{x_1(n),...,x_l(n),y_1(n),...,y_m(n)\}$ denote any positive solution of system (5.4.2). Assume*

$$\min_{1\leq i\leq l;\, 1\leq j\leq m}\left\{\frac{\overline{a}_{ii}M_i'}{\underline{r}_i - \sum_{k=1,k\neq i}^{l} \overline{a}_{ik}M_k' - \sum_{k=1}^{m} \overline{e}_{ik}W_k'}, \frac{\overline{c}_{jj}W_j'}{\sum_{k=1}^{l} \underline{d}_{jk}m_k' - \overline{b}_j - \sum_{k=1,k\neq j}^{m} \overline{c}_{jk}W_k'}\right\} > 1$$

holds. Then there exist positive constants M'_i, W'_j, m'_i and w'_j such that

$$m'_i \leq \lim_{n\to\infty} \inf x_i(n) \leq \lim_{n\to\infty} \sup x_i(n) \leq M'_i \quad (i = 1, 2, ..., l),$$

$$w'_j \leq \lim_{n\to\infty} \inf y_j(n) \leq \lim_{n\to\infty} \sup y_j(n) \leq W'_j \quad (j = 1, 2, ..., m),$$

where

$$M'_i = \frac{\exp(\bar{r}_i - 1)}{a_{ii}}, \qquad W'_j = \frac{\exp(\sum_{k=1}^{l} \bar{d}_{jk} M'_k - \underline{b}_j - 1)}{\underline{c}_{jj}},$$

$$m'_i = \frac{\underline{r}_1 - \sum_{k=1}^{l} \bar{a}_{ik} M'_k - \sum_{k=1}^{m} \bar{e}_{ik} W'_k}{\bar{a}_{ii}} \exp\{(\bar{r}_i - \sum_{k=1}^{l} a_{ik} M'_k)$$

$$\times (1 - \frac{\bar{a}_{ii} M'_i}{\underline{r}_i - \sum_{k=1}^{l} \bar{a}_{ik} M'_k - \sum_{k=1}^{m} \bar{e}_{ik} W'_k})\},$$

$$w'_j = \frac{\sum_{k=1}^{l} \underline{d}_{jk} m'_k - \bar{b}_j - \sum_{k=1}^{m} \bar{c}_{jk} W'_k}{\bar{c}_{jj}} \exp\{(-\underline{b}_j + \sum_{k=1}^{l} \bar{d}_{jk} m'_k - \sum_{k=1}^{m} \underline{c}_{jk} W'_k)$$

$$\times (1 - \frac{\bar{c}_{jj} W'_j}{\sum_{k=1}^{l} \underline{d}_{jk} m'_k - \bar{b}_j - \sum_{k=1}^{m} \bar{c}_{jk} W'_k})\}.$$

Finally, we give a suitable example to illustrate the feasibility of Theorem 5.4.1.

Example 5.1. We consider the following system:

$$x(n+1) = x(n) \exp\{1 - x(n-1) - \tfrac{1}{60}(3 + \sin n) y_1(n)\},$$

$$y_1(n+1) = y_1(n) \exp\{-1 + \tfrac{2}{9}(8 + \cos n) x(n) - y_1(n-1) - \tfrac{1}{80}(3 + \sin n) y_2(n)\},$$

$$y_2(n+1) = y_2(n) \exp\{-1 - \cos n + \tfrac{2}{9}(8 + \cos n) x(n) - y_2(n-1)\}.$$

It is easy to verify that the system satisfies the condition (H). Therefore, by Theorem 5.4.1 the system is permanent.

5.5 Open Problems

Open Problem 1.
In this section we consider the Lotka-Volterra predator-prey model given by Elaydi et al. [90] and extend it to predator-prey model with ratio dependence. Let x and y represent population densities of a prey and a predator, respectively. In [70], Fan and Wang discretized a continuous model with ratio dependence and obtained the Lotka-Volterra discrete predator-prey model with ratio dependence

$$\begin{cases} x(n+1) = x(n)exp\left\{a(n) - b(n)x(n) - \dfrac{c(n)y(n)}{m(n)y(n) + x(n)}\right\} \\[4mm] \quad y(n+1) = y(n)exp\left\{-d(n) + \dfrac{f(n)x(n)}{m(n)y(n) + x(n)}\right\} \end{cases}$$

$$(5.5.1)$$

We refer to [70] for the specific interpretation of the coefficients. In [70] the authors proved the existence of positive periodic solution of (5.5.1) by using Coincidence Theory or Degree Theory.

We propose the reader uses the idea of [176] and show the positive periodic solution is actually globally asymptotically stable by constructing a suitable Lyapunov function.

Chapter 6
Exponential and l_p-Stability in Volterra Equations

This chapter is devoted primarily to the exponential and l_p-stability of Volterra difference equations. Lyapunov functionals are the main tools in the analysis. It is pointed out that in the case of exponential stability, Lyapunov functionals are hard to extend to vector Volterra difference equations or to Volterra difference equations with infinite delay. In addition, we use nonstandard discretization scheme due to Mickens [122] and apply them to continuous Volterra integro-differential equations. We will show that under the discretization scheme the stability of the zero solution of the continuous dynamical system is preserved. Also, under the same discretization, using a combination of Lyapunov functionals, Laplace transforms, and z-transforms, we show that the boundedness of solutions of the continuous dynamical system is preserved. We end the chapter with a brief section introducing semigroup, which should stir up some curiosity in the application of semigroup to Volterra difference equations. The chapter concludes with multiple open problems. The work of this chapter heavily depends on the materials in [9, 51, 59, 76, 91], and [98].

6.1 Exponential Stability

We consider the scalar linear difference equation with multiple delays

$$x(t+1) = a(t)x(t) + \sum_{s=t-r}^{t-1} b(t,s)x(s), \ t \geq 0, \tag{6.1.1}$$

where $r \in \mathbb{Z}^+$, $a : \mathbb{Z}^+ \to \mathbb{R}$ and $b : \mathbb{Z}^+ \times [-r, \infty) \to \mathbb{R}$. We will use Lyapunov functionals and obtain some inequalities regarding the solutions of (6.1.1) from which we can deduce exponential asymptotic stability of the zero solution. Also, we will provide a criteria for the unboundedness of solutions and the instability of the zero solution of (6.1.1) by means of Lyapunov type functionals.

© Springer Nature Switzerland AG 2018
Y. N. Raffoul, *Qualitative Theory of Volterra Difference Equations*,
https://doi.org/10.1007/978-3-319-97190-2_6

Consider the kth-order scalar difference equation

$$x(t+k) + p_1 x(t+k-1) + p_2 x(t+k-2) + \cdots + p_k x(t) = 0, \qquad (6.1.2)$$

where the p_i's are real numbers. It is well known that the zero solution of (6.1.2) is asymptotically stable if and only if $|\lambda| < 1$ for every characteristic root λ of (6.1.2). There are no easy criteria to test for exponential stability of the zero solution of equations that are similar to (6.1.2) for variable coefficients. This itself highlights the importance of the creativity of constructing a suitable Lyapunov functional that leads to the exponential stability. When using Lyapunov functionals, one faces the difficulties of relating the constructed Lyapunov functional back to the solution x so that stability can be deduced. This task is tedious and we did overcome it. The authors have done an extensive literature search and could not find any work that dealt with exponential stability of Volterra equations of the form of (6.1.1). This research offers easily verifiable conditions that guarantee exponential stability. Moreover, we give criteria for the instability of the zero solution. Most importantly, our results will hold for $|a(t)| \geq 1$. We will illustrate our theory with several examples and numerical simulations. It is scarce to find results concerning the use of Lyapunov functionals in the stability of finite delay difference equations due to the unforeseen difficulties in constructing such functions. This section intends to fill some of the gap and moreover, we will compare the results obtained in this section to known ones where other methods are used, such as operator theory.

In Chapter 2, we looked at the system of functional difference equation of the form

$$x(n+1) = G(n, x_n), \ x \in \mathbb{R}^k \qquad (6.1.3)$$

where $G : \mathbb{Z}^+ \times \mathbb{R} \to \mathbb{R}$ is continuous in x. Let x be any solution of (6.1.3). Quite often when using Lyapunov functional to study system (6.1.3) we encounter pair of inequalities in the form of

$$W_1(x(n)) \leq V(n, x(\cdot)) = W_2(x(n)) + \sum_{s=0}^{n-1} K(n,s) W_3(x(s)),$$

$$\triangle V(n, x(\cdot)) \leq -W_4(x(n)) + F(n)$$

where V is a Lyapunov functional bounded below, x is the unknown solution of the functional difference equation, and K, F, and $W_i, i = 1,2,3,4$ are scalar positive functions. The wedge W_1 is mandatory in order to relate the solutions x back to V. Hence, identifying such a W_1 is not an easy job, as we shall see later on in this section. It is even more difficult when using Lyapunov functionals to obtain exponential stability, since it requires that along the solutions of (6.1.3) we have for some $\alpha > 0$

$$\triangle V(n, x(\cdot)) \leq -\alpha V(n, x(\cdot)).$$

The above inequality presents us with formidable challenges as maybe seen later in the section. However, a simple but clever rewriting of the difference equation points us in the right direction in constructing the appropriate Lyapunov functional, as we shall see from (6.1.5).

Let $\psi : [-h,0] \to (-\infty, \infty)$ be a given bounded initial function with

$$||\psi|| = \max_{-h \le s \le 0} |\psi(s)|.$$

It should cause no confusion to denote the norm of a function $\varphi : [-h, \infty) \to (-\infty, \infty)$ with

$$||\varphi|| = \sup_{-h \le s < \infty} |\varphi(s)|.$$

The notation x_t means that $x_t(\tau) = x(t + \tau), \tau \in [-h, 0]$ as long as $x(t + \tau)$ is defined. Thus, x_t is a function mapping an interval $[-h, 0]$ into $\mathbb{R}$. We say $x(t) \equiv x(t, t_0, \psi)$ is a solution of (6.1.1) if $x(t)$ satisfies (6.1.1) for $t \ge t_0$ and $x_{t_0} = x(t_0 + s) = \psi(s)$, $s \in [-h, 0]$.

In preparation for our main results, we let

$$A(t, s) = \sum_{u=t-s}^{r} b(u+s, s). \tag{6.1.4}$$

By noting that

$$A(t, t-r-1) = 0,$$

we have that (6.1.1) is equivalent to

$$\triangle x(t) = \big(a(t) + A(t+1, t) - 1\big)x(t) - \triangle_t \sum_{s=t-r-1}^{t-1} A(t,s)x(s). \tag{6.1.5}$$

In [138], the author used the same method and studied the exponential stability and instability of the zero solution of

$$x(t+1) = a(t)x(t) - b(t)x(t-h).$$

One of the novelty of rewriting (6.1.1) in the form of (6.1.5) is that it allows us to obtain stability results concerning the totally delayed Volterra difference equation

$$x(t+1) = \sum_{s=t-r}^{t-1} b(t,s)x(s), \ t \ge 0. \tag{6.1.6}$$

We have the following definition.

Definition 6.1.1. The zero solution of (6.1.1) is said to be exponentially stable if any solution $x(t, t_0, \psi)$ of (6.1.1) satisfies

$$|x(t, t_0, \psi)| \le C\big(||\psi||, t_0\big)\zeta^{\gamma(t-t_0)}, \quad \text{for all } t \ge t_0,$$

where ζ is constant with $0 < \zeta < 1$, $C : \mathbb{R}^+ \times \mathbb{Z}^+ \to \mathbb{R}^+$, and γ is a positive constant. The zero solution of (6.1.1) is said to be uniformly exponentially stable if C is independent of t_0.

For simplicity we let

$$Q(t) = a(t) + A(t+1,t) - 1.$$

Assume

$$\triangle_t A^2(t,z) \le 0, \quad \text{for all } t+s+1 \le z \le t-1. \tag{6.1.7}$$

Lemma 6.1 ([98]). *Let $A(t,s)$ be given by (6.1.4) and that for $\delta > 0$ the inequality*

$$-\frac{\delta}{(\delta+1)r} \le Q(t) \le -r\delta A^2(t+1,t) - Q^2(t) \tag{6.1.8}$$

holds. If

$$V(t) = \left[x(t) + \sum_{s=t-r-1}^{t-1} A(t,s)x(s) \right]^2$$

$$+ \delta \sum_{s=-r}^{-1} \sum_{z=t+s}^{t-1} A^2(t,z)x^2(z), \tag{6.1.9}$$

then along the solutions of (6.1.1) we have

$$\triangle V(t) \le Q(t)V(t).$$

Proof. First we note that due to condition (6.1.8), $Q(t) < 0$ for all $t \ge 0$. Also, we use the fact that if $u(t)$ is a sequence, then $\triangle u^2(t) = u(t+1)\triangle u(t) + u(t)\triangle u(t)$. Let $x(t) = x(t,t_0,\psi)$ be a solution of (6.1.1) and define $V(t)$ by (6.1.9). Then along solutions of (6.1.5) we have

$$\triangle V(t) = \left[x(t+1) + \sum_{s=t-r}^{t} A(t+1,s)x(s) \right] \triangle_t \left[x(t) + \sum_{s=t-r-1}^{t-1} A(t,s)x(s) \right]$$

$$+ \left[x(t) + \sum_{s=t-r-1}^{t-1} A(t,s)x(s) \right] \triangle_t \left[x(t) + \sum_{s=t-r-1}^{t-1} A(t,s)x(s) \right]$$

$$+ \delta \triangle_t \sum_{s=-r}^{-1} \sum_{z=t+s}^{t-1} A^2(t,z)x^2(z). \tag{6.1.10}$$

We note that

$$x(t+1) + \sum_{s=t-r}^{t} A(t+1,s)x(s) = (Q(t)+1)x(t) - \triangle_t \sum_{s=t-r-1}^{t-1} A(t,s)x(s)$$

$$+ \sum_{s=t-r}^{t} A(t+1,s)x(s)$$

$$= (Q(t)+1)x(t) + \sum_{s=t-r-1}^{t-1} A(t,s)x(s)$$

$$= (Q(t)+1)x(t) + \sum_{s=t-r}^{t-1} A(t,s)x(s),$$

since $A(t,t-r-1) = 0$. With this in mind, (6.1.10) reduces to

$$\triangle V(t) = \left[(Q(t)+1)x(t) + \sum_{s=t-r}^{t-1} A(t,s)x(s) \right] Q(t)x(t)$$

$$+ \left[x(t) + \sum_{s=t-r}^{t-1} A(t,s)x(s) \right] Q(t)x(t)$$

$$+ \delta\triangle_t \sum_{s=-r}^{-1} \sum_{z=t+s}^{t-1} A^2(t,z)x^2(z)$$

$$= Q(t)V(t) + \left(Q^2(t) + Q(t) \right))x^2(t)$$

$$- \delta Q(t) \sum_{s=-r}^{-1} \sum_{z=t+s}^{t-1} A^2(t,z)x^2(z)$$

$$+ \delta\triangle_t \sum_{s=-r}^{-1} \sum_{z=t+s}^{t-1} A^2(t,z)x^2(z)$$

$$- Q(t) \left(\sum_{s=t-r}^{t-1} A(t,s)x(s) \right)^2. \tag{6.1.11}$$

Also, using (6.1.4), we arrive at

$$\triangle_t \sum_{s=-r}^{-1} \sum_{z=t+s}^{t-1} A^2(t,z)x^2(z) = \sum_{s=-r}^{-1} \sum_{z=t+s+1}^{t} A^2(t+1,z)x^2(z)$$

$$- \sum_{s=-r}^{-1} \sum_{z=t+s}^{t-1} A^2(t,z)x^2(z)$$

$$= \sum_{s=-r}^{-1} \left[A^2(t+1,t)x^2(t) + \sum_{z=t+s+1}^{t-1} A^2(t+1,z)x^2(z) \right.$$

$$\left. - \sum_{z=t+s+1}^{t-1} A^2(t,z)x^2(z) - A^2(t,t+s)x^2(t+s) \right]$$

$$= \sum_{s=-r}^{-1} \left(A^2(t+1,t)x^2(t) - A^2(t,t+s)x^2(t+s) \right)$$

$$+ \sum_{s=-r}^{-2} \sum_{z=t+s+1}^{t-1} \triangle_t A^2(t,z)x^2(z)$$

$$= rA^2(t+1,t)x^2(t) - \sum_{s=-r}^{-1} A^2(t,t+s)x^2(t+s)$$

$$+ \sum_{s=-r}^{-2} \sum_{z=t+s+1}^{t-1} \triangle_t A^2(t,z)x^2(z)$$

$$\leq rA^2(t+1,t)x^2(t)$$

$$- \sum_{s=-r}^{-1} A^2(t,t+s)x^2(t+s). \tag{6.1.12}$$

With the aid of Hölder's inequality, we have

$$\left(\sum_{s=t-r}^{t-1} A(t,s)x(s) \right)^2 \leq r \sum_{s=t-r}^{t-1} A^2(t,s))x^2(s). \tag{6.1.13}$$

Also,

$$\sum_{s=-r}^{-1} \sum_{z=t+s}^{t-1} A^2(t,z)x^2(z) \leq r \sum_{s=t-r}^{t-1} A^2(t,s)x^2(s). \tag{6.1.14}$$

By invoking (6.1.8) and substituting expressions (6.1.12)–(6.1.14) into (6.1.11), we obtain

$$\triangle V(t) \leq Q(t)V(t) + \left(Q^2(t) + Q(t) + r\delta A^2(t+1,t) \right)x^2(t)$$

$$+ [-(\delta+1)rQ(t) - \delta] \sum_{s=t-h}^{t-1} A^2(t,s)x^2(s)$$

$$\leq Q(t)V(t). \tag{6.1.15}$$

This completes the proof.

Theorem 6.1.1 ([98]). *Assume the hypothesis of Lemma 6.1 holds and suppose there exists a number $\alpha < 1$ such that $0 < a(t) + A(t+1,t)) \leq \alpha$. Then any solution $x(t) = x(t,t_0,\psi)$ of (6.1.1) satisfies the exponential inequality*

$$|x(t)| \leq \sqrt{\frac{r+\delta}{\delta} V(t_0) \prod_{s=t_0}^{t-1} (a(s) + A(s+1,s))} \tag{6.1.16}$$

for $t \geq t_0$.

Proof. First we note that condition (6.1.8) implies that there exists some positive number $\alpha < 1$ such that $|a(t) + A(t+1,t))| < \alpha$. Now by changing the order of summation we have

$$\delta \sum_{s=-r}^{-1} \sum_{z=t+s}^{t-1} A^2(t,z)x^2(z) = \delta \sum_{z=t-r}^{t-1} \sum_{s=-r}^{z-1} A^2(t,z)x^2(z)$$

$$= \delta \sum_{z=t-r}^{t-1} A^2(t,z)x^2(z)(z-t+r+1)$$

$$\geq \delta \sum_{z=t-r}^{t-1} A^2(t,z)x^2(z),$$

where we have used the fact that $t - h \leq z \leq t - 1 \implies 1 \leq z - t + h + 1 \leq h$. Also

$$\left(\sum_{z=t-r}^{t-1} A(t,s)x(s) \right)^2 \leq r \sum_{z=t-r}^{t-1} A^2(t,s)x^2(s),$$

and hence,

$$\delta \sum_{s=-r}^{-1} \sum_{z=t+s}^{t-1} A^2(t,z)x^2(z) \geq \frac{\delta}{r} \left(\sum_{z=t-r}^{t-1} A(t,z)x(z) \right)^2.$$

Let $V(t)$ be given by (6.1.9). Then

$$V(t) = \left[x(t) + \sum_{s=t-r-1}^{t-1} A(t,s)x(s) \right]^2 + \delta \sum_{s=-r}^{-1} \sum_{z=t+s}^{t-1} A^2(t,z)x^2(z)$$

$$\geq \left[x(t) + \sum_{s=t-r-1}^{t-1} A(t,s)x(s) \right]^2 + \frac{\delta}{r} \left(\sum_{z=t-r}^{t-1} A(t,z)x(z) \right)^2$$

$$\geq \frac{\delta}{r+\delta} x^2(t) + \left[\sqrt{\frac{r}{r+\delta}} x(t) + \sqrt{\frac{r+\delta}{r}} \sum_{z=t-r}^{t-1} A(t,z)x(z) \right]^2$$

$$\geq \frac{\delta}{r+\delta} x^2(t).$$

Consequently,

$$\frac{\delta}{r+\delta} x^2(t) \leq V(t).$$

From (6.1.15) we get

$$V(t) \leq V(t_0) \prod_{s=t_0}^{t-1} (a(s) + A(s+1,s)).$$

Thus we arrive at

$$|x(t)| \leq \sqrt{\frac{r+\delta}{\delta} V(t_0) \prod_{s=t_0}^{t-1} \left(a(s) + A(s+1,s) \right)}$$

for $t \geq t_0$. This completes the proof.

Corollary 6.1 ([98]). *Assume the hypothesis of Theorem 6.1.1 holds. Then the zero solution of* (6.1.1) *is exponentially stable.*

Proof. From inequality (6.1.16) we have that

$$|x(t)| \leq \sqrt{\frac{r+\delta}{\delta} V(t_0) \prod_{s=t_0}^{t-1} \left(a(s) + A(s+1,s) \right)}$$

$$\leq \sqrt{\frac{r+\delta}{\delta} V(t_0) \alpha^{t-t_0}}$$

for $t \geq t_0$. The proof is complete since $\alpha \in (0,1)$.

Now we state a corollary regarding the exponential stability of the zero solution of (6.1.6).

Corollary 6.2 ([98]). *Assume the hypothesis of Theorem 6.1.1 holds with $Q(t) = A(t+1,t) - 1$. Then the zero solution of* (6.1.6) *is exponentially stable.*

Remark 6.1. If for a positive constant M we have

$$V(t_0) \leq M, \text{ for all } t_0 \geq 0,$$

then the zero solution of (6.1.1) is uniformly exponentially stable. This follows from the exponential inequality (6.1.16).

6.2 Criterion for Instability

In this section we use a nonnegative definite Lyapunov functional and obtain criteria that can be easily applied to test for instability of the zero solution of (6.1.1).

Theorem 6.2.1 ([98]). *Let $H > r$ be a constant. Assume $Q(t) > 0$ such that*

$$Q^2(t) + Q(t) - HA^2(t+1,t) \geq 0. \tag{6.2.1}$$

If

$$V(t) = \left[x(t) + \sum_{s=t-r-1}^{t-1} A(t,s)x(s) \right]^2$$

$$- H \sum_{s=t-r-1}^{t-1} A^2(t,s)x^2(s), \tag{6.2.2}$$

then along the solutions of (6.1.1) we have

$$\Delta V(t) \geq Q(t)V(t).$$

Proof. Let $x(t) = x(t,t_0,\psi)$ be a solution of (6.1.1) and assume $V(t)$ is given by (6.2.2). Since the calculation is similar to the one in Lemma 6.1 we have that

$$\Delta V(t) = Q(t)V(t) + (Q^2(t) + Q(t) - HA^2(t+1,t))x^2(t)$$

$$+ Q(t)(H-r) \left(\sum_{s=t-r-1}^{t-1} A^2(t,s)x^2(s) \right)^2$$

$$\geq Q(t)V(t), \tag{6.2.3}$$

where we have used

$$\left(\sum_{s=t-r}^{t-1} A(t,s)x(s) \right)^2 \leq r \sum_{s=t-r}^{t-1} A^2(t,s)x^2(s)$$

and (6.2.1). This completes the proof.

We remark that condition (6.2.1) is satisfied for

$$Q(t) \geq \frac{-1 + \sqrt{1 + 4HA^2(t+1,t)}}{2}.$$

Theorem 6.2.2 ([98]). *Suppose the hypothesis of Theorem 6.2.1 holds. Then all solutions of (6.1.1) are unbounded and its zero solution is unstable, provided that*

$$\prod^{\infty} (a(s) + A(s+1,s)) = \infty. \tag{6.2.4}$$

Proof. From (6.2.3) we have

$$V(t) \geq V(t_0) \prod_{s=t_0}^{t-1} (a(s) + A(s+1,s)). \tag{6.2.5}$$

Let $V(t)$ be given by (6.2.2). Then

$$V(t) = x^2(t) + 2x(t) \sum_{t-r-1}^{t-1} A(t,s)x(s) + \left[\sum_{t-r-1}^{t-1} A(t,s)x(s) \right]^2$$

$$- H \sum_{t-r-1}^{t-1} A^2(t,s)x^2(s). \tag{6.2.6}$$

Let $\beta = H - r$. Then from

$$\left(\frac{\sqrt{r}}{\sqrt{\beta}} a - \frac{\sqrt{\beta}}{\sqrt{r}} b \right)^2 \geq 0,$$

we have

$$2ab \leq \frac{r}{\beta} a^2 + \frac{\beta}{r} b^2.$$

With this in mind we arrive at

$$2x(t) \sum_{t-r-1}^{t-1} A(t,s)x(s) \leq 2|x(t)| \left| \sum_{t-r-1}^{t-1} A(t,s)x(s) \right|$$

$$\leq \frac{r}{\beta} x^2(t) + \frac{\beta}{r} \left[\sum_{t-r-1}^{t-1} A(t,s)x(s) \right]^2$$

$$\leq \frac{r}{\beta} x^2(t) + \frac{\beta}{r} r \sum_{t-r}^{t-1} A^2(t,s)x^2(s).$$

A substitution of the above inequality into (6.2.6) yields

$$V(t) \leq x^2(t) + \frac{r}{\beta} x^2(t) + (\beta + r - H) \sum_{t-r-1}^{t-1} A^2(t,s)x^2(s)$$

$$= \frac{\beta + r}{\beta} x^2(t)$$

$$= \frac{H}{H - r} x^2(t).$$

Using inequality (6.2.5), we get

$$|x(t)| \geq \sqrt{\frac{H - r}{H}} V^{1/2}(t)$$

$$= \sqrt{\frac{H - r}{H}} V^{1/2}(t_0) \left(\prod_{s=t_0}^{t-1} (a(s) + + A(s+1,s)) \right)^{\frac{1}{2}}.$$

This completes the proof.

We have the following corollary regarding the unboundedness and instability of (6.1.6).

Corollary 6.3 ([98]). *Suppose the hypothesis of Theorem 6.2.1 holds with $Q(t) = A(t+1,t) - 1$. Then all solutions of (6.1.6) are unbounded and its zero solution is unstable, provided that*

$$\prod^{\infty} (A(s+1,s)) = \infty.$$

6.2.1 Applications and Numerical Evidence

In this section we provide examples that illustrate our theoretical results in two instances: when the coefficients $a(t)$ and $b(t,s)$ are constants, and when they are functions.

First, if $a(t) = a$ and $b(t,s) = b$ $(a,b \in \mathbb{R})$ we have $A(t,s) = \sum\limits_{u=t-s}^{r} b$. Then $A(t,s) = b(r+1-t+s)$. Hence, $\triangle_t A^2(t,s) = b^2(r-t+s)^2 - b^2(r+1-t+s)^2 \leq 0$ and thus condition (6.1.7) holds. Also $A(t+1,t) = br$, and hence condition (6.1.8) becomes

$$-\frac{\delta}{(\delta+1)r} \leq a+br-1 \leq -\left[\delta b^2 r^3 + (a+br-1)^2\right]. \qquad (6.2.7)$$

It is obvious from (6.2.7) that when $a = 1$, b has to be negative.
Next we give four examples where the emphasis is on $|a| \geq 1$.

Example 6.1 ([98]). Let $a = r = 1, b = -0.3$ and $\delta = 0.5$. Then one can easily verify that (6.2.7) is satisfied. Hence the zero solution of the delay difference equation

$$x(t+1) = x(t) - 0.3x(t-1) \qquad (6.2.8)$$

is exponentially stable.

Example 6.2 ([98]). Let $a = 1.2, b = -0.3, r = 1$, and $\delta = 0.5$. Then one can easily verify that (6.2.7) is satisfied. Hence the zero solution of the delay difference equation

$$x(t+1) = 1.2x(t) - 0.3x(t-1) \qquad (6.2.9)$$

is exponentially stable as illustrated in Figure 6.1.

Example 6.3 ([98]). Let $a = 1.29, b = -0.6, r = 1$, and $\delta = 0.5$. With these values (6.2.7) is satisfied, and therefore the zero solution of the delay difference equation

$$x(t+1) = 1.29x(t) - 0.6x(t-1)$$

is exponentially stable as shown in Figure 6.2.

Example 6.4 ([98]). $a = 1.125, b = -0.15, r = 2$, and $\delta = \frac{2}{3}$. Then one can easily verify that (6.2.7) is satisfied. Hence the zero solution of the delay difference equation

$$x(t+1) = 1.125x(t) - 0.15\big(x(t-1) + x(t-2)\big)$$

is exponentially stable as shown in Figure 6.3.

It is worth mentioning that in both papers [87] and [136], in which fixed point theory was used, it was assumed that

$$\prod_{s=0}^{t-1} a(s) \to 0, \text{ as } t \to \infty$$

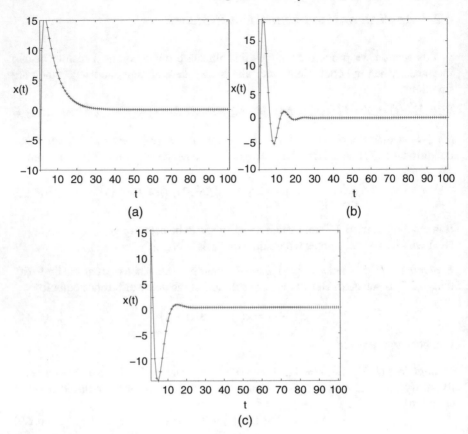

Fig. 6.1 Trajectories of (6.1.1) when $a(t)$ and $b(t,s)$ are constant. Figure 6.1 refers to Example 6.2 where $a = 1.2, b = -0.3$, and $r = 1$ with initial condition $x(0) = -10$ and $x(1) = 10.3$. Figure 6.2 refers to Example 6.3 where $a = 1.29, b = -0.6$, and $r = 1$ with initial condition $x(0) = -10$ and $x(1) = 10.3$. Figure 6.3 refers to Example 6.4 where $a = 1.125, b = -0.6$, and $r = 2$ with initial condition $x(0) = 15, x(1) = 2$, and $x(3) = -10$.

for the asymptotic stability.

Example 6.5 ([98]). Let $a = 1.3, b = -0.2, r = 1$ and $H = 1.1$. Then $Q(t) = 0.1 >$ 0. Moreover $Q(t) \geq \dfrac{-1 + \sqrt{1 + 4HA^2(t+1,t)}}{2} = 0.0422$. Thus conditions (6.2.1) and (6.2.4) are satisfied and the zero solution of

$$x(t+1) = 1.3x(t) - 0.2x(t-1) \tag{6.2.10}$$

is unstable. Actually, all its solutions become unbounded for large t. Figure 6.2 shows a trajectory for the above equation with initial condition $x(0) = -10$ and $x(1) = -1.3$.

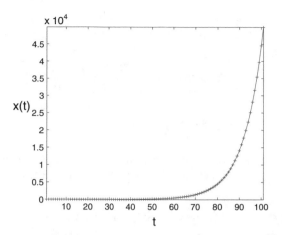

Fig. 6.2 Trajectories of (6.1.1) when $a(t)$ and $b(t,s)$ are constant. This graph corresponds to Example 6.5 where $a = 1.3, b = -0.2$, and $r = 1$ with initial condition $x(0) = -10$ and $x(1) = -1.3$.

Remark 6.2. When $a(t)$ and $b(t,s)$ are constant the solution $x(t)$ of the delay difference equation (6.1.1) is the same as the sequence $(x_n)_{n\in\mathbb{N}_0}$ defined recursively as

$$x_{n+r+1} = ax_{n+r} + b(x_{n+r-1} + \cdots + x_n), n \in \mathbb{N}_0, \qquad (6.2.11)$$

and for which the general solution can be obtained analytically. For $r = 1$ in particular, the general solution to (6.2.11) is easily calculated. For instance, the exact solution to (6.2.8) in Example 6.1 is

$$x(t) = \left(\frac{\sqrt{30}}{10}\right)^t \left(x(0)\cos(t\theta) + \frac{\frac{10x(1)}{\sqrt{30}} - x(0)\cos\theta}{\sin\theta}\sin(t\theta)\right),$$

where $\theta = \arctan\left(\frac{\sqrt{5}}{5}\right)$. Since $\left|\frac{\sqrt{30}}{10}\right| < 1$ we see that $\lim_{t\to+\infty}|x(t)| = 0$ with an exponential convergence. The exact solution to (6.2.9) in Example 6.2 is

$$x(t) = \frac{1}{2}\left[\left(x(0) + 10\frac{x(1) - \frac{6x(0)}{10}}{\sqrt{6}}\right)\left(\frac{6+\sqrt{6}}{10}\right)^t\right.$$
$$\left. + \left(x(0) - 10\frac{x(1) - \frac{6x(0)}{10}}{\sqrt{6}}\right)\left(\frac{6-\sqrt{6}}{10}\right)^t\right].$$

Since $\left|\frac{6\pm\sqrt{6}}{10}\right| < 1$, we see that the solution $x(t)$ of Example 6.2 converges exponentially to zero. Similar calculations can be done for Examples 6.3 and 6.4. Finally, the exact solution to (6.2.10) in Example 6.5 is

$$x(t) = \frac{1}{2}\left[\left(x(0) + 20\frac{x(1) - \frac{13x(0)}{20}}{\sqrt{89}}\right)\left(\frac{13 + \sqrt{89}}{20}\right)^t\right.$$

$$\left. + \left(x(0) - 20\frac{x(1) - \frac{13x(0)}{20}}{\sqrt{89}}\right)\left(\frac{13 - \sqrt{89}}{20}\right)^t\right].$$

Since $\left|\frac{13+\sqrt{89}}{20}\right| > 1$, we see that $\lim_{t\to+\infty} |x(t)| = +\infty$.

We now give two examples that illustrate the exponentially stable and unstable case when $a(t)$ and $b(t,s)$ are functions. We corroborate our results with numerical simulations.

Example 6.6 ([98]). Let $a(t) = d^{2t+1} + \frac{2}{3}$ and $b(t,s) = -d^{t+s}$ for $d \in \mathbb{R}$. Then $A(t,s) = -d^{2s}\sum_{u=t-s}^{r} d^u$, and therefore $A(t+1,t) = -d^{2t}\sum_{u=1}^{r} d^u = -d^{2t+1}$ for $r = 1$. We can show that $\triangle_t A^2(t,z) \le 0$ for all $t+s+1 \le z \le t-1$. If we take $r = 1$ and $\delta = 1$, we obtain $Q(t) = -\frac{1}{3}$. With these choices we see that the left inequality of condition (6.1.8) is trivially satisfied. To obtain the right inequality, we need to choose d such that $\left(d^2(d^4)^t + \frac{1}{9}\right) \le -Q(t) = \frac{1}{3}$ for t large enough. It is therefore sufficient to choose $d \in (0,1)$. In that case, $\lim_{t\to+\infty} (d^4)^t = 0$ which implies that the right inequality of condition (6.1.8) will eventually be satisfied. Note that condition (6.1.8) is satisfied for all $t \ge 0$ when $d \in (0, \frac{\sqrt{2}}{3}]$. With these choices for the parameters d, δ, and r, we can conclude that the zero solution of the delay difference equation

$$x(t+1) = \left(d^{2t+1} + \frac{2}{3}\right)x(t) - d^{2t+1}x(t-1)$$

is exponentially stable. We plotted two of its trajectories in Figure 6.3.

Example 6.7 ([98]). Let $a(t) = d^{2t+1} + 1.1$ and $b(t,s) = -d^{t+s}$. Then from Example 6.6 we have $A(t+1,t) = -d^{2t+1}$ when $r = 1$. In that case choosing $H = 1$ yields $Q(t) = 0.1 > 0$. With these choices we see that condition (6.2.1) is satisfied if $d \in (0,1)$ and hence the zero solution of the delay difference equation

$$x(t+1) = \left(d^{2t+1} + 1.1\right)x(t) - d^{2t+1}x(t-1)$$

is unstable as illustrated in Figure 6.4. In fact, the zero solution is unstable for all choices of $a(t) = d^{2t+1} + v$ with $v > 1$. We note that with these choices of $a(t)$ and $b(t,s)$ we have

$$\prod^{\infty} (a(s) + A(s+1,s)) = \prod^{\infty} v = +\infty,$$

and hence (6.2.4) is verified.

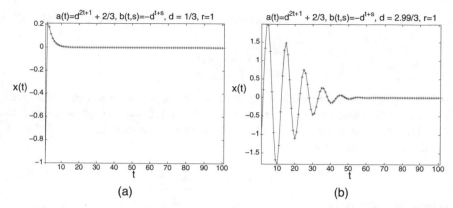

Fig. 6.3 Trajectories of (6.1.1) when $a(t) = d^{2t+1} + \frac{2}{3}$ and $b(t,s) = -d^{t+s}$. These plots refer to Example 6.6 with $r = 1$. The initial condition was taken to be $x(0) = -1$ and $x(1) = 0.21$. In Figure 6.3(a) we plotted the trajectory obtained with $d = \frac{2}{3}$, and in Figure 6.3(b) we plotted the trajectory with $d = \frac{2.99}{3}$. In the latter case, since condition (6.1.8) is verified only after a certain value of t, the first few terms of the trajectory $x(t)$ are not converging to zero until condition (6.1.8) is satisfied.

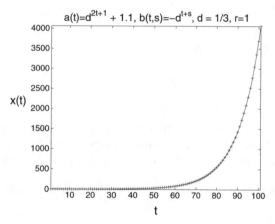

Fig. 6.4 Trajectories of (6.1.1) when $a(t) = d^{2t+1} + 1.1$ and $b(t,s) = -d^{t+s}$. This graph corresponds to Example 6.7 with $r = 1$ and initial condition $x(0) = -1$ and $x(1) = 0.21$.

Next we compare our results to existing ones. Let $a = 1.2, b_1 = -0.2, b_2 = -0.088, h^* = 2$, and $\delta = 0.5$. Then one can easily verify that (6.1.8) is satisfied. Hence the zero solution of the difference equation with multiple delays

$$x(t+1) = 1.2x(t) - 0.2x(t-1) - 0.088x(t-2). \tag{6.2.12}$$

is exponentially stable.

It is worth mentioning that in both papers [87] and [136] it was assumed that

$$\prod_{s=0}^{t-1} a(s) \to 0, \text{ as } t \to \infty$$

for the asymptotic stability. Of course our $a = 1.2$ does not satisfy such a condition, and yet we concluded exponential stability. Let $a = 1.2, b = -0.3, h = 1$, and $\delta = 0.5$. Then one can easily verify that (6.1.8) is satisfied. Hence the zero solution of the delay difference equation

$$x(t+1) = 1.2x(t) - 0.3x(t-1) \tag{6.2.13}$$

is exponentially stable.

Moreover the above condition of [87] and [136] cannot be satisfied since our $a = 1.2$. Next we compare our results with the results obtained in [125] by El-Morshedy. Hence, we begin with the statement of the following.

Lemma 6.2 ([125]). *If there exists $\lambda \in (0,1)$ such that*

$$\left| \prod_{j=0}^{N} a(n-j) + b(n) \right| + \sum_{s=1}^{N} \left| \prod_{j=0}^{s-1} a(n-j) \right| |b(n-s)| \leq \lambda, \tag{6.2.14}$$

for large n, then the zero solution of

$$x(n+1) = a(n)x(n) + b(n)x(n-N) \tag{6.2.15}$$

is globally exponentially stable.

It can be easily seen that condition (6.2.14) cannot be satisfied for the data given in the above (6.2.12). Next we state two major results from [125], by Berezansky and Braverman, so we can compare with equation (6.2.12) and (6.2.13).

Lemma 6.3 ([16]). *Let $0 < \gamma < 1$ and $\sum_{l=2}^{m} |a_l(n)| + |1 - a_1(n)| \leq \gamma$ for n large enough.*

Then the equation

$$x(n+1) - x(n) = -a_1(n)x(n) - \sum_{l=2}^{m} a_l(n)x(g_l(n)) \tag{6.2.16}$$

is exponentially stable. Here $n - T \leq g_l(n) \leq n$ for some integer $T > 0$.

Lemma 6.4 ([16]). *Suppose that for some $\gamma \in (0,1)$ the following inequality is satisfied for n large enough:*

$$\sum_{k=2}^{m} |a_k(n)| \sum_{j=g_k(n)}^{n-1} \sum_{l=1}^{m} |a_l(j)| + |1 - \sum_{k=1}^{m} a_k(n)| \leq \gamma. \tag{6.2.17}$$

Then (6.2.16) is exponentially stable.

In the spirit of (6.2.16) we rewrite (6.2.13) as

$$x(t+1) - x(n) = .2x(t) - 0.3x(t-1).$$

Then the condition in Lemma 6.3 is equivalent to

$$|a_2(n)| + |1 - a_1(n)| = 0.3 + |1 + 0.2| > 1,$$

is not satisfied. Also, condition (6.2.17) is equivalent to

$$|a_2(n)||a_1(n)| + |1 - a_1(n)| = 0.3(0.2) + |1 + 0.2| > 1,$$

is not satisfied. Thus, we have demonstrated that Lyapunov functionals yield better results as seen from the improvement over the results of [16] and [125]. For more comparison with existing literature, we have the following theorem by Berezansky, Braverman, and Karabash.

Theorem 6.2.3 ([17]). *Consider the two delays difference equation*

$$x(n+1) - x(n) = -a_0 x(n) - a_1 x(n - h_1) - a_2 x(n - h_2), \ h_1, \ h_2 > 0. \qquad (6.2.18)$$

Suppose at least one of the following conditions hold:
1) $1 > a_0 > 0$, $|a_1| + |a_2| < a_0$;
2) $0 < a_0 + a_1 + a_2 < 1$, $|a_1|h_1 + |a_2|h_2 < \dfrac{a_0 + a_1 + a_2}{|a_0| + |a_1| + |a_2|}$;

3) $0 < a_0 + a_2 < 1$, $|a_2|h_2 < \dfrac{a_0 + a_2 - |a_1|}{|a_0| + |a_1| + |a_2|}$.
Then Equation (6.2.18) is exponentially stable.

The next theorem is due to Cooke and Győri.

Theorem 6.2.4 ([42]). *The multiple delays difference equation*

$$x(n+1) - x(n) = -\sum_{k=1}^{N} a_k x(n - h_k), \ a_k \geq 0, \ h_k \geq 0, \qquad (6.2.19)$$

is asymptotically stable if

$$\sum_{k=1}^{N} a_k h_k < 1. \qquad (6.2.20)$$

The next theorem is due to Elaydi (1994) and Kocić and Laddas (1993).

Theorem 6.2.5 ([63, 94]). *The multiple delays difference equation*

$$x(n+1) - x(n) = a_0(n)x(n) - \sum_{k=1}^{N} a_k(n)x(g_k(n)), \ g_k(n) \leq n \qquad (6.2.21)$$

is asymptotically stable if

$$\sum_{k=1}^{N} |a_k(n)| = \begin{cases} a_0(n) - \varepsilon & \text{for } 0 < a_0(n) < 1 \\ 2 - a_0(n) - \varepsilon & \text{for } 1 \le a_0(n) < 2 \end{cases} \qquad (6.2.22)$$

The next theorem is due to Hartung and Győri.

Theorem 6.2.6 ([77]). *The multiple delays difference equation*

$$x(n+1) - x(n) = -\sum_{k=1}^{N} a_k x(g_k(n)), \; a_k \ge 0, \; g_k(n) \le n, \qquad (6.2.23)$$

is exponentially stable if

$$\limsup_{n \to \infty} (n - g_k(n)) < \infty, \qquad (6.2.24)$$

and

$$\sum_{k=1}^{N} a_k \limsup_{n \to \infty} (n - g_k(n)) < 1 + \frac{1}{e} - \sum_{k=1}^{N} a_k. \qquad (6.2.25)$$

We remark that Theorems 6.2.4 and 6.2.5 only give results regarding asymptotic stability.

Consider the difference equation in Example 6.4, where we have shown its zero solution is exponentially stable.

$$x(n+1) - x(n) = -(-.125)x(n) - 0.15x(n-1) - 0.15x(n-2). \qquad (6.2.26)$$

Then $a_0 = -.125$, $a_1 = 0.15$, and, $a_1 = 0.15$. It is clear that 1) of Theorem 6.2.3 cannot hold since our a_0 is negative. Similarly, condition (6.2.22) cannot hold since it requires that $a_0 > 0$. Notice that Theorems 6.2.4 and 6.2.6 are not applicable to the results of this section since the coefficients (6.1.1) depend on time. In Example 6.6 we showed the zero solution is exponentially stable. Also we remark that our theorems do not require sign conditions on the coefficients and the fact that if we rewrite the equations of Theorems 6.2.4, 6.2.5, and 6.2.6, in the form of

$$x(n+1) = (a_0 + 1)x(n) + a_1 x(n - h_1) + a_2 x(n - h_2)$$

then their first coefficient, $a_0 + 1 > 1$, unlike our theorems that are applicable to $|a_0 + 1| > 1$ as it was demonstrated in the examples.

In general, it is a major problem to find an appropriate Lyapunov functional and hence, the dependence of the quality of the corresponding results on such functional. However, once a suitable Lyapunov functional is found, investigators may continue for decades deriving more and more information from that Lyapunov functional. It is a common knowledge among researchers that stability and boundedness results go hand in hand with the type of the Lyapunov functional that is being used. To illustrate our concern, we consider the delay difference equation

$$x(t+1) = a(t)x(t) + b(t)x(t-\tau) + p(t), \; t \in \mathbb{Z}^+, \tag{6.2.27}$$

where $a, b, p : \mathbb{Z}^+ \to \mathbb{R}$, τ is a positive integer. We have the following theorem.

Theorem 6.2.7. *Assume*

$$|a(t)| < 1, \text{ for all } t \in \mathbb{Z} \tag{6.2.28}$$

and there is a $\delta > 0$ such that

$$|b(t)| + \delta < 1, \tag{6.2.29}$$

and

$$|a(t)| \le \delta, \text{ and } |p(t)| \le K, \text{ for some positive constant } K. \tag{6.2.30}$$

Then all solutions of (6.2.27) are bounded. If $p(t) = 0$ for all t, then the zero solution of (6.2.27) is (UAS).

Proof. Consider the Lyapunov functional

$$V(t, x(\cdot)) = |x(t)| + \delta \sum_{s=t-\tau}^{t-1} |x(s)|.$$

Then along solutions of (6.2.27) we have

$$\Delta V = |x(t+1)| - |x(t)| + \delta \sum_{s=t+1-\tau}^{t} |x(s)| - \delta \sum_{s=t-\tau}^{t-1} |x(s)|$$

$$\le |a(t)||x(t)| - |x(t)| + |b(t)||x(t-\tau)| + \delta \sum_{s=t+1-\tau}^{t} |x(s)| - \delta \sum_{s=t-\tau}^{t-1} |x(s)| + |p(t)|$$

$$= \big(|a(t)| + \delta - 1\big)|x(t)| + \big(|b(t)| - \delta\big)|x(t-\tau)| + |p(t)|$$

$$\le \big(|a(t)| + \delta - 1\big)|x(t)| + |p(t)|$$

$$\le -\gamma|x(t)| + |p(t)|, \text{ for some positive constant } \gamma.$$

The results follow from either [133] or Theorem 2.2.4.

It is evident from Example 6.2 that Theorem 6.2.7 does not give any result concerning the exponential stability of the single delay difference equation

$$x(t+1) = 1.2x(t) - 0.3x(t-1).$$

This illustrates the uncertainty we face when using Lyapunov functionals. On the other hand, it is tricky to construct a Lyapunov functional that deals with multiple delays.

As we indicated before, there is always a price to pay. By using Lyapunov functionals, our method relaxed the stringent conditions on the size of the coefficients. On the other hand, it puts a severe demand on the size of the delay h. The next theorem, which is due to Clark [35], does exactly the opposite; however, it asks for the coefficients to be constants.

Theorem 6.2.8 ([35]). *Suppose the coefficients a and b of (6.2.15) are constants. Then Equation (6.2.15) is asymptotically stable provided that*

$$|a| + |b| < 1.$$

6.3 Vector Equation

In this section we try to extend the concept of exponential stability of finite delay scalar Volterra equation to the finite delay vector Volterra equation

$$x(t+1) = Px(t) + \sum_{s=t-r}^{t-1} C(t,s)g(x(s)), \qquad (6.3.1)$$

where r is a positive integer, P is a constant $n \times n$ matrix, and C is an $n \times n$ matrix of functions that are defined on $-r \le t \le s < \infty$, where $t, s \in [-r, \infty) \cap \mathbb{Z}$. The nonlinear function $g : \mathbb{R}^n \to \mathbb{R}^n$ is continuous in x. Throughout this paper it is understood that if $x \in \mathbb{R}^n$, then $|x|$ is taken to be the Euclidean norm. Obtaining exponential stability through the method of Lyapunov functional V requires that along the solutions, we have $\triangle V(t,x) \le -\alpha V(t,x)$, something that is almost impossible to obtain in vector equations. Materials of this section can be found in [51] and the references therein. Let $U = (u)_{ij}$ be an $n \times n$ matrix. Then we define the norm $|U|$ to be

$$|U| = \max_{1 \le j \le n} \sum_{i=1}^{n} |u_{ij}|.$$

It should cause no confusion to denote the norm of a sequence function $\varphi : [-r, \infty) \cap \mathbb{Z} \to \mathbb{R}^n$ with

$$\|\varphi\| = \sup_{-r \le s < \infty} |\varphi(s)|.$$

The notation x_t means that $x_t(\tau) = x(t+\tau), \tau \in [-r, 0] \cap \mathbb{Z}$ as long as $x(t+\tau)$ is defined. Thus, x_t is a function mapping an interval $[-r, 0] \cap \mathbb{Z}$ into $\mathbb{R}^n$. We say $x(t) \equiv x(t, t_0, \psi)$ is a solution of (6.3.1) if $x(t)$ satisfies (6.3.1) for $t \ge t_0$ and $x_{t_0} = x(t_0 + s) = \psi(s)$, $s \in [-r, 0] \cap \mathbb{Z}$. Throughout this paper it is to be understood that when a function is written without its argument, then the argument is t. We begin with a stability definition. For $t_0 \ge 0$ we define

$$E_{t_0} = [-r, t_0] \cap \mathbb{Z}.$$

Let $C(t)$ denote the set of sequences $\phi : [-r, \infty) \cap \mathbb{Z} \to \mathbb{R}^n$ and $\|\phi\| = \sup\{|\phi(s)| : s \in [-r, t] \cap \mathbb{Z}\}$.

Definition 6.3.1. The zero solution of (6.3.1) is stable if for each $\varepsilon > 0$ and each $t_0 \ge -r$, there exists a $\delta = \delta(\varepsilon, t_0) > 0$ such that $[\phi \in E_{t_0} \to \mathbb{R}^n, \phi \in C(t) : \|\phi\| < \delta]$ implies $|x(t, t_0, \phi)| < \varepsilon$ for all $t_0 \ge 0$.

In order to be able to handle the calculations for $\triangle V(t)$ along the solutions of (6.3.1), we let

$$A(t,s) := \sum_{u=t-s}^{r} C(u+s,s), \, t,s \in [0, \infty) \cap \mathbb{Z}\}.$$

Clearly $A(t,t-r-1) = 0$, and as a consequence, one can easily verify that (6.3.1) is equivalent to the new system

$$\triangle x(t) = Qx(t) + A(t+1,t)g(x(t)) - \triangle_t \sum_{s=t-r-1}^{t-1} A(t,s)g(x(s)), \qquad (6.3.2)$$

where the matrix Q is given by $Q = P - I$ and I is the identity $n \times n$ matrix.

Remark 6.3. Writing (6.3.1) in the form of (6.3.2) allows us to obtain stability result regarding the nonlinear Volterra difference equation

$$x(t+1) = \sum_{s=t-r}^{t-1} C(t,s)g(x(s)). \qquad (6.3.3)$$

This is remarkable since (6.3.1) is considered as the perturbed form of $x(t+1) = Px(t)$, which implies that the stability of the zero solution of (6.3.1) depends on the stability of linear part; that is, one must require that the magnitude of all the eigenvalues of the matrix A be inside the unit circle.

Before we state and prove our next theorem, we assume there exists a positive definite symmetric and constant $n \times n$ matrix D such that for positive constants λ, μ_1, and μ_2 we have

$$P^T DQ + Q^T D = -\mu_1 I. \qquad (6.3.4)$$

Due to the nonlinearity of the function g, we require that

$$x^T \left(P^T DA(t+1,t) + DA(t+1,t) \right) g(x) \leq -\mu_2 |x|^2 \text{ if } x \neq 0, \qquad (6.3.5)$$

and

$$|g(x)| \leq \lambda |x|. \qquad (6.3.6)$$

It is clear that conditions (6.3.5) and (6.3.6) imply that $g(0) = 0$ and hence $x = 0$ is a solution for system (6.3.1). We note that since D is a positive definite symmetric matrix, there exists a positive constant k such that

$$k|x|^2 \leq x^T Dx, \text{ for all } x. \qquad (6.3.7)$$

If $W(t)$ and $Z(t)$ are two sequences, then $\triangle W(t)Z(t) = W(t+1)\triangle Z(t) + (\triangle W(t)) Z(t)$.

Theorem 6.3.1 ([51]). *Let (6.3.4)–(6.3.6) hold, and suppose there are constants $\gamma > 0$ and $\alpha > 0$ so that*

$$-\mu_1 - 2\mu_2 + \gamma r\lambda^2 |A(t+1,t)| + \left(\lambda |A^T(t+1,t)D| + |Q^T D| \right) \sum_{s=t-r}^{t-1} |A(t,s)| \leq -\alpha,$$

$$(6.3.8)$$

$$-\gamma + \lambda |A^T(t+1,t)D| + |Q^T D| \leq 0, \tag{6.3.9}$$

and

$$1 - \lambda \sum_{s=t-r-1}^{t-1} |A(t,s)| > 0 \tag{6.3.10}$$

then the zero solution of (6.3.1) is stable.

Proof. Define the Lyapunov functional $V(t) = V(t,x)$ by

$$V(t) = \left(x(t) + \sum_{s=t-r-1}^{t-1} A(t,s)g(x(s))\right)^T D\left(x(t) + \sum_{s=t-r-1}^{t-1} A(t,s)g(x(s))\right)$$

$$+ \gamma \sum_{s=-r}^{-1} \sum_{z=t+s}^{t-1} |A(t,z)| \|g(x(z))\|^2. \tag{6.3.11}$$

First we note that the right side of (6.3.11) is scalar. Let $x(t) = x(t,t_0,\psi)$ be a solution of (6.3.1) and define $V(t)$ by (6.3.11). Then along solutions of (6.3.1) we have

$$\triangle V(t) = \left(x(t+1) + \sum_{s=t-r}^{t} A(t+1,s)g(x(s))\right)^T D$$

$$\times \triangle\left(x(t) + \sum_{s=t-r-1}^{t-1} A(t,s)g(x(s))\right)$$

$$+ \triangle\left(x(t) + \sum_{s=t-r-1}^{t-1} A(t,s)g(x(s))\right)^T$$

$$\times D\left(x(t) + \sum_{s=t-r-1}^{t-1} A(t,s)g(x(s))\right)$$

$$+ \gamma r |A(t+1,t)| \|g(x(t))\|^2 - \gamma \sum_{s=-r}^{-1} |A(t,t+s)| \|g(x(t+s))\|^2.$$

Using (6.3.2) one can easily show that

$$x(t+1) + \sum_{s=t-r}^{t} A(t+1,s)g(x(s)) = Px(t) + \sum_{s=t-r}^{t-1} A(t,s)g(x(s)).$$

With this in mind $\triangle V$ becomes

$$\triangle V(t) = \left(Px(t) + \sum_{s=t-r}^{t-1} A(t,s)g(x(s))\right)^T D\left(Qx(t) + A(t+1,t)g(x(t))\right)$$

$$+ \left(Qx(t) + A(t+1,t)g(x(t))\right)^T D\left(x(t) + \sum_{s=t-r-1}^{t-1} A(t,s)g(x(s))\right)$$

$$+ \gamma r \lambda^2 |A(t+1,t)| \|g(x(t))\|^2 - \gamma \sum_{s=-r}^{-1} |A(t,t+s)| \|g(x(t+s))\|^2.$$

After rearranging terms, the above expression simplifies to

$$\triangle V(t) = x^T(t)\left(P^T DQ + Q^T D\right)x(t) + x^T(t)P^T DA(t+1,t))g(x(t))$$
$$+ \left(\sum_{s=t-r}^{t-1} A(t,s)g(x(s))\right)^T DQx(t)$$
$$+ \left(\sum_{s=t-r}^{t-1} A(t,s)g(x(s))\right)^T DA(t+1,t)g(x(t))$$
$$+ x^T(t)Q^T D \sum_{s=t-r-1}^{t-1} A(t,s)g(x(s)) + g^T(x(t))A^T(t+1,t)Dx(t)$$
$$+ g^T(x(t))A^T(t+1,t)D \sum_{s=t-r-1}^{t-1} A(t,s)g(x(s)) + \gamma r|A(t+1,t)||g(x(t))|^2$$
$$- \gamma \sum_{s=-r}^{-1} |A(t,t+s)||g(x(t+s))|^2. \tag{6.3.12}$$

Next we try to simplify (6.3.12) by combining likewise terms. We begin by noting that $g^T(x)A^T(t+1,t)Dx = \left[x^T DA(t+1,t))g(x)\right]^T$, and hence we have

$$x^T P^T DA(t+1,t))g(x) + g^T(x)A^T(t+1,t)Dx$$
$$= x^T\left(P^T DA(t+1,t) + DA(t+1,t)\right)g(x).$$

$$\left(\sum_{s=t-r}^{t-1} A(t,s)g(x(s))\right)^T DQx(t) + x^T(t)Q^T D \sum_{s=t-r-1}^{t-1} A(t,s)g(x(s))$$
$$= x^T(t)Q^T D \sum_{s=t-r-1}^{t-1} A(t,s)g(x(s))$$
$$+ \left[\left(\sum_{s=t-r}^{t-1} A(t,s)g(x(s))\right)^T DQx(t)\right]^T$$
$$= 2x^T(t)Q^T D \sum_{s=t-r-1}^{t-1} A(t,s)g(x(s))$$
$$\le 2|x^T(t)||Q^T D| \sum_{s=t-r}^{t-1} |A(t,s)||g(x(s))|$$
$$\le |Q^T D| \sum_{s=t-r}^{t-1} |A(t,s)|(|x(t)|^2 + |g(x(s))|^2),$$

where we have used the inequality $2ab \leq a^2 + b^2$. Similarly,

$$\sum_{s=t-r}^{t-1} A(t,s)g(x(s)) \Big)^T DA(t+1,t)g(x(t))$$

$$+ g^T(x(t))A^T(t+1,t)D \sum_{s=t-r-1}^{t-1} A(t,s)g(x(s))$$

$$= g^T(x(t))A^T(t+1,t)D \sum_{s=t-r-1}^{t-1} A(t,s)g(x(s))$$

$$+ \Big[\Big(\sum_{s=t-r}^{t-1} A(t,s)g(x(s)) \Big)^T DA(t+1,t)g(x(t)) \Big]^T$$

$$= 2g^T(x(t))A^T(t+1,t)D \sum_{s=t-r-1}^{t-1} A(t,s)g(x(s))$$

$$\leq 2\lambda |x(t)||A^T(t+1,t)D| \sum_{s=t-r}^{t-1} |A(t,s)||g(x(s))|$$

$$\leq \lambda |A^T(t+1,t)D| \sum_{s=t-r}^{t-1} |A(t,s)| \big(|x(t)|^2 + |g(x(s))|^2 \big).$$

Let $u = t + s$, then

$$\gamma \sum_{s=-r}^{-1} |A(t,t+s)||g(x(t+s))|^2 = -\gamma \sum_{s=t-r}^{t-1} |A(t,s)||g(x(s))|^2.$$

By substituting the above four simplified expressions into (6.3.12) yields

$$\triangle V(t) \leq \Big[-\mu_1 - \mu_2 + \gamma r \lambda^2 |A(t+1,t)|$$

$$+ \big(\lambda |A^T(t+1,t)D| + |Q^T D| \big) \sum_{t-r}^{t-1} |A(t,s)| \Big] |x(t)|^2$$

$$+ \Big[-\gamma + \lambda |A^T(t+1,t)D| + |Q^T D| \Big] \sum_{t-r}^{t-1} |A(t,s)||g(x(s))|^2.$$

$$\leq -\alpha |x(t)|^2. \tag{6.3.13}$$

Let $\varepsilon > 0$ be given, we will find $\delta > 0$ so that $|x(t,t_0,\phi)| < \varepsilon$ as long as $[\phi \in E_{t_0} \to \mathbb{R} : \|\phi\| < \delta]$. Let

$$L^2 = |D| \Big(1 + \lambda \sum_{t_0-r}^{t_0-1} |A(t_0,s)| \Big)^2 + \lambda^2 v \sum_{s=-r}^{-1} \sum_{z=t_0+s}^{t_0-1} |A(t_0,z)|.$$

By (6.3.13) we have V is decreasing and hence for $t \geq t_0 \geq 0$ we have that

$$V(t,x) \leq V(t_0,\phi)$$

$$\leq |D|\Big(\phi(t_0) + \sum_{t_0-r}^{t_0-1} A(t_0,s)g(\phi(s))\Big)^2$$

$$+ v\lambda^2 \sum_{s=-r}^{-1} \sum_{z=t_0+s}^{t_0-1} |A(t_0,z)||\phi(z)|^2$$

$$= \delta^2 |D|\Big(1 + \lambda \sum_{t_0-r}^{t_0-1} |A(t_0,s)|\Big)^2$$

$$+ v\lambda^2 \delta^2 \sum_{s=-r}^{-1} \sum_{z=t_0+s}^{t_0-1} |A(t_0,z)|$$

$$\leq \delta^2 L^2. \tag{6.3.14}$$

By (6.3.11), we have

$$V(t,x) \geq \Big(x(t) + \sum_{s=t-r-1}^{t-1} A(t,s)g(x(s))\Big)^T$$

$$\times D\Big(x(t) + \sum_{s=t-r-1}^{t-1} A(t,s)g(x(s))\Big)$$

$$\geq k^2\Big(|x| - |\sum_{s=t-r-1}^{t-1} A(t,s)g(x(s))|\Big)^2. \tag{6.3.15}$$

Combining the two inequalities (6.3.14) and (6.3.15) we arrive at

$$|x(t)| \leq \frac{\delta L}{k} + \lambda \sum_{s=t-r-1}^{t-1} |A(t,s)||x(s)|.$$

So as long as $|x(t)| < \varepsilon$, we have

$$|x(t)| < \frac{\delta L}{k} + \varepsilon\lambda \sum_{s=t-r-1}^{t-1} |A(t,s)|, \text{ for all } t \geq t_0.$$

Thus, we have from the above inequality

$$|x(t)| < \varepsilon \text{ for } \delta < \frac{k}{L}\Big(1 - \lambda \sum_{s=t-r-1}^{t-1} |A(t,s)|\Big)\varepsilon.$$

Note that by (6.3.10), the above inequality regarding δ is valid. This completes the proof.

We have the following corollary.

Corollary 6.4 ([51]). *Assume all the conditions of Theorem 6.3.1 hold. Let $x(t)$ be any solution of (6.3.1). Then $|x(t)|^2 \in l\big([t_0,\infty) \cap \mathbb{Z}\big)$.*

Proof. We know from Theorem 6.3.1 that the zero solution is stable. Thus, for the same δ of stability, we take $|x(t,t_0,\phi)| < 1$. Since V is decreasing, we have by summing (6.3.13) from t_0 to $t-1$ and using (6.3.14) that,

$$V(t,x) \le V(t_0,\phi) \le \delta^2 L^2 - \alpha \sum_{t_0}^{t-1} |x(s)|^2.$$

Since,

$$V(t,x) \ge \Big(x + \sum_{s=t-r-1}^{t-1} A(t,s)g(x(s))\Big)^T D\Big(x + \sum_{s=t-r-1}^{t-1} A(t,s)g(x(s))\Big),$$

we have that

$$\Big(x + \sum_{s=t-r-1}^{t-1} A(t,s)g(x(s))\Big)^T D\Big(x + \sum_{s=t-r-1}^{t-1} A(t,s)g(x(s))\Big)$$

$$\le \delta^2 L^2 - \alpha \sum_{t_0}^{t-1} |x(s)|^2. \tag{6.3.16}$$

Also, using Schwarz inequality one obtains

$$\Big(\sum_{s=t-r-1}^{t-1} |A(t,s)||g(x(s))|\Big)^2 = \Big(\sum_{s=t-r-1}^{t-1} |A(t,s)|^{1/2}|A(t,s)|^{1/2}|g(x(s))|\Big)^2$$

$$\le \lambda^2 \sum_{s=t-r-1}^{t-1} |A(t,s)| \sum_{s=t-r-1}^{t-1} |A(t,s)||x(s)|^2.$$

As $\sum_{s=t-r-1}^{t-1} |A(t,s)|$ is bounded by (6.3.10) and $|x|^2 < 1$, we have $\sum_{s=t-r-1}^{t-1} |A(t,s)|$ $|x(s)|^2$ is bounded and hence $\sum_{s=t-r-1}^{t-1} |A(t,s)||g(x(s))|$ is bounded. Therefore, from (6.3.16), we arrive at

$$\alpha \sum_{s=t_0}^{t-1} |x(s)|^2 \le \delta^2 L^2 - \Big(x + \sum_{s=t-r-1}^{t-1} A(t,s)g(x(s))\Big)^T$$

$$\times D\Big(x + \sum_{s=t-r-1}^{t-1} A(t,s)g(x(s))\Big)$$

$$\le \delta^2 L^2 + |D|\Big(|x| + \sum_{s=t-r-1}^{t-1} A(t,s)g(x(s))|\Big)^2 \le K,$$

from which we deduce that $|x(t)|^2 \in l\big([t_0,\infty) \cap \mathbb{Z}\big)$.

Due to our previous remark, it is straightforward to extend Theorem 6.3.1 and Corollary 6.4 to (6.3.3) by setting the coefficient matrix $P = 0$.

Theorem 6.3.2 ([51]). *Let (6.3.4) and (6.3.5) hold for $P = 0$ matrix. Assume (6.3.6) and suppose there are constants $\gamma > 0$ and $\alpha > 0$ so that*

$$-\mu_1 - \mu_2 + \gamma r \lambda^2 |A(t+1,t)| + \left(\lambda |A^T(t+1,t)D|\right.$$
$$\left. + |D|\right) \sum_{t-r}^{t-1} |A(t,s)| \le -\alpha,$$

$$-\gamma + \lambda |A^T(t+1,t)D| + |D| \le 0,$$

and

$$1 - \lambda \sum_{s=t-r-1}^{t-1} |A(t,s)| > 0$$

then the zero solution of (6.3.3) is stable and $|x(t)|^2 \in l\big([t_0,\infty)\cap\mathbb{Z}\big)$.

Proof. The proof is immediate consequence of Theorem 6.3.1 and Corollary 6.4 by taking the matrix P to be the zero matrix which implies that $Q = I$.

Next, we resort to fundamental matrix solution to characterize solutions of (6.3.1) and then compare both results. We begin by considering the homogenous system,

$$x(t+1) = Ax(t) \tag{6.3.17}$$

where $A = (a_{ij})$ is constant $n \times n$ nonsingular matrix. Then if $\Phi(t)$ is a matrix that is nonsingular for all $t \ge t_0$ and satisfies (6.3.17), then it is said to be a fundamental matrix for (6.3.17). Also, it is known that if all eigenvalues of A reside inside the unit circle, then there exist positive constants l and $\eta \in (0,1)$ such that $|\Phi(t)\Phi^{-1}(t_0)| \le l\eta^{t-t_0}$. For more on Linearization of systems of the form of (6.3.17), we refer the reader to [57]. Suppose the function g is Lipschitz. That is, there exists a positive constant L such that

$$|g(x) - g(y)| \le L|x - y| \tag{6.3.18}$$

for all x and y. Then (6.3.18) along with $g(0) = 0$ imply that $|g(x)| \le L|x|$.

Theorem 6.3.3 ([51]). *Assume all eigenvalues of A of system (6.3.17) reside inside the unit circle. Also, assume (6.3.18) along with $g(0) = 0$. In addition we ask that for some positive constant R*

$$\sum_{s=-r}^{\infty} |C(u,s)| \le R, \tag{6.3.19}$$

then the zero solution of (6.3.1) is exponentially stable provided that $RL < \frac{1-\eta}{l}$.

Proof. Let $\Phi(t)$ be the fundamental matrix for (6.3.17). For a given initial function $\phi : [-r,\infty)\cap\mathbb{Z} \to \mathbb{R}^n$, by using the variation of parameters, we have that

$$x(t) = \Phi(t)\Phi^{-1}(t_0)\phi(t_0) + \sum_{u=t_0}^{t-1} \Phi(t)\Phi^{-1}(u+1) \sum_{s=u-r}^{u-1} C(u,s)g(x(s)). \quad (6.3.20)$$

Then $x(t)$ given by (6.3.20) is a solution of (6.3.1) (see [57]). Hence, for $t \geq t_0$ we have

$$|x(t)| \leq l\eta^{t-t_0}|\phi(t_0)| + RLl\eta^{t-1} \sum_{u=t_0}^{t-1} \eta^{-u}|x(u)|.$$

The rest of the proof follows along the lines of Theorem 4.35 of [57], by invoking Gronwall's inequality (see Corollary 1.1).

Theorem 6.3.3 gives stronger type of stability since it requires the zero solution of (6.3.17) to be exponentially stable. We end this section with an example.

Example 6.8. Let $P = \begin{pmatrix} 1/2 & 0 \\ 0 & 1/2 \end{pmatrix}$ and $C(t,s) = \begin{pmatrix} 1/3 & 0 \\ 0 & 1/3 \end{pmatrix}$,

then $A(t,s) = \begin{pmatrix} \frac{1}{3}(r-t+s+1) & 0 \\ 0 & \frac{1}{3}(r-t+s+1) \end{pmatrix}$

and $A(t+1,t) = \begin{pmatrix} \frac{1}{3}r & 0 \\ 0 & \frac{1}{3}r \end{pmatrix}$.

From $P^T DQ + Q^T D = -\mu_1 I$, we obtain

$$D = \begin{pmatrix} \frac{4}{3}\mu_1 & 0 \\ 0 & \frac{4}{3}\mu_1 \end{pmatrix}. \text{ Let } g(x) = \begin{pmatrix} \frac{-9\mu_2}{8\mu_1 r}x_1 \\ \frac{-9\mu_2}{8\mu_1 r}x_2 \end{pmatrix}.$$

Then

$$x^T \left(P^T DA(t+1,t) + DA(t+1,t)\right)g(x) = -\mu_2 \left(x_1^2 + x_2^2\right).$$

Hence (6.3.5) is satisfied. By letting $\frac{9\mu_2}{8r\mu_1} \leq \lambda < \frac{3}{r(r+1)}$ we have that $|g(x)| \leq \lambda |x|$. For the sake of verifying (6.3.10), we note that

$$|A(t,s)| \leq \frac{1}{3}|r-t+s+1| \leq \frac{r}{3}, \text{ for } s \in [t-r, t-1].$$

Thus,

$$\sum_{s=t-r-1}^{t-1} |A(t,s)| \leq \sum_{s=t-r-1}^{t-1} \frac{r}{3} \leq \frac{r(r+1)}{3}.$$

Thus, $1 - \lambda \sum_{s=t-r-1}^{t-1} |A(t,s)| > 0$ for $\lambda < \frac{3}{r(r+1)}$. Left to verify conditions (6.3.8) and (6.3.9). As before, by simple calculations one can easily show that (6.3.8) and (6.3.9) correspond to

$$-\mu_1 - 2\mu_2 + \gamma r\lambda^2\frac{r}{3} + \frac{4\lambda r\mu_1}{9} + \frac{2}{3}\mu_1\left(\frac{r}{3}\right) \leq -\alpha, \quad (6.3.21)$$

and

$$-\gamma + \frac{4\lambda r\mu_1}{9} + \frac{2}{3}\mu_1 \leq 0, \quad (6.3.22)$$

respectively. Now inequalities (6.3.21) and (6.3.22) can be satisfied by the choice of appropriate μ_1, μ_2, and r. Thus we have shown that the zero solution of

$$x(t+1) = \begin{pmatrix} 1/2 & 0 \\ 0 & 1/2 \end{pmatrix} x(t) - \sum_{s=t-r}^{t-1} \begin{pmatrix} 1/3 & 0 \\ 0 & 1/3 \end{pmatrix} \begin{pmatrix} \frac{-9\mu_2}{8\mu_1 r} x_1 \\ \frac{-9\mu_2}{8\mu_1 r} x_2 \end{pmatrix}$$

is stable by invoking Theorem 6.3.1.

Next we consider the nonlinear Volterra difference equation

$$y(n+1) = f(y(n)) + \sum_{s=0}^{n} C(n,s)h(y(s)) + g(n) \tag{6.3.23}$$

where f and h are $k \times 1$ vector functions that are continuous in x and g is $k \times 1$ vector sequence. In addition C is $k \times k$ matrix functions on $\mathbb{Z}^+$ and $\mathbb{Z}^+ \times \mathbb{Z}^+$. Note that (6.3.23) has no delay. We are mainly interested in the Uniform boundedness on the solutions of (6.3.23) and its exponential stability when $g(n) = 0$ for all $n \in \mathbb{Z}^+$. We make the assumptions that for positive constants λ_1, λ_2, and M that

$$|f(y)| \leq \lambda_1 |y|, \ |h(y)| \leq \lambda_2 |y|, \ \text{and} \ |g| \leq M. \tag{6.3.24}$$

If $A = (a_{ij})$ is a $k \times k$ real matrix, then we define the norm of A by

$$|A| = \max_{1 \leq i \leq k} \left\{ \sum_{j=1}^{k} |a_{ij}| \right\}.$$

Similarly, for $x \in \mathbb{R}^k$, $|x|$ denotes the maximum norm of x. In the next theorem we construct a Lyapunov functional to obtain uniform boundedness and the exponential stability of the zero solution.

Theorem 6.3.4. *Assume* (6.3.24). *Also, we assume that*

$$\sum_{s=0}^{n-1} \sum_{j=n}^{\infty} |C(j,s)| < \infty, \tag{6.3.25}$$

$$\lambda_1 + \lambda_2 |C(n,n)| + K \sum_{j=n}^{\infty} |C(j,n)| \leq 1 - \alpha, \tag{6.3.26}$$

$$|C(n,s)| \geq \lambda \sum_{j=n}^{\infty} |C(j,s)| \ \text{where} \ \lambda = \frac{K\alpha}{\varepsilon}, \tag{6.3.27}$$

where ε, α, and K are positive constants with $\alpha \in (0,1)$ and $K = \varepsilon + \lambda_2$. Then every solution $y(n)$ of (6.3.23) is uniformly bounded and $\limsup_{x \to \infty} |y(n)| \leq \frac{M}{\alpha}$. Moreover, if $g(n) = 0$ for all $n \in \mathbb{Z}^+$, then the zero solution of (6.3.23) is exponentially stable.

Proof. Let's begin by defining the Lyapunov functional V by

$$V(n) = |y(n)| + K \sum_{s=0}^{n-1} \sum_{j=n}^{\infty} |C(j,s)||y(s)|. \tag{6.3.28}$$

Then using (6.3.28) we have along the solutions of (6.3.23) that

$$\triangle V(n) = |y(n+1)| - |y(n)| + K \left(\sum_{s=0}^{n} \sum_{j=n}^{\infty} |C(j,s)||y(s)| - \sum_{s=0}^{n-1} \sum_{j=n}^{\infty} |C(j,s)||y(s)| \right).$$

Or

$$\triangle V(n) = (|y(n+1)| - |y(n)|) + K \left(\sum_{j=n}^{\infty} |C(j,n)||y(n)| - \sum_{s=0}^{n-1} |C(n,s)||y(s)| \right).$$

Substitute $y(n+1)$ and use (6.3.26) to obtain

$$\triangle V(n) \le \left(|f(y(n))| + |g(n)| + \sum_{s=0}^{n} |C(n,s)||h(y(s))| - |y(n)| \right)$$

$$+ K \left(\sum_{j=n}^{\infty} |C(j,n)||y(n)| - \sum_{s=0}^{n-1} |C(n,s)||y(s)| \right).$$

Or

$$\triangle V(n) \le \left(\lambda_1 |(y(n))| + M + \lambda_2 \sum_{s=0}^{n} |C(n,s)||(y(s))| - |y(n)| \right)$$

$$+ K \left(\sum_{j=n}^{\infty} |C(j,n)||y(n)| - \sum_{s=0}^{n-1} |C(n,s)||y(s)| \right).$$

After simplification we arrive at

$$\triangle V(n) \le \left(\lambda_1 - 1 + \lambda_2 |C(n,n)| + K \sum_{j=n}^{\infty} |C(j,n)| \right) |y(n)|$$

$$+ M + \sum_{s=0}^{n-1} (\lambda_2 - K)|C(n,s)||y(s)|,$$

which reduces to

$$\triangle V(n) \le -\alpha |y(n)| + M + (\lambda_2 - K) \sum_{s=0}^{n-1} |C(n,s)||y(s)|.$$

Using (6.3.27) we get

$$\triangle V(n) \le -\alpha \left(|y(n)| - \frac{M}{\alpha} - \frac{\lambda(\lambda_2 - K)}{\alpha} \sum_{s=0}^{n-1} \sum_{j=n}^{\infty} |C(j,s)||y(s)| \right).$$

Now use the fact that $\lambda := \frac{\alpha K}{\varepsilon}$ and $\varepsilon := K - \lambda_2 \Rightarrow -\frac{\lambda(\lambda_2 - K)}{\alpha} = K$ to simplify to

$$\Delta V(n) \leq -\alpha \left(|y(n)| + K \sum_{s=0}^{n-1} \sum_{j=n}^{\infty} |C(j,s)||y(s)| \right) + M$$

$$= -\alpha V(n) + M.$$

Now, by applying the variations of parameters formula we get:

$$V(n) \leq (1-\alpha)^n V(0) + M \sum_{s=0}^{n-1} (1-\alpha)^{(n-s-1)},$$

which simplifies to

$$V(n) \leq (1-\alpha)^n V(0) + \frac{M}{\alpha}.$$

Using (6.3.28) we arrive at

$$|y(n)| \leq (1-\alpha)^n |y(0)| + \frac{M}{\alpha} \tag{6.3.29}$$

$$\leq |y(0)| + \frac{M}{\alpha}.$$

Hence we have uniform boundedness. If $g(n) - 0$ for all $n \in \mathbb{Z}^+$, then from (6.3.29) we have

$$|y(n)| \leq (1-\alpha)^n |y(0)|,$$

which implies the exponential stability. This completes the proof.
For the next theorem we consider the scalar Volterra difference equation

$$x(n+1) = \mu(n)x(n) + \sum_{s=0}^{n-1} h(n,s)x(s) + f(n), \tag{6.3.30}$$

and show, under suitable conditions, all its solutions are uniformly bounded and its zero solution is uniformly exponentially stable when $f(n)$ is identically zero. We assume the existence of an initial sequence $\phi : \mathbb{Z}^+ \to [0, \infty)$, that is bounded and $||\phi|| = \max_{0 \leq s \leq n_0} |\phi(s)|$, $n_0 \geq 0$.

Theorem 6.3.5 (Raffoul). *Suppose there is a scalar sequence $\alpha : \mathbb{Z}^+ \to [0, \infty)$. Assume there are positive constants $a > 1$ and b such that*

$$\alpha(s)a^{-b(n-s-1)} - \sum_{u=s}^{n-1} a^{-b(n-s-1)} |h(u,s)| > 0, \tag{6.3.31}$$

$$|\mu(n)| + |\alpha(n)| - |h(n,n)| - 1 \leq -(1 - a^{-b}), \tag{6.3.32}$$

and for some positive constant M

$$\sum_{s=0}^{n-1}(1-a^{-b})^{(n-s-1)}|f(s)| \le M, \ \ for \ 0 \le n < \infty.$$

(i) If

$$\max_{n \ge n_0} \sum_{s=0}^{n} \left(\alpha(s)a^{-b(n-s-1)} - \sum_{u=s}^{n} a^{-b(n-s-1)}|h(u,s)| \right) < \infty$$

then all solutions of (6.3.30) are uniformly bounded and its zero solution is uniformly exponentially stable when $f(n)$ is identically zero.

(ii) If for every $n_0 \ge 0$, there is a constant $M(n_0)$ depending on n_0 such that

$$\sum_{s=0}^{n_0-1} \alpha(s)a^{-b(n_0-s-1)} - \sum_{u=s}^{n_0-1} a^{-b(n_0-s-1)}|h(u,s)| < M(n_0),$$

then all solutions of (6.3.30) are bounded and its zero solution is exponentially stable when $f(n)$ is identically zero.

Proof. Consider the Lyapunov functional

$$V(n,x) = |x(n)|$$
$$+ \sum_{s=0}^{n-1} \left[\alpha(s)a^{-b(n-s-1)} - \sum_{u=s}^{n-1} a^{-b(n-u-1)}|h(u,s)| \right] |x(s)|. \tag{6.3.33}$$

Then along the solutions of (6.3.30) we have

$$\Delta V(n,x) \le |\mu(n)||x(n)| + \sum_{s=0}^{n-1} |h(n,s)||x(s)| + |f(n)|$$

$$+ \sum_{s=0}^{n} \left[\alpha(s)a^{-b(n-s)} - \sum_{u=s}^{n} a^{-b(n-u)}|h(u,s)| \right] |x(s)|$$

$$- \sum_{s=0}^{n-1} \left[\alpha(s)a^{-b(n-s-1)} - \sum_{u=s}^{n-1} a^{-b(n-u-1)}|h(u,s)| \right] |x(s)|.$$

Next we try to simplify $\Delta V(n,x)$.

$$\sum_{s=0}^{n} \left[\alpha(s)a^{-b(n-s)} - \sum_{u=s}^{n} a^{-b(n-u)}|h(u,s)| \right] |x(s)|$$

$$= \sum_{s=0}^{n} \left[\alpha(s)a^{-b(n-s)} - \sum_{u=s}^{n-1} a^{-b(n-u)}|h(u,s)| - |h(n,s)| \right] |x(s)|$$

$$= \sum_{s=0}^{n-1} \left[\alpha(s)a^{-b(n-s)} - \sum_{u=s}^{n-1} a^{-b(n-u)}|h(u,s)| - |h(n,s)| \right] |x(s)|$$

$$+ \alpha(n)|x(n)| - |h(n,n)||x(n)|$$

$$= a^{-b} \sum_{s=0}^{n-1} [\alpha(s)a^{-b(n-s-1)} - \sum_{u=s}^{n-1} a^{-b(n-u-1)}|h(u,s)|] |x(s)|$$

$$- \sum_{s=0}^{n-1} |h(n,s)||x(s)| + \alpha(n)|x(n)| - |h(n,n)||x(n)|.$$

Substituting the above expression into (6.3.34) and making use of (6.3.32) yield

$$\triangle V(n,x) \le [|\mu(n)| + |\alpha(n)| - |h(n,n)| - 1]|x(n)|$$

$$- (1 - a^{-b}) \sum_{s=0}^{n-1} [\alpha(s)a^{-b(n-s-1)} - \sum_{u=s}^{n-1} a^{-b(n-u-1)}|h(u,s)|] |x(s)| + |f(n)|$$

$$\le -(1 - a^{-b}) [|x(n)|$$

$$+ \sum_{s=0}^{n-1} [\alpha(s)a^{-b(n-s)} - \sum_{u=s}^{n-1} a^{-b(n-u)}|h(u,s)|] |x(s)| + |f(n)|$$

$$= -(1 - a^{-b})V(n,x) + |f(n)|. \tag{6.3.34}$$

Set $\beta = (1 - a^{-b}) \in (0,1)$ and apply the variation of parameters formula to get

$$V(n,x(n)) \le (1 - \beta)^{n-n_0} V(n_0, \phi) + \sum_{s=n_0}^{n-1} (1 - \alpha)^{(n-s-1)}|f(s)|$$

$$\le (1 - \beta)^{n-n_0} ||\phi|| [1 +$$

$$+ \sum_{s=0}^{n_0-1} [\alpha(s)a^{-b(n_0-s-1)} - \sum_{u=s}^{n_0-1} a^{-b(n_0-u-1)}|h(u,s)|]$$

$$+ \sum_{s=n_0}^{n-1} (1 - \alpha)^{(n-s-1)}|f(s)|. \tag{6.3.35}$$

The results readily follow from (6.3.35) and the fact that $|x(n)| \le V(n,x)$. This completes the proof.

6.4 z-Transform and Lyapunov Functionals

Next we use combination of Lyapunov functionals and z-transform to obtain boundedness and stability of Equation (6.3.23).

We assume the existence of a sequence $\varphi(n)$ such that

$$\varphi(n) \ge 0, \ \triangle \varphi(n) \le 0 \text{ and } \sum_{n=0}^{\infty} \varphi(n) < \infty. \tag{6.4.1}$$

Lemma 6.5. *Assume* (6.4.1) *and if*

$$H(n) = \beta(n) + \lambda \sum_{s=0}^{n-1} \varphi(n-s-1)|\beta(s)|, \tag{6.4.2}$$

$$\triangle H(n) = -\alpha\beta(n) \quad \beta(0) = 1, \tag{6.4.3}$$

where $\beta(n)$ *and* $H(n)$ *are scalar sequences, then*

$$\beta(n) + \sum_{s=0}^{n-1} \{\alpha + \lambda\varphi(n-s-1)\}\beta(s) = 1 \text{ for all } n = 1, 2, 3, \ldots, \tag{6.4.4}$$

$$\beta(n) > 0 \quad \text{for all } n = 1, 2, 3, \ldots, \tag{6.4.5}$$

$$\sum_{n=0}^{\infty} \beta(n) < \infty, \tag{6.4.6}$$

and

$$\tilde{\beta}(z) = \left[1 + \frac{\alpha}{z-1} + \lambda\frac{\tilde{\varphi}}{z} \right]^{-1} \left(\frac{z}{z-1} \right), \tag{6.4.7}$$

where $\tilde{\beta}(z)$, $\tilde{\varphi}(z)$ *are Z-transforms of* β *and* φ.

Proof. By (6.4.3) we obtain

$$H(n) = H(0) - \alpha \sum_{s=0}^{n-1} \beta(s),$$

and hence

$$H(n) = H(0) - \alpha \sum_{s=0}^{n-1} \beta(s)$$

$$= \beta(n) + \sum_{s=0}^{n-1} \{\alpha + \lambda\varphi(n-s-1)\}\beta(s).$$

Since $\beta(0) = H(0)$, we have (6.4.4). The proof of (6.4.5) follows by an induction argument on (6.4.4) and by noting that the summation term is positive and $\beta(0) = 1$. For the proof of (6.4.6) we sum (6.4.3) from $s = 0$ to $s = n-1$ and get

$$\alpha \sum_{s=0}^{n-1} \beta(s) = H(0) - H(n).$$

Since $\beta(n) > 0 \ \forall n \geq 0$, we have that H is monotonically decreasing. Therefore

$$0 < \sum_{s=0}^{n-1} \beta(s) < \frac{H(0)}{\alpha} \qquad \text{for every } n,$$

which proves (6.4.6). Left to prove (6.4.7). The z-transforms of φ and β exist for some $|z| > d$, where $d > 0$. Therefore, replacing n by $n+1$ in equation (6.4.4) gives

$$\beta(n+1) + \sum_{s=0}^{n} \{\alpha + \lambda \varphi(n-s)\} \beta(s) = 1.$$

Taking the z-transform of both sides and using the fact that $\beta(0) = 1$ give

$$z\tilde{\beta}(z) - z\beta(0) + \alpha \frac{z}{z-1} \tilde{\beta}(z) + \lambda \tilde{\varphi}(z) \tilde{\beta}(z) = \frac{z}{z-1},$$

or

$$\left\{ z + \alpha \frac{z}{z-1} + \lambda \tilde{\varphi}(z) \right\} \tilde{\beta}(z) = \frac{z^2}{z-1}.$$

Since $z > 0$ we can divide through by z and get,

$$\left\{ 1 + \alpha \frac{1}{z-1} + \frac{\lambda \tilde{\varphi}(z)}{z} \right\} \tilde{\beta}(z) = \frac{z}{z-1}.$$

Finally solving for $\tilde{\beta}(z)$ gives

$$\tilde{\beta}(z) = \left\{ 1 + \alpha \frac{1}{z-1} + \frac{\lambda \tilde{\varphi}(z)}{z} \right\}^{-1} \left(\frac{z}{z-1} \right),$$

which proves (6.4.7).

Theorem 6.4.1. *Assume the hypothesis of Lemma 6.5. Assume there is $\lambda > 0$ such that*

$$\lambda \triangle \varphi(n-s-1) + \lambda_2 |C(n,s)| \leq 0 \qquad \text{for } 0 \leq s < n \text{ for } n \in \mathbb{Z}^+, \qquad (6.4.8)$$

and

$$\lambda_1 + \lambda_2 |C(n,n)| + \lambda \varphi(0) \leq 1 - \alpha, \qquad (6.4.9)$$

where $\alpha \in (0,1)$, then for every solution $y(n)$ of (6.3.23), $|y(n)|$ is uniformly bounded and

$$\limsup_{n \to \infty} |y(n)| \leq \frac{M}{\alpha}.$$

Proof. Define the Lyapunov functional V by

$$V(n) \equiv |y(n)| + \lambda \sum_{s=0}^{n-1} \varphi(n-s-1) |y(s)|, \qquad n \geq 0. \qquad (6.4.10)$$

Then, using (6.4.10) we have along the solutions of (6.3.23) that

$$\triangle V(n) = |y(n+1)| + \lambda \sum_{s=0}^{n} \varphi(n-s) |y(s)| - |y(n)| - \lambda \sum_{s=0}^{n-1} \varphi(n-s-1) |y(s)|,$$

which simplifies to

$$\triangle V(n) = \left\{ |y(n+1)| - |y(n)| \right\}$$
$$+ \lambda \left\{ \sum_{s=0}^{n} \varphi(n-s) \, \| \, y(s) \, \| \, - \sum_{s=0}^{n-1} \varphi(n-s-1) \, \| \, y(s) \, \| \, \right\}$$
$$= \left\{ |y(n+1)| - |y(n)| \right\}$$
$$+ \lambda \left\{ \sum_{s=0}^{n-1} \triangle_n \varphi(n-s-1)|y(s)| + \varphi(0)|y(n)| \right\}.$$

Along the solutions of (6.3.23) we have

$$\triangle V(n) \le \left\{ |f(y(n))| + |g(n)| + \sum_{s=0}^{n} |C(n,s)||h(y(s))| - |y(n)| \right\}$$
$$+ \lambda \left\{ \sum_{s=0}^{n-1} \triangle_n \varphi(n-s-1)|y(s)| + \varphi(0)|y(n)| \right\}$$
$$\le \left\{ \lambda_1 |y(n)| + M + \lambda_2 \sum_{s=0}^{n} |C(n,s)||y(s)| - |y(n)| \right\}$$
$$+ \lambda \left\{ \sum_{s=0}^{n-1} \triangle_n \varphi(n-s-1)|y(s)| + \varphi(0)|y(n)| \right\}.$$

After some algebra, we arrive at the simplified expression,

$$\triangle V(n) \le \left\{ [\lambda_1 + \lambda_2 |C(n,n)| - 1 + \lambda \varphi(0)] \right\} |y(n)| + M$$
$$+ \left\{ \sum_{s=0}^{n-1} [\lambda_2 |C(n,s)| + \lambda \triangle_n \varphi(n-s-1)]|y(s)| \right\}.$$

Using (6.4.8) and (6.4.9) we arrive at

$$\triangle V(n) \le -\alpha |y(n)| + M, \qquad M > 0 \tag{6.4.11}$$

Due to (6.4.11), there is a nonnegative sequence $\eta(n) : \mathbb{Z}^+ \to \mathbb{R}$ such that

$$\triangle V(n) = -\alpha |y(n)| + M - \eta(n).$$

Since η is a linear combination of functions of exponential order, η is also of exponential order and so we can take Z transform and have

$$z\tilde{V}(z) - zV(0) - \tilde{V}(z) = -\alpha |\tilde{y}(z)| + M\frac{z}{z-1} - \tilde{\eta}(z).$$

We solve for $\tilde{V}$ and get

$$\tilde{V}(z) = \frac{z}{z-1}V(0) - \frac{\alpha}{z-1}|\tilde{y}(z)| + M\frac{z}{(z-1)^2} - \frac{\tilde{\eta}(z)}{z-1}.$$

To derive the second expression for $\tilde{V}$, use the fact that

$$Z\left[\sum_{s=0}^{n-1}x(n-s-1)g(s)\right] = \frac{1}{z}\tilde{x}(z)\tilde{g}(z).$$

Taking the Z-transform in (6.4.10) we arrive at

$$\tilde{V}(z) = \left\{1 + \lambda\frac{\tilde{\varphi}(z)}{z}\right\}|\tilde{y}(z)|.$$

Substituting it into

$$\tilde{V}(z) = \frac{z}{z-1}V(0) - \frac{\alpha}{z-1}|\tilde{y}(z)| + M\frac{z}{(z-1)^2} - \frac{\tilde{\eta}(z)}{z-1}$$

gives

$$\frac{z}{z-1}V(0) - \frac{\alpha}{z-1}|\tilde{y}(z)| + M\frac{z}{(z-1)^2} - \frac{\tilde{\eta}(z)}{z-1} = |\tilde{y}(z)| + \lambda\frac{\tilde{\varphi}(z)|\tilde{y}(z)|}{z}.$$

Solving for $|\tilde{y}|$ gives

$$\begin{aligned}
|\tilde{y}(z)| &= \left[M\frac{1}{(z-1)} + V(0) - \frac{\tilde{\eta}(z)}{z}\right]\left\{\frac{\alpha}{z-1} + 1 + \lambda\frac{\tilde{\varphi}(z)}{z}\right\}^{-1}\left(\frac{z}{z-1}\right) \\
&= \left[M\frac{1}{(z-1)} + V(0) - \frac{\tilde{\eta}(z)}{z}\right]\tilde{\beta}(z) \\
&= \left[M\frac{1}{(z-1)}\tilde{\beta}(z) + V(0)\tilde{\beta}(z) - \frac{\tilde{\eta}(z)}{z}\tilde{\beta}(z)\right].
\end{aligned} \tag{6.4.12}$$

Taking the z inverse in (6.4.12) gives

$$\begin{aligned}
|y(n)| &= V(0)\beta(n) - M\beta(n) + M\sum_{s=0}^{n}\beta(s) - \sum_{s=0}^{n-1}\eta(n-s-1)\beta(s) \\
&= V(0)\beta(n) + M\sum_{s=0}^{n-1}\beta(s) - \sum_{s=0}^{n-1}\eta(n-s-1)\beta(s) \\
&\leq V(0)\beta(n) + M\sum_{s=0}^{n-1}\beta(s).
\end{aligned}$$

Since $\beta(n)$ is bounded, there exists a positive constant R such that $\beta(n) \leq R$ for all $n \geq 0$. Hence, the above inequality gives

$$|y(n)| \leq V(0)R + \frac{M}{\alpha}.$$

This shows that all solutions $y(n)$ of (6.3.23) are uniformly bounded.
Note that since $\sum_{n=0}^{\infty} \beta(n) < \infty$, we have that $\beta(n) \to 0$, as $n \to \infty$ and hence

$$\limsup_{n \to \infty} |y(n)| \leq M \sum_{n=0}^{\infty} \beta(n) \leq \frac{M}{\alpha}.$$

This completes the proof.

We end the section with the following example.

Example 6.9. Let

$$f(y(n)) := \left(\frac{1}{16\sqrt{2}} \right) \begin{bmatrix} \frac{|y_1(n)|}{1+|y_1(n)|} \\ \frac{|y_2(n)|}{1+|y_2(n)|} \end{bmatrix}$$

$$h(y(n)) := \begin{pmatrix} y_1(n) \\ 0 \end{pmatrix}$$

$$C(n,s) := \frac{1}{2^{(n+4-s)}} \begin{bmatrix} 1 & 0 \\ 0 & \frac{1}{2} \end{bmatrix}$$

and

$$g(n) := \cos\left(\frac{n\pi}{4} \right) \begin{bmatrix} 1 \\ 0 \end{bmatrix}.$$

Then we have, $|f(y(n))| \leq \frac{1}{16}|y(n)|$, $|h(y(n))| \leq |y(n)|, |C(n,s)| \leq \frac{1}{2^{(n+4-s)}}$ and $|g(n)| \leq 1$. Let

$$\text{Let } \phi(n) = \frac{1}{2^{(n+3)}}$$

and define the Lyapunov functional V by

$$V(n) = |y(n)| + \lambda \sum_{s=0}^{n-1} \phi(n-s-1)|y(s)|.$$

Then we have $\lambda = \frac{1}{2}$, $\lambda_1 = \frac{1}{16}$, and $\lambda_2 = 1$. Then all conditions of Theorem 6.4.1 are satisfied since $\triangle\phi(n) \leq 0$, $\lambda\triangle\phi(n-s-1) + \lambda_2|C(n,s)| \leq 0$. Moreover, condition (6.4.9) is satisfied for $\alpha = \frac{13}{16}$. Thus, by Theorem 6.4.1 all solutions are uniformly bounded and satisfy

$$\limsup_{x \to \infty} |y(n)| \leq \frac{M}{\alpha} = \frac{16}{13}.$$

6.5 l_p-Stability

In this section we state the definition of l_p-stability and prove theorems under which it occurs. We begin by considering the nonautonomous nonlinear discrete system

$$x(n+1) = G(n,x(s); \ 0 \le s \le n) \overset{def}{=} G(n,x(\cdot)) \qquad (6.5.1)$$

where $G : \mathbb{Z}^+ \times \mathbb{R}^k \to \mathbb{R}^k$ is continuous in x and $G(n,0) = 0$. Let $C(n)$ denote the set of functions $\phi : [0,n] \to \mathbb{R}$ and $\|\phi\| = \sup\{|\phi(s)| : 0 \le s \le n\}$.
We say that $x(n) = x(n,n_0,\phi)$ is a solution of (6.5.1) with a bounded initial function $\phi : [0,n_0] \to \mathbb{R}^k$ if it satisfies (6.5.1) for $n > n_0$ and $x(j) = \phi(j)$ for $j \le n_0$.

Definition 6.5.1. The zero solution of (6.5.1) is stable (S) if for each $\varepsilon > 0$, there is a $\delta = \delta(n_0,\varepsilon) > 0$ such that $[n_0 \ge 0, \phi \in C(n_0), \|\phi\| < \delta]$ imply $|x(n,n_0,\phi)| < \varepsilon$ for all $n \ge n_0$. It is uniformly stable (US) if it is stable and δ is independent of n_0. It is asymptotically stable (AS) if it is (S) and $|x(n,n_0,\phi)| \to 0$, as $n \to \infty$.

Definition 6.5.2. The zero solution of system (6.5.1) is said to be l_p-stable if it is stable and if $\displaystyle\sum_{n=n_0}^{\infty} \|x(n,n_0,\phi)\|^p < \infty$ for positive p.

We have the following elementary theorem.

Theorem 6.5.1. *If the zero solution of* (6.5.1) *is exponentially stable, then it is also l_p-stable.*

Proof. Since the zero solution of (6.5.1) is exponentially stable, we have by the above definition that

$$\sum_{n=n_0}^{\infty} \|x(n,n_0,\phi)\| \le [C(\|\phi\|,n_0)]^p \sum_{n=n_0}^{\infty} a^{p\eta(n-n_0)}$$

$$= [C(\|\phi\|,n_0)]^p a^{-n_0 p\eta} \sum_{n=n_0}^{\infty} a^{p\eta n}$$

$$= [C(\|\phi\|,n_0)]^p / (1 - a^{p\eta}),$$

which is finite. This completes the proof.

We caution that l_p-stability is not uniform with respect to p, as the next example shows. Also, it shows that (AS) does not imply l_p-stability for all p. In Chapter 1, we considered the difference equation

$$x(n+1) = \frac{n}{n+1}x(n), \ x(n_0) = x_0 \ne 0, \ n_0 \ge 1$$

and showed its solution is given by

$$x(n) := x(n, n_0, x_0) = \frac{x_0 n_0}{n}.$$

Clearly the zero solution is (US) and (AS). However, for $n_0 = n$, we have

$$x(2n, n, x_0) = \frac{x_0 n}{2n} \to \frac{x_0}{2} \neq 0$$

which implies that the zero solution is not (UAS). Moreover,

$$\sum_{n=n_0}^{\infty} ||x(n, n_0, x_0)||^p \leq \sum_{n=n_0}^{\infty} |(\frac{x_0 n_0}{n})|^p = |x_0|^p (n_0)^p \sum_{n=n_0}^{\infty} (\frac{1}{n})^p,$$

which diverges for $0 < p \leq 1$ and converges for $p > 1$.

The next example shows that asymptotic stability does not necessarily imply l_p-stability for any $p > 0$. Let $g : [0, \infty) \to (0, \infty)$ with $\lim_{n \to \infty} g(n) = \infty$. Consider the nonautonomous difference equation

$$x(n+1) = [g(n)/g(n+1)]x(n), \ x(n_0) = x_0, \tag{6.5.2}$$

which has the solution $x(n, n_0, x_0) = \frac{g(n_0)}{g(n)} x_0$. It is obvious that as $n \to \infty$ the solution tends to zero, for fixed initial n_0 and the zero solution is indeed asymptotically stable. On the other hand

$$\sum_{n=n_0}^{\infty} ||x(n, n_0, x_0)||^p = [g(n_0)x_0]^p \sum_{n=n_0}^{\infty} \left(\frac{1}{g(n)} \right)^p, \tag{6.5.3}$$

which may not converge for any $p > 0$. For example, if we take

$$g(n) = log(n+2),$$

then from (6.5.3) we have

$$\sum_{n=n_0}^{\infty} ||x(n, n_0, x_0)||^p = [log(n_0 + 2)]^p ||x_0||^p \sum_{n=n_0}^{\infty} \left(\frac{1}{log(n+2)} \right)^p,$$

which is known to diverge for all $p \geq 0$.

The next theorem relates l_p-stability to Lyapunov functionals.

Theorem 6.5.2. *If there exists a positive definite V (see Definition 1.2.1) and along the solutions of (6.5.1), V satisfies $\triangle V \leq -c||x||^p$, for some positive constants c and p, then the zero solution of (6.5.1) is l_p-stable.*

Proof. Set the solution $x(n) := x(n, n_0, \phi)$. The hypothesis of the theorem implies the zero solution is stable. Thus, for $n \geq n_0$ there is a positive constant M such that $||x(n, n_0, \phi)|| \leq M$. For $n \geq n_0$ we set

$$L(n) = V(n,x(n)) + c \sum_{s=n_0}^{n-1} ||x(s)||^p.$$

Then for all $n \geq n_0$ we have

$$\triangle L(n) = \triangle V(n,x) + ||x||^p$$
$$\leq -c||x||^p + c||x||^p = 0.$$

Therefore, $L(n)$ is decreasing and hence $0 \leq L(n) \leq L(n_0) = V(n_0, \phi)$, $n \geq n_0$. This implies that $0 \leq L(n) = V(n,x) + c \sum_{s=n_0}^{n-1} ||x(s)||^p \leq V(n_0, \phi)$, $n \geq n_0$ so that

$$0 \leq V(n,x) \leq -c \sum_{s=n_0}^{n-1} ||x(s)||^p + V(n_0, \phi).$$

As a consequence,

$$\sum_{s=n_0}^{n-1} ||x(s,n_0,\phi)||^p \leq V(n_0,\phi)/c, \ n \geq n_0.$$

Letting $n \to \infty$ on both sides of the above inequality gives

$$\sum_{n=n_0}^{\infty} ||x(n,n_0,\phi)||^p \leq V(n_0,\phi)/c < \infty.$$

This completes the proof.

In the next two examples we show that the l_p-stability depends on the type of Lyapunov functional that is being used. Moreover, there will be a price to pay if you want to obtain l_p-stability for higher values of p.

Example 6.10. Consider the scalar Volterra difference equation

$$x(n+1) = a(n)x(n) + \sum_{s=0}^{n-1} b(n,s)f(s,x(s)) \tag{6.5.4}$$

with f being continuous and there exists a constant λ_1 such that $f(n,x)| \leq \lambda_1|x|$. Assume there exists a positive α such that

$$|a(n)| + \lambda \sum_{s=n+1}^{\infty} |b(s,n)| + \lambda_1|b(n,n)| - 1 \leq -\alpha, \tag{6.5.5}$$

and for some positive constant λ which is to be specified later, we have

$$\lambda_1 \leq \lambda, \tag{6.5.6}$$

then the zero solution of (6.5.4) is l_1-stable.

Proof. Define the Lyapunov functional V by

$$V(n,x) = |x(n)| + \lambda \sum_{j=0}^{n-1} \sum_{s=n}^{\infty} |b(s,j)||x(j)|.$$

We have along the solutions of (6.5.4) that

$$\triangle V(t) \le \left(|a(n)| + \lambda \sum_{s=n+1}^{\infty} |b(s,n)| + \lambda_1|b(n,n)| - 1\right)|x(n)|$$

$$+ (\lambda_1 - \lambda) \sum_{s=0}^{n-1} |b(n,s)||x(s)|$$

$$\le -\alpha|x(n)|.$$

This implies the zero solution is stable and l_1-stable by Theorem 6.5.2. This completes the proof.

Example 6.11. Consider (6.5.4) and assume f is continuous with $|f(n,x)| \le \lambda_1 x^2$. Assume there exists a positive constant α such that

$$a^2(n) + \lambda \sum_{s=n+1}^{\infty} |b(s,n)| + \lambda_1|a(n)| \sum_{s=0}^{n} |b(n,s)| - 1 \le -\alpha, \qquad (6.5.7)$$

and for some positive constant λ which is to be specified later, we have

$$\lambda_1|a(n)| + \lambda_1^2 \sum_{s=0}^{n-1} |b(n,s)| - \lambda \le 0. \qquad (6.5.8)$$

Then the zero solution of (6.5.4) is l_2-stable.

Proof. define the Lyapunov functional V by

$$V(n,x) = x^2(n) + \lambda \sum_{j=0}^{n-1} \sum_{s=n}^{\infty} |b(s,j)|x^2(j).$$

We have along the solutions of (6.5.4) that

$$\triangle V(t) = \left(a(n)x(n) + \sum_{s=0}^{n-1} b(n,s)f(s,x(s))\right)^2 - x^2(n)$$

$$+ \lambda x^2(n) \sum_{s=n+1}^{\infty} |b(s,n)| - \lambda \sum_{s=0}^{n-1} |b(n,s)|x^2(s) - x^2(n)$$

$$\le a^2(n)x^2(n) + 2\lambda_1|a(n)||x(n)| \sum_{s=0}^{n-1} |b(n,s)||x(s)| + \left(\sum_{s=0}^{n-1} b(n,s)f(s,x(s))\right)^2$$

$$+ \lambda x^2(n) \sum_{s=n+1}^{\infty} |b(s,n)| - \lambda \sum_{s=0}^{n-1} |b(n,s)|x^2(s) - x^2(n).$$

As a consequence of $2zw \leq z^2 + w^2$, for any real numbers z and w we have

$$2\lambda_1 |a(n)||x(n)| \sum_{s=0}^{n-1} |b(n,s)||x(s)| \leq \lambda_1 |a(n)| \sum_{s=0}^{n-1} |b(n,s)|(x^2(n) + x^2(s)).$$

Also, using Schwartz inequality we obtain

$$\left(\sum_{s=0}^{n-1} b(n,s)f(s,x(s)) \right)^2 = \sum_{s=0}^{n-1} |b(n,s)|^{1/2} |b(n,s)|^{1/2} |f(s,x(s))|$$

$$\leq \sum_{s=0}^{n-1} |b(n,s)| \sum_{s=0}^{n-1} |b(n,s)||f^2(s,x(s))$$

$$\leq \lambda_1^2 \sum_{s=0}^{n-1} |b(n,s)| \sum_{s=0}^{n-1} |b(n,s)||x^2(s).$$

Putting all together, we get

$$\triangle V(t) \leq \left(a^2(n) + \lambda \sum_{s=n+1}^{\infty} |b(s,n)| + \lambda_1 |a(n)| \sum_{s=0}^{n} |b(n,s)| - 1 \right) x^2(n)$$

$$+ \left(\lambda_1 |a(n)| + \lambda_1^2 \sum_{s=0}^{n-1} |b(n,s)| - \lambda \right) \sum_{s=0}^{n-1} |b(n,s)|x^2$$

$$\leq -\alpha x^2(n).$$

This implies the zero solution is stable and l_2-stable by Theorem 6.5.2. This completes the proof.

A quick comparison of (6.5.5) with (6.5.7) and (6.5.6) with (6.5.8) reveals that the conditions for the l_2-stability are more stringent than of the conditions for l_1-stability.

6.6 Discretization Scheme Preserving Stability and Boundedness

In Chapter 1, we briefly discussed the notion that Volterra discrete equations play major role in numerical solutions of Volterra integro-differential equations. In this section we apply a nonstandard discretization scheme due to Mickens (see [119]) to a Volterra integro-differential equation, to form a Volterra discrete system. By displaying suitable Lyapunov functionals, one for the Volterra integro-differential equation and another for the Volterra discrete system, we will show that under the same conditions on some of the coefficients, the stability of the zero solution and boundedness of solutions are preserved in both systems.

This section is intended to give a brief introduction to the subject of discretization, although by no means, should it be considered a complete study of the subject. The author is not claiming that the discretization scheme used here is the most general

nor it is the most efficient. The sole purpose of this section is to introduce the reader
to the effectiveness of Lyapunov functionals when dealing with preserving the qualitative behaviors of solutions. However, this section should set the stage for future
research in preserving the characteristics of Volterra integro-differential equations
when nonstandard discretization schemes are used in obtaining the corresponding
Volterra discrete systems. For comprehensive treatment of the subject of nonstandard discretization we refer to [119] and [120].

For motivational purpose, consider the differential equation

$$x'(t) = ax(t), \text{ for some constant } a < 0, \tag{6.6.1}$$

which has the solution $x(t) = e^{a(t-t_0)}$ and $x(t) \to 0$ as $t \to \infty$.

On the other hand, if we consider the difference equation

$$x(t+1) = ax(t), \ x(t_0) = x_0, \tag{6.6.2}$$

then the unique solution of (6.6.2) is

$$x(t) = x_0 a^{t-t_0}$$

and

$$x(t) \to 0 \text{ as } t \to \infty$$

provided that $|a| < 1$. We see that the stability is not preserved. Applying the ap-
proximations

$$x'(t) = \frac{x(t+h) - x(t)}{h}, \ x(t) = \frac{x(t+h) + x(t)}{2} \tag{6.6.3}$$

to equation (6.6.1) we have the analogous discrete system

$$x(n+1) = \frac{2+ah}{2-ah}x(n), \tag{6.6.4}$$

where $x(n+1) = x(t+h)$ and $x(n) = x(t)$. All solutions $x(n)$ of (6.6.4) satisfy
$x(n) \to 0$ as $n \to \infty$, provided that

$$\left|\frac{2+ah}{2-ah}\right| < 1. \tag{6.6.5}$$

Clearly, inequality (6.6.5) is satisfied for $a < 0$ and $0 < h < 1$. Thus, we see that the
discretization scheme defined by (6.6.3) preserved the stability of the zero solution.
It is noted that the result holds under the restriction that the step-size $\triangle t = h$ satisfies
the restriction

$$0 < h < 1. \tag{6.6.6}$$

Restriction (6.6.6) is a direct consequence of how we discretize equation (6.6.1). To
ease the restriction given by (6.6.6), we use a nonstandard discretization scheme due
to Mickens [122]; that is we let

$$x'(t) = \frac{x(t+h) - x(t)}{\Phi(a,h)}, \quad \Phi(a,h) = \frac{e^{ah} - 1}{a}. \tag{6.6.7}$$

We note that this scheme holds for all $h > 0$. For more on the use of nonstandard discretization, we refer the reader to [119, 120, 121, 122]. Under discretization (6.6.7), equation (6.6.1) becomes

$$x(n+1) = (1 + a\Phi(a,h))x(n) = e^{ah}x(n). \tag{6.6.8}$$

Since $a < 0$, we have that $e^{ah} < 1$, and hence all solutions of (6.6.8) go to zero asymptotically without any restriction on the step-size h. Thus, we see that the discretization scheme defined by (6.6.7) preserved the stability of the zero solution.

Definition 6.6.1. A resulting difference equation is said to be *consistent with respect to property P* under a given discretization scheme with its continuous counterpart if they both exhibit property P under equivalent conditions.

Based on Definition 6.6.1, we see that (6.6.5) is consistent with respect to asymptotic stability with (6.6.1) under discretization (6.6.3) provided that (6.6.6) holds. The same is true for (6.6.7) but without further restriction on the size h.

Next we discuss the stability, uniform asymptotic stability, and exponential stability of Volterra integro-differential equations and their corresponding discrete systems with respect to certain discretization schemes. Consider the scalar Volterra integro-differential equation

$$x'(t) = ax(t) + \int_0^t B(t,s) f(s,x(s))ds, \ t \geq 0. \tag{6.6.9}$$

We assume $f(t,x)$ is continuous in x and t and satisfy

$$|f(t,x)| \leq \gamma|x|, \tag{6.6.10}$$

where γ is a positive constant. The kernel $B : \mathbb{R}^2 \to \mathbb{R}$ is continuous in both arguments. By considering the discretization scheme (6.6.3) for

$$x'(t) = ax(t)$$

and by approximating the integral term with

$$\int_0^t B(t,s)f(s,x(s)) \ ds = h \sum_{s=0}^t B(t,s)f(s,x(s)), \tag{6.6.11}$$

we arrive at the corresponding discrete Volterra equation,

$$x(n+1) = \frac{2+ah}{2-ah}x(n) + \frac{2h^2}{2-ah} \sum_{s=0}^n B(n,s)f(s,x(s)), \ n \geq 0, \tag{6.6.12}$$

where $x(n+1) = x(t+h)$, $x(n) = x(t)$ and $0 < h < 1$. Similarly by considering discretizations (6.6.7) and (6.6.11) we arrive at the corresponding discrete Volterra equation,

$$x(n+1) = e^{ah}x(n) + h\Phi(a,h)\sum_{s=0}^{n}B(n,s)f(s,x(s)), \quad n \ge 0, \qquad (6.6.13)$$

The study of Volterra discrete systems is important since they play a major role in the fields of numerical analysis, control theory, and computer science. Thus, finding a discretization scheme under which Equation (6.6.12) is consistent with (6.6.9) is important. Throughout this section it is assumed that the step size h satisfies $0 < h < 1$. In preparation for the next theorem we make the following assumptions.

$$|B(t,s)| \text{ is monotonically decreasing in } t \qquad (6.6.14)$$

and there exists a constant $\alpha > 0$ such that $\forall t \ge 0$

$$a + \gamma\int_{t}^{\infty}|B(u,t)|du \le -\alpha. \qquad (6.6.15)$$

Theorem 6.6.1. *Assume conditions* (6.6.14) *and* (6.6.15) *hold. Then* (6.6.13) *is consistent with respect to uniform asymptotic stability under the discretization scheme* (6.7.7) *and* (6.6.11) *with its continuous counterpart* (6.6.9).

Proof. Define the Lyapunov functional V by

$$V(t) = |x(t)| + \gamma\int_{0}^{t}\int_{t}^{\infty}|B(u,s)||x(s)|du\,ds.$$

Then by making use of (6.6.15), we have along the solutions (6.6.9) that

$$\begin{aligned}
V'(t) &= \frac{x(t)}{|x(t)|}x'(t) + \gamma\int_{t}^{\infty}|B(u,t)||x(t)|du - \gamma\int_{0}^{t}|B(t,s)||x(s)|ds \\
&\le a|x(t)| + \gamma\int_{0}^{t}|B(t,s)||x(s)|ds \\
&\quad + \gamma\int_{t}^{\infty}|B(u,t)||x(t)|du - \gamma\int_{0}^{t}|B(t,s)||x(s)|ds \\
&\le \left[a + \gamma\int_{t}^{\infty}|B(u,t)|du\right]|x(t)| \\
&\le -\alpha|x(t)|.
\end{aligned}$$

Then by Theorem 2.6.1 of [22], the zero solution of (6.6.9) is (UAS). Now we turn our attention to (6.6.12). Define V by

$$V(n) = |x(n)| + \gamma h\Phi(a,h)\sum_{s=0}^{n-1}\sum_{u=n}^{\infty}|B(u,s)||x(s)|.$$

It can be easily shown that along the solutions of (6.6.13)

$$\Delta V(n) \leq \left[e^{ah} + \gamma h \Phi(a,h) \sum_{u=n}^{\infty} |B(u,n)| - 1 \right] |x(n)|.$$

Due to condition (6.6.15) there exists a positive constant β such that

$$\gamma \int_t^{\infty} |B(u,t)| du \leq \beta.$$

We can choose h small enough so that the above inequality combined with (6.6.14) and the fact that $a < 0$ to imply that there exists a positive constant η such that

$$e^{ah} + \gamma h \Phi(a,h) \sum_{u=n}^{\infty} |B(u,n)| - 1 \leq -\eta.$$

Therefore,

$$\Delta V(n) \leq -\eta |x(n)|.$$

By setting $\alpha = 0$ in Theorem 2.2.4 we have the zero solution of (6.6.13) is (UAS). The proof is complete.

In Theorem 6.6.1 we showed that the discretization scheme given by (6.6.3) and (6.6.11) preserved the uniform asymptotic stability of the zero solutions of Equations (6.6.9) and (6.6.12). In the next theorem we will show that the discretization scheme given by (6.6.3) and (6.6.11) preserves the exponential asymptotic stability of the zero solutions of Equations (6.6.9) and (6.6.12) under more stringent conditions on the kernel $B(t,s)$. For the next theorem we make the following assumptions.

$$|B(t,s)| \text{ is monotonically decreasing in } t \text{ and } s. \tag{6.6.16}$$

Suppose there exist constants $k > 1$ and $\alpha > 0$ such that

$$a + \gamma k \int_t^{\infty} |B(u,t)| du \leq -\alpha < 0 \tag{6.6.17}$$

where $k = 1 + \varepsilon$ for some $\varepsilon > 0$. Suppose

$$|B(t,s)| \geq \lambda \int_t^{\infty} |B(u,s)| du \tag{6.6.18}$$

where $\lambda \geq \frac{k\alpha}{\varepsilon} > 0, 0 \leq s < t \leq u < \infty$, and

$$\gamma \int_0^{t_0} \int_{t_0}^{\infty} |B(u,s)| du \, ds \leq \rho < \infty \quad \text{for all } t_0 \geq 0. \tag{6.6.19}$$

Remark 6.4. Due to conditions (6.6.16) and (6.6.17) there exists a positive constant β such that

$$\gamma k \int_t^\infty |B(u,t)|\, du \le \beta.$$

Similarly, by conditions (6.6.16) and (6.6.19) there is a constant such that

$$h\Phi(a,h)\gamma k \sum_{s=0}^{n_0-1} \sum_{j=n_0}^\infty |B(j,s)| \le \Gamma_1.$$

Finally, as a consequence of (6.6.16) and (6.6.18) we have

$$|B(n,s)| \ge \lambda \sum_{j=n}^\infty |B(j,s)|.$$

Theorem 6.6.2 ([91]). *Assume conditions (6.6.16)–(6.6.19) hold. Then (6.6.13) is consistent with respect to uniform exponential stability under the discretization scheme (6.6.3) and (6.6.11) with its continuous counterpart (6.6.9).*

Proof. Define

$$V(t,x) = |x(t)| + k \int_0^t \int_t^\infty |B(u,s)|\,du|f(s,x(s))|\,ds. \qquad (6.6.20)$$

Let $V'(t,x) = \frac{d}{dt}V(t,x(t))$. Then along the solutions of (6.6.9) we have,

$$
\begin{aligned}
V'(t,x) &= \frac{x(t)}{|x(t)|}x'(t) + k\int_t^\infty |B(u,t)|\,du|f(t,x(t))| - k\int_0^t |B(t,s)||f(s,x(s))|\,ds \\
&\le a|x(t)| + \int_0^t |B(t,s)||f(s,x(s))|\,ds \\
&\quad + k\int_t^\infty |B(u,t)|\,du|f(t,x(t))| - k\int_0^t |B(t,s)||f(s,x(s))|\,ds \\
&\le \left[a + k\int_t^\infty |B(u,t)|\,du\gamma\right]|x(t)| + (1-k)\int_0^t |B(t,s)||f(s,x(s))|\,ds \\
&\le -\alpha|x(t)| - \varepsilon\int_0^t |B(t,s)||f(s,x(s))|\,ds \\
&\le -\alpha|x(t)| - \varepsilon\lambda \int_0^t \int_t^\infty |B(u,s)|\,du|f(s,x(s))|\,ds \\
&\le -\alpha\left[|x(t)| + k\int_0^t \int_t^\infty |B(u,s)|\,du|f(s,x(s))|\,ds\right] \\
&\le -\alpha V(t,x). \qquad (6.6.21)
\end{aligned}
$$

Hence inequality (6.6.21) yields

$$V(t,x) \le V(t_0,\phi(.))e^{-\alpha(t-t_0)}.$$

As a consequence, we have

$$|x(t)| \le V(t_0,\phi(.))e^{\ \alpha(t-t_0)} \quad \text{for } t \ge t_0$$

$$\leq \|\phi\| \left[1 + k\gamma \int_0^{t_0} \int_{t_0}^{\infty} |B(u,s)| du\, ds \right] e^{-\alpha(t-t_0)} \quad \text{for } t \geq t_0.$$

Hence, the zero solution of (6.6.9) is uniformly exponentially stable.

Remark 6.5. Suppose $\rho(t_0)$ is a constant depending on t_0. If condition (6.6.19) is substituted with

$$\int_0^{t_0} \int_{t_0}^{\infty} |B(u,s)| du \gamma(s) ds \leq \rho(t_0), \quad \text{for } t_0 \geq 0,$$

then a slight modification of the proceeding paragraph shows that the zero solution of (6.6.9) is exponentially stable.

To show the zero solution of (6.6.12) is uniformly exponentially stable, we define $V(n) = V(n,x)$ by

$$V(n) = |x(n)| + kh\Phi(a,h) \sum_{s=0}^{n-1} \sum_{j=n}^{\infty} |B(j,s)||f(s,x(s))|.$$

Then along solutions of (6.6.12), we have

$$\Delta V(n) = |x(n+1)| - |x(n)| + kh\Phi(a,h) \sum_{s=0}^{n} \sum_{j=n+1}^{\infty} |B(j,s)||f(s,x(s))|$$

$$-kh\Phi(a,h) \sum_{s=0}^{n-1} \sum_{j=n}^{\infty} |B(j,s)||f(s,x(s))|$$

$$= \left| e^{ah}x(n) + h\Phi(a,h) \sum_{s=0}^{n} B(n,s)f(s,x(s)) \right|$$

$$-|x(n)| + kh\Phi(a,h) \sum_{s=0}^{n} \left[\sum_{j=n}^{\infty} |B(j,s)||f(s,x(s))| \right.$$

$$\left. -|B(n,s)||f(s,x(s))| \right] - kh\Phi(a,h) \sum_{s=0}^{n-1} \sum_{j=n}^{\infty} |B(j,s)||f(s,x(s))|$$

$$\leq \left[e^{ah} + \gamma kh\Phi(a,h) \sum_{j=n}^{\infty} |B(j,n)| - 1 \right] |x(n)|$$

$$+ h\Phi(a,h)(1-k) \sum_{s=0}^{n-1} |B(n,s)||f(s,x(s))|.$$

Let α be defined by (6.6.17). Then by (6.6.16) and (6.6.17), we can choose an appropriate h so that

$$e^{ah} + \gamma kh\Phi(a,h) \sum_{j=n}^{\infty} |B(j,n)| - 1 \leq -\alpha.$$

As a consequence,

$$\triangle V(n) \leq -\alpha |x(n)| + h\Phi(a,h)(1-k)\sum_{s=0}^{n-1} |B(n,s)||f(s,x(s))|$$

$$\leq -\alpha |x(n)| - \varepsilon\lambda h\Phi(a,h)\sum_{s=0}^{n-1}\sum_{j=n}^{\infty} |B(j,s)||f(s,x(s))|$$

$$\leq -\alpha \left[|x(n)| + h\Phi(a,h)k\sum_{s=0}^{n-1}\sum_{j=n}^{\infty} |B(j,s)||f(s,x(s))| \right]$$

$$= -\alpha V(n).$$

The above inequality implies that

$$V(n) \leq (1-\alpha)^{n-n_0}V(n_0), \ n \geq n_0 \geq 0.$$

Or

$$|x(n)| \leq (1-\alpha)^{n-n_0}V(n_0)$$

$$\leq ||\phi|| \left[1 + h\Phi(a,h)\sum_{s=0}^{n_0-1}\sum_{j=n_0}^{\infty} |B(j,s)| \right] (1-\alpha)^{n-n_0}, \ n \geq n_0 \geq 0.$$

This completes the proof.

Now we turn our attention to the preservation of boundedness. Consider the Volterra linear integro-differential equation

$$x'(t) = ax(t) + \int_0^t B(t,s)x(s)ds + g(t), \ t \geq 0 \tag{6.6.22}$$

and its analogous discrete Volterra equation, under discretizations (6.6.7) and (6.6.11)

$$x(n+1) = e^{ah}x(n) + h\Phi(a,h)\sum_{s=0}^{n} B(n,s)x(s) + h\Phi(a,h)g(n), \ n \geq 0 \tag{6.6.23}$$

where a, B are as defined before and g is continuous and uniformly bounded. Thus, there exists a positive constant M such that

$$h\Phi(a,h)|g(t)| \leq M, \text{ for all } t \geq 0. \tag{6.6.24}$$

Theorem 6.6.3. *Suppose there is a continuous function $\psi : [0,\infty) \to [0,\infty)$ with $\psi' \leq 0$ for $t \geq 0$, $\int_0^t \psi(u)du < \infty$, and $\frac{\partial}{\partial t}\psi(t-s) + |B(t,s)| \leq 0$ for $0 \leq s < t < \infty$, where $|B(t,t)|$ is uniformly bounded. If for $t \geq 0$, $a + \psi(0) \leq -\alpha < 0$, for some positive constant α, then (6.6.23) is consistent with respect to boundedness under the discretization scheme (6.6.7) and (6.6.11) with its continuous counterpart (6.6.22).*

Proof. Define a Lyapunov functional

$$V(t,x) = |x(t)| + \int_0^t \psi(t-s)|x(s)|ds.$$

Along the solutions of (6.6.22) we have,

$$
\begin{aligned}
V'(t,x) &= \frac{x(t)}{|x(t)|}x'(t) + \psi(0)|x(t)| + \int_0^t \frac{\partial}{\partial t}\psi(t-s)|x(s)|ds \\
&\le a|x(t)| + \int_0^t |B(t,s)||x(s)|ds \\
&\quad + |g(t)| + \psi(0)|x(t)| + \int_0^t \frac{\partial}{\partial t}\psi(t-s)|x(s)|ds \\
&\le [a + \psi(0)]|x(t)| + M + \int_0^t \left[\frac{\partial}{\partial t}\psi(t-s) + |B(t,s)|\right]|x(s)|ds \\
&\le -\alpha|x(t)| + M.
\end{aligned}
$$

By ([23], pp. 109–111) we have all solutions of (6.6.22) are bounded. With respect to (6.6.23) we consider the Lyapunov functional

$$V(n) = |x(n)| + h\Phi(a,h)\sum_{s=0}^{n-1}\psi(n-s-1)|x(s)|.$$

Then along solutions of (6.6.23), we have

$$
\begin{aligned}
\triangle V(n) &\le \left[e^{ah} + h\Phi(a,h)\left(|B(n,n)| + \psi(0)\right) - 1\right]|x(n)| \\
&\quad + h\Phi(a,h)\sum_{s=0}^{n-1}\left[\triangle_n\psi(n-s-1) + |B(n,s)|\right]|x(s)| + |g(n)|.
\end{aligned}
$$

Due to condition $\frac{\partial}{\partial t}\psi(t-s) + |B(t,s)| \le 0$ for $0 \le s < t < \infty$, we have $\triangle_n\psi(n-s-1) + |B(n,s)| \le 0$ for $0 \le s < n < \infty$.
Also, due to condition $a + \psi(0) \le -\alpha < 0$ we have $a < 0$. Moreover, since $|B(t,t)|$ is uniformly bounded we arrive at the fact that we can choose h small enough so that

$$e^{ah} + h\Phi(a,h)\left(|B(n,n)| + \psi(0)\right) - 1 \le -\alpha,$$

for some positive constant α. As a consequence,

$$\triangle V(n) \le -\alpha|x(n)| + M.$$

By Theorem 2.1.1 we have all solutions of (6.6.23) are bounded. Thus, (6.6.23) is consistent with respect to boundedness under the discretization scheme (6.6.7) and (6.6.11) with its continuous counterpart (6.6.22).
Next we state the following corollaries using discretization (6.6.3).

Corollary 6.5 ([91]). *Assume conditions (6.6.14) and (6.6.15) hold. Then (6.6.12) is consistent with respect to uniform asymptotic stability under the discretization scheme (6.6.3) and (6.6.11) with its continuous counterpart (6.6.9).*

The proof follows along the lines of the proof of Theorem 6.6.1 by taking

$$V(n) = |x(n)| + \gamma \frac{2h^2}{2-ah} \sum_{s=0}^{n-1} \sum_{u=n}^{\infty} |B(u,s)||x(s)|.$$

Corollary 6.6 ([91]). *Assume conditions (6.6.16)–(6.6.19) hold. Then (6.6.12) is consistent with respect to uniform exponential stability under the discretization scheme (6.6.3) and (6.6.11) with its continuous counterpart (6.6.9).*

The proof follows along the lines of the proof of Theorem 6.6.2 by taking

$$V(n) = |x(n)| + \frac{2h^2}{2-ah} k \sum_{s=0}^{n-1} \sum_{j=n}^{\infty} |B(j,s)||f(s,x(s))|.$$

For the next corollary we consider (6.6.22) and its analogous discrete Volterra difference equation

$$x(n+1) = \frac{2+ah}{2-ah} x(n) + \frac{2h^2}{2-ah} \sum_{s=0}^{n} B(n,s)x(s) + g(n), \ n \geq 0 \qquad (6.6.25)$$

under discretization scheme (6.6.3) and (6.6.11).

Corollary 6.7 ([91]). *Suppose there is a continuous function $\psi : [0,\infty) \to [0,\infty)$ with $\psi' \leq 0$ for $t \geq 0$, $\int_0^t \psi(u)du < \infty$, and $\frac{\partial}{\partial t}\psi(t-s) + |B(t,s)| \leq 0$ for $0 \leq s < t < \infty$, where $|B(t,t)|$ is uniformly bounded. If for $t \geq 0$, $a + \psi(0) \leq -\alpha < 0$, for some positive constant α, then (6.6.25) is consistent with respect to boundedness under the discretization scheme (6.6.3) and (6.6.11) with its continuous counterpart (6.6.22).*

The proof follows along the lines of the proof of Theorem 6.6.3 by taking

$$V(n) = |x(n)| + \frac{2h^2}{2-ah} \sum_{s=0}^{n-1} \psi(n-s-1)|x(s)|.$$

6.7 Semigroup

We end the book with a brief introduction of the concept of semigroup. The notion of semigroup falls under the umbrella of fixed point theory. In continuous dynamical systems, including partial differential equations, semigroup has been the main tools in studying boundedness, uniform exponential stability, strong stability, weak stability, almost weak stability, the existence of weak solutions, and almost periodic solutions. The theory of semigroup has been well developed for continuous dynamical systems, which is not the case for discrete dynamical systems. This section is only intended to raise curiosity about the subject of semigroup and how it can be effectively used to qualitatively study solutions of discrete dynamical systems, and in particular, Volterra difference equations.

Let X be a Banach space and $\mathscr{B}(X)$ the Banach algebra of all linear and bounded operators acting on X.

Definition 6.7.1. The subset $\mathbb{T} = \{T(n)\}_{n \in \mathbb{Z}}$ of $\mathscr{B}(X)$ is called discrete semigroup if it satisfies the following conditions:
(i) $T(0) = I$, where I is the identity operator on X.
(ii) $T(n+m) = T(n)T(m)$, for all $n, m \in \mathbb{Z}^+$.

Definition 6.7.2. A linear operator A is called the generator of semigroup T if

$$\lim_{s \to 1} \frac{T(s)x - T(1)x}{s - 1}, \; x \in D(A),$$

where the domain $D(A)$ of A is the set of all $x \in X$ for which the above limit exists.

Next we consider the discrete initial value problem

$$x(t+1) = Ax(t), \; x(t_0) = x_0 \in D(A), \; t \geq t_0, \; t, t_0 \in \mathbb{Z}^+, \tag{6.7.1}$$

where A is the generator of T. By [76] the initial value problem (6.7.1) has the unique solution

$$x(t) = T(t - t_0)x_0. \tag{6.7.2}$$

Denote the norms in X and $\mathscr{B}(X)$ by $\|\cdot\|$. We have the following theorems. First, for concise definitions and terminology regarding stability and boundedness we refer to [76].

Theorem 6.7.1 ([76]). *The following statements are equivalent:*
(i) Equation (6.7.1) is stable;
(ii) $\{T(t) : t \in \mathbb{Z}\}$ is bounded;
(iii) Equation (6.7.1) is uniformly stable.

Theorem 6.7.2 ([76]). *The following statements are equivalent:*
(i) Equation (6.7.1) is asymptotically stable;
(ii) $\lim_{t \to \infty} \|T(t)x\| = 0$, for every $x \in X$;
(iii) Equation (6.7.1) is globally asymptotically stable;
(iv) Equation (6.7.1) is uniformly asymptotically stable.

Next we turn our attention to using semigroup in Volterra difference equations. Thus, we consider the linear convolution Volterra difference equations with infinite delays

$$x(n+1) = \sum_{s=-\infty}^{n} C(n-s)x(s), \; n \geq n_0 \geq 0, \; n, n_0 \in \mathbb{Z}^+, \tag{6.7.3}$$

and

$$x(n+1) = \sum_{s=-\infty}^{n} \{C(n-s) + G(n,s)\}x(s), \; n \geq n_0 \geq 0, n, n_0 \in \mathbb{Z}^+. \qquad (6.7.4)$$

Our intention is to write (6.7.3) as a functional difference equation so that semigroup can be used to derive conditions that relate solutions of (6.7.3) and (6.7.4). Let γ be a positive constant. Define the set

$$B^\gamma = \{\varphi : \mathbb{Z}^- \to \mathbb{C}^k : \sup_{t \in \mathbb{Z}^-} |\varphi(t)|e^{\gamma t} < \infty\},$$

where $\mathbb{C}$ is the set of complex numbers. Then B^γ is a Banach space when endowed with the norm

$$||\varphi|| = \sup_{t \in \mathbb{Z}^-} |\varphi(t)|e^{\gamma t} < \infty, \; \varphi \in B^\gamma.$$

As we have done before, for $x_n \in B^\gamma$, we set

$$x_n(s) = x(n+s), \; s \in \mathbb{Z}^+.$$

Then we may write (6.7.3) as

$$x(n+1) = L(x_n), \qquad (6.7.5)$$

where $L(\cdot) : B^\gamma \to \mathbb{C}^k$ is a functional given by

$$L(\varphi) = \sum_{j=0}^{\infty} C(j)\varphi(-j), \; \varphi \in B^\gamma.$$

Let $T(n)$ denote the solution of (6.7.5). Then $T(n)\varphi = x_n(\varphi)$, for $\varphi \in B^\gamma$. Moreover, we denote by $x(\cdot, \varphi)$ the solution of (6.7.5) satisfying $x(s, \varphi) = \varphi(s)$, for $s \in \mathbb{Z}^-$. Then it can be easily shown that $T(n)$ is a bounded linear operator on B^γ and satisfies the semigroup property

$$T(n+m) = T(n)T(m).$$

We have the following theorems.

Theorem 6.7.3 ([59]). *Suppose system (6.7.5) possesses an ordinary dichotomy with dichotomy constant M (see [59]). Assume*

$$\sum_{n=0}^{\infty} |C(n)|e^{\gamma n} < \infty \text{ and } \sum_{s=-\infty}^{n} \sup_{n \geq n_0} |G(n,s)|e^{\gamma(n-s)} < \infty, \qquad (6.7.6)$$

$$\sum_{s=n_0}^{\infty} \sum_{j=-\infty}^{n_0-1} |G(s,j)|e^{\gamma(n_0-j)} + \sum_{s=n_0}^{\infty} \sum_{j=n_0}^{s} |G(s,j)| < 1/M. \qquad (6.7.7)$$

Then for any bounded solution $x(n)$ of (6.7.3) on $[n_0, \infty)$ there exists a unique bounded solution $y(n)$ of (6.7.4) on $[n_0, \infty)$ such that

$$y(n) = x(n) + \sum_{s=n_0}^{n-1} T(n-s-1)PE^0\left(\sum_{j=-\infty}^{s} |G(s,j)y(j)| \right)$$

$$- \sum_{s=n}^{\infty} T(n-s-1)(I-P)E^0\left(\sum_{j=-\infty}^{s} |G(s,j)y(j)| \right), \; n \geq n_0, \quad (6.7.8)$$

where $E^0(t) = I$ if $t = 0$ and $E^0(t) = 0$ (matrix) if $t \neq 0$.

Theorem 6.7.4 ([59]). *Assume* (6.7.6) *and* (6.7.7) *Suppose system* (6.7.5) *possesses an ordinary dichotomy with dichotomy constant M and related projection P (see [59]) such that*

$$||T(n)P|| \leq Ma^n \text{ for some } a, 0 < a < 1.$$

Then there is a one-to-one correspondence between bounded solutions $x(n)$ of (6.7.3) *on $[n_0,\infty)$ and bounded solutions $y(n)$ of* (6.7.4) *on $[n_0,\infty)$, and the asymptotic relation*

$$y(n) = x(n) + o(1) \; (n \to \infty)$$

holds.

Naturally, the resolvent operator that was developed in Chapter 1, Section 1.3, might be used to define a semigroup for Volterra difference equations. To see this, we consider Volterra difference equation of convolution type

$$x(n+1) = Ax(n) + \sum_{s=0}^{n} B(n-s)x(s) \tag{6.7.9}$$

for all integers $n \geq 0$ and for integers, $0 \leq s \leq n$, where A, B are $k \times k$ matrix functions, and x is a $k \times 1$ unknown vector. Then, we saw that the resolvent matrix equation of (6.7.9) takes the form

$$R(n+1) = AR(n) + \sum_{u=0}^{n} B(n-u)R(u), \; R(0) = I, \; n \in \mathbb{Z}^+. \tag{6.7.10}$$

Let A and $B(\cdot)$ be closed operators in X. Hence $D(A)$ endowed with the graph norm $|x| = ||x|| + ||Ax||$ is a Banach space denoted by Y. Next we use the resolvent operator to define the solution of the nonhomogenous Volterra difference equation of convolution type

$$x(n+1) = Ax(n) + \sum_{s=0}^{n} B(n-s)x(s) + f(n), \tag{6.7.11}$$

where f is a $k \times 1$ given vector. First, we have the following definition.

Definition 6.7.3. $R(\cdot)$ is a resolvent of (6.7.11) if $R(n) \in \mathscr{B}(X)$ for $n \in \mathbb{Z}^+$ and satisfies

1. $R(0) = I$ (the identity operator on X).
2. $R(n) \in \mathscr{B}(Y)$ for $n \in \mathbb{Z}^+$ and for $y \in Y$, we have

$$R(n+1)y = AR(n)y + \sum_{u=0}^{n} B(n-u)R(u)y$$

$$= R(n)Ay + \sum_{u=0}^{n} R(n-u)B(u)y. \qquad (6.7.12)$$

We note that item 2. of Definition 6.7.3 is needed for (ii) of Definition 6.7.1. Suppose $R(n)$ is the resolvent operator of (6.7.11). Then, it can be easily shown using the results of Section 1.3 that the solution of (6.7.11) is given by

$$x(n) = R(n)x_0 + \sum_{u=0}^{n} R(n-u-1)f(u), \ n \in \mathbb{Z}^+. \qquad (6.7.13)$$

Now, one can use the concept of the resolvent operator given by (6.7.12) to obtain various results concerning the qualitative analysis of solutions of Volterra difference equations.

It is worth noting, however, that using the resolvent operator in Volterra difference equations to define a semigroup and obtain a meaningful result is in dire need for further development.

6.8 Open Problems

Open Problem 1
Prove a parallel theorem to Theorem 6.3.5 by considering (6.3.30) as a vector equation.

Open Problem 2
Extend Theorem 6.3.5 to the delay Volterra difference equation

$$x(n+1) = \mu(n)x(n) + \sum_{s=n-h}^{n-1} h(n,s)x(s) + f(n),$$

where h is a positive integer.

Open Problem 3 (Extremely Hard)
Extend the results of Section 6.1 to the following Volterra difference equations

$$x(n+1) = Px(n) + \sum_{s=-\infty}^{n-1} H(n,s)g(x(s)), \ (vector)$$

$$x(n+1) = a(n)x(n) + \sum_{s=-\infty}^{n-1} h(n,s)g(x(s)), \quad (scalar)$$

and

$$x(n+1) = Px(n) + \sum_{s=n-h}^{n-1} H(n,s)g(x(s)), \quad (vector)$$

where h is a positive integer.

References

1. Adivar M., Koyuncuoğlu H. and Raffoul Y., *Existence of periodic and non-periodic solutions of systems of nonlinear Volterra difference equations with Infinite delay*, Journal of Difference Equations and Applications, Volume 19, Issue 12, December 2013, pages 1927–1939.

2. Adivar, M., and Raffoul, Y., *Existence of for Volterra integral equations on time scales,* Bull. Aust. Math. Soc. 82 (2010), 139–155.

3. Adivar, M., and Raffoul, Y., *Qualitative analysis of nonlinear Volterra integral equations on time scales using and Lyapunov functionals,* Applied Mathematics and Computation 273 (2016) 258–266.

4. Adivar, M,. Islam, M. and Raffoul, Y., *Separate contraction and existence of periodic solutions in totally nonlinear delay differential equations,* Hacettepe Journal of Mathematics and Statistics, 41 (1) (2012), 1–13.

5. Agarwal, R., and Pang, P.Y., *On a generalized difference system*, Nonlinear Anal., TM and Appl., 30(1997), 365–376.

6. Agarwal, R., *Difference Equations and Inequalities. Theory, Methods and Applications, second ed., in: Monographs and Textbooks in Pure and Applied Mathematics*, Marcel Decker, Inc., New York, 2000.

7. Agarwal, R., and O'Regan, D., *Infinite Interval Problems for Differential, Difference, and Integral Equations. Kluwer Academic*, 2001.

8. Agarwal R. and Wong, P.J.Y., On the existence of positive solutions of higher order difference equations, *Topological Methods in Nonlinear Analysis* **10** (1997) 2, 339–351.

9. Alsahafi, S., Raffoul, Y., and Sanbo, A., *Qualitative analysis of solutions in Volterra nonlinear systems of difference equations,* Int. Journal of Math. Analysis, Vol. 8, 2014, no. 31.

10. Ashegi, R., *Bifurcations and dynamics of discrete predator-prey system,* Journal of Biological Dynamics, Vol.8, No.1, 161–186.

11. Appleby, J.D., and Rodkina, A., *Stability of nonlinear stochastic Volterra difference equations with respect to a fading perturbation*, Int. J. Difference Equ., 4 (2009), no.2, pp. 165–184.

© Springer Nature Switzerland AG 2018
Y. N. Raffoul, *Qualitative Theory of Volterra Difference Equations*,
https://doi.org/10.1007/978-3-319-97190-2

12. Bakke, V.L., and Jackiewicz, Z., *Boundedness of solutions of difference equations and applications to numerical solutions of Volterra difference equations of the second kind,* J. Math. Anal. Appl. 115 (1986), pp. 592–605.

13. Bailey, A., and Nicholson, J., *The balance of animal populations,* I Proc. Zool. Soc. Lon. 3 (1935), pp. 551–598.

14. Becker, L.C., *Stability Consideration for Volterra Integro-differential Equations,* Ph.D. Dissertation, Southern Illinois University, 1979.

15. Beddington, R., Free, C, and Lawton, J., *Dynamic complexity in predator-prey models framed in difference equations,* Nature 225, (1975), pp. 58–60.

16. Berezansky, L., and Braverman, E., *Exponential stability of difference equations with several delays: Recursive approach,* Adv. Difference. Equ. Vol. 2009, Article ID 104310, 13, pages.

17. Berezansky, L., and Braverman, E., *On exponential dichotomy, Bohl–Perron type theorems and stability of difference equations,* Journal of Mathematical Analysis and Applications 304 (2), 511–530.

18. Bohner, M., and Peterson, A., *Dynamic Equations on Time Scales, An Introduction with Applications,* Birkhäuser, Boston, 2001.

19. Brunner, H., *The numerical analysis of functional integral and integro-differential equations of Volterra type,* Acta Numerica, 13 (2004), pp.55–145.

20. Brunner, H., and Van der Houwen, P.J., *The Numerical Solution of Volterra Equations,* SIAM, Philadelphia (1985)

21. Burton, T. A., *Uniform asymptotic stability in functional differential equations,* Proc. Amer. Math. Soc. 68(1978), 195–199.

22. Burton, T. A., *Volterra Integral and Differential Equations,* Academic Press, New York, 1983.

23. Burton, T. A., *Stability and Periodic Solutions of Ordinary and Functional Differential Equations,* Academic Press, New York, 1985.

24. Burton, T.A. *Integral equations, implicit functions, and fixed points,* Proc. Amer. Math. Soc. 124 (1996), 2383–2390.

25. Burton, T.A. and Kirk, C., *A fixed point theorem of Krasnoselskii-Schaefer type,* Math. Nachr., 189 (1998), 23–31.

26. Burton, T.A., *fixed points and differential equations with asymptotically constant or periodic solutions,* Electron. J. Qual. Theory Differ. Equ. 2004 (2004), no.11, 1–31.

27. Burton, T.A. *fixed points, Volterra equations, and Becker's resolvent,* Acta. Math. Hungar. 108(2005), 261–281.

28. Burton, T.A. *Integral equations, periodicity, and fixed points,* fixed point theory, 9 (2008), no. 1, 47–65.

29. Burton, T.A. *Integral equations, Volterra equations, and the remarkable resolvent,* E. J. Qualitative Theory of Diff. Equ. (2006) No. 2, 1–17.

30. Burton, T.A. *Integral equations, L^p-forcing, remarkable resolvent: Lyapunov functionals,* Nonlinear Anal. 68, 35–46.

31. Chen, L., Liujuan Chen, L., and Li, Z., *Permanence of a delayed discrete mutualism model with feedback controls,* Mathematical and Computer Modelling 50 (2009) 1083–089.

32. Chen, F.D., Liao, X.Y., and Huang, Z.K. *The dynamic behavior of n-species cooperation system with continuous time delays and feedback controls,* Appl. Math. Comput. 181(2006) 803–815.

33. Chen, F.D., *Permanence and global stability of nonautonomous Lotka-Volterra system with predator-prey and deviating arguments,* Appl. Math. Comput. 173(2006) 1082–1100.

34. Cheng, S., and Zhang, G. *Existence of positive periodic solutions for non-autonomous functional differential equations,* Electronic Journal of Differential Equations. **59** (2001), 1–8.

35. Clark, C.W., *A delay-recruitment model of population dynamics, with application to baleen whale populations,* J. Math. Biol. 3 (1976), 381–391.

36. Crisci, M.R., Kolmanovskii, V.B., and Vecchio. A., *Boundedness of discrete Volterra equations,* J. Math. Analy. Appl. 211(1997), 106–130.

37. Crisci, M.R., Kolmanovskii, V.B., and Vecchio. A., *Stability of difference Volterra equations: direct Lyapunov method and numerical procedure,* Advances in Difference Equations, II. Comput. Math. Appl. 36 (1998) 10–12, 77–97.

38. Crisci, M.R., Kolmanovskii, V.B., and Vecchio. A., *On the exponential stability of discrete Volterra systems,* J. Differ. Equations Appl. 6 (2000) 6, 667–680.

39. Cooke L.K., *Functional-differential equations; some models and perturbation problems,* Differential and Dynamical Systems, Academic Press, New York (1967), 167–183.

40. Cooke L.K., *An epidemic equation with immigration,* Math. Biosciences 29 (1976), 135–158.

41. Cooke L.K., and Yorke J.A.,*Some equations modeling growth process and gonorrhea epidemics,* Math. Biosciences 16 (1973), 75–101.

42. Cooke, K.L., and Győri, N., *Numerical approximation of the solutions of delay differential equations on an infinite interval using piecewise constant arguments,* Computers & Mathematics with Applications 28 (1–3), 81–92.

43. Cushing, J.M., *An operator equation and bounded solutions of integro-differential systems,* SIAM J. Math. Anal. 6 (1975) 433–445.

44. Cushing, J.M., *Integro-differential Equations and Delay Models in Population Dynamics,* Lecture Notes in Biomathematics, Vol. 20, Springer, Berlin, New York, 1977.

45. Datta, A. and Henderson, J. *Differences and smoothness of solutions for functional difference equations,* Proceedings Difference Equations **1** (1995), 133–142.

46. Dannan, F., Elaydi, S., and Liu, P., *Periodic solutions of difference equations,* J. Difference Equ. Appl., 6 (2000), no. 2, pp. 203–232.

47. Diblík, J., and Schmeidel, E., *On the existence of solutions of linear Volterra difference equations asymptotically equivalent to a given sequence,* Appl. Math. Comput. 218(18), 9310–9320 (2012).

48. Diblik, J., and Ruzickova, M., *Existence of asymptotically periodic solutions of system of Volterra difference equations,* J. Difference Equ. Appl., 15 (2009), no. 11–12, pp. 1165–1177.

49. Diblik, J., and Ruzickova, M., *Asymptotically periodic solutions of Volterra system of difference equations*, Comput. Math. Appl., 59 (2010), no.8, pp. 2854–2867.

50. Ding, D., Fang, K., and Zhao, Y., *Mathematical analysis for a discrete predator-prey model with time delay and holding II functional response*, Discrete Dynamics in Nature and Society, Vol. 2015, Article ID 797542, 8 pages.

51. Eid, G., Ghalayani, B., and Raffoul, Y., *Lyapunov functional and stability in nonlinear finite delay Volterra discrete systems*, International Journal of Difference Equations, 2015, 10(1), 77–90.

52. Elaydi, S.E., *Periodicity and stability of linear Volterra difference systems*, J. Math Anal. Appl. 181(1994), 483–492.

53. Elaydi, S.E. and Zhang, S., *Periodic solutions Volterra difference equations with Infinite delay I: The linear case,* Proceedings of the First International Conference on Difference Equations. Gorden and Breach (1994), 163–174.

54. Elaydi, S.E. and Zhang, S., *Periodic solutions Volterra difference equations with Infinite delay II: The nonlinear case,* Proceedings of the First International Conference on Difference Equations. Gorden and Breach,(1994), 175–183.

55. Elaydi, S.E., *Periodicity and Stability of linear Volterra difference systems*, J. Math. Anal. Appl., 181 (1994), no. 2, pp. 483–492.

56. Elaydi, S.E., *Stability of Volterra difference equations of convolution type*, Dynamical Systems. Nankai Ser. Pure Appl. Math. Theoret. Phys., 4 (1993), 66–72.

57. Elaydi, S.E., *An Introduction to Difference Equations,* Second edition. Undergraduate Texts in Mathematics. Springer-Verlag, New York, 1999.

58. Elaydi, S.E., *Stability and asymptoticity of Volterra difference equations, A progress report*, J. Comput. Appl. Math. 228 (2009) 2, 504?513.

59. Elaydi, S.E., *stability and asymptotocity of Volterra difference equations: A progress report*, J. Compu. and Appl. Math. 228 (2009) 504–513.

60. Elaydi, S.E., and Murakami, S., *Asymptotic stability versus exponential stability in linear Volterra difference equations of convolution type*, Journal of Difference Equations 2(1996), 401–410.

61. Elaydi, S.E., and Murakami, S., *Uniform asymptotic stability in linear Volterra difference equations*, Journal of Difference Equations 3(1998), 203–218.

62. Elaydi, S.E., Murakami, S. and Kamiyama, E., *Asymptotic equivalence for difference equations with infinite delay*, J. Difference Equ. Appl., 5 (1999), no. 1, pp. 1–23.

63. Elaydi, S.E., and Zhang, S., *Stability and periodicity of difference equations with finite delay,* Funkcial. Ekvac 37 (3), 401–413

64. Eloe, P., and Islam, M., *Stability properties and integrability of the resolvent of linear Volterra equations*, Tohoku Math. J., 47(1995), 263–269.

65. Eloe, P., Islam, M., and Raffoul, Y., *Uniform asymptotic stability in nonlinear Volterra discrete systems,* Special Issue on Advances in Difference Equations IV, Computers and Mathematics with Applications 45 (2003), pp. 1033–1039.

66. Eloe, P., Islam, M., and Zhang, B., *Uniform asymptotic stability in linear Volterra integro-differential equations with applications to delay systems*, Dynamic Systems and Applications, 9 (2000), 331–344.

67. Eloe, P., Raffoul, Y., Reid, D., and Yin, K., *Positive solutions of nonlinear functional difference equations,* Computers and Mathematics With applications. **42** (2001) , 639–646.

68. Fan, M., Wang, K., and Jiang, D., *Existence and global attractivity of positive periodic solutions of periodic n-species Lotka-Volterra competition systems with several deviating argunents,* Math. Biosci. 160(1999) 47–61.

69. Fan, Y., and Li, W.T., *Permanence for a delayed discrete ratio-dependent predator-prey system with Holling type functional response,* J. Math. Anal. Appl. 299(2004) 357–374.

70. Fan, M., and Wang, K., *Periodic solutions of a discrete time nonautonomous ratio-dependent predator-prey system. Math. Comput. Modelling,* 35 (2002), 951–961.

71. Gaines, R., and Mawhin, J., *Coincidence Degree, and Nonlinear Differential Equations*, Lecture Notes in Mathematics 568, Springer, Berlin, 1977.

72. Gronek, T., and Schmeidel, E., *Existence of bounded solution of Volterra difference equations via Darbos fixed-point theorem,* J. Differ. Equ. Appl. 19(10), (2013) 1645–1653.

73. Hakim, N., Kaufmann, J., Cerf, G., and Meadows, H., *Nonlinear time series prediction with a discrete-time recurrent neural network model,* Neural Networks, 1991., IJCNN-91-Seattle International Joint Conference on Neural Networks.

74. Hale, J., and Verduyn L., *Introduction to Functional Differential Equations,* Springer Verlag, New York,1993.

75. Hamaya, Y., *Existence of an almost periodic solution in a difference equation with Infinite delay*, J. Difference Equ. Appl., 9 (2003), no. 2, pp. 227–237.

76. Hamza, A., and Oraby, M. K., *Stability of abstract dynamic equations on time scales*, Advances in Difference Equations, 2012, 2012:143.

77. Hartung, F., and Győri, I., *Preservation of stability in a linear neutral differential equation under delay perturbations,* Dynamic Systems and Applications 10 (2), 225–242

78. Henderson, J., and A. Peterson, A., *Properties of delay variation in solutions of delay difference equations,* Journal of Differential Equations **1** (1995), 29–38.

79. Henderson, J., and Lauer, S. *Existence of a positive solution for an nth order boundary value problem for nonlinear difference equations,* Applied and Abstract Analysis, (1997), 271–279.

80. Henderson J., and Hudson, W.N., *Eigenvalue problems for nonlinear differential equations,* Communications on Applied Nonlinear Analysis **3** (1996), 51–58.

81. Hilker, F., *Strange periodic attractors in a prey-predator system with infected prey,* Math. Popul. Stud. 13 (2006), 119–134.

82. Hino, Y., and Murakami, S., *Stabilities in linear integro-differential equations,* Lecture Notes in Numerical and Applied Analysis, Kinokuniya, Tokyo, 15(1996), 31–46.

83. Islam, M., and Raffoul, Y., *Uniform asymptotic stability in linear Volterra difference equations*, PanAmerican Mathematical Journal, 11 (2001), No. 1, pp. 61–73.

84. Islam, M., and Raffoul, Y., *Stability properties of linear Volterra integrodifferential equations with nonlinear perturbation*, Communication of Applied Analysis, Vol. 7, No. 3 (2003) 406–416.

85. Islam, M., and Raffoul, Y., *Exponential stability in nonlinear difference equations,* Journal of Difference Equations and Applications, (2003), Vol. 9, No. 9, pp. 819–825.

86. Islam, M., and Raffoul, Y., *Periodic and asymptotically periodic solutions in coupled nonlinear systems of Volterra integro-differential equations,* Dynamic Systems and Applications, 23 (2014) 235–244.

87. Islam, M., and Yankson, E., *Boundedness and stability in nonlinear delay difference equations employing fixed point theory, Electron. J. Qual. Theory Differ. Equ.* 26 (2005).

88. Jiang, D. Wei, J., and Zhang, B., *Positive periodic solutions of functional differential equations and population models,* Electronic Journal of Differential equations, Vol. 2002(2002), No. 71, pp. 1–13.

89. Jirsa, V.K., and Haken H. *Field theory of electromagnetic brain activity,* Phys Rev Lett 1996; 77: 960–963.

90. Kapçak, S., Ufuktepe, U., and Elaydi, S., *Stability and invariant manifolds of a generalized Beddington host-parasitoid model,* Journal of Biological Dynamics, 2013 Vol. 7, No. 1, 233–253.

91. Kaufmann, E., and Raffoul, Y., *Discretization scheme in Volterra integrodifferential equations that Preserves stability and boundedness,* Journal of Difference Equations and Applications, 12 (2006), No. 7, 731–740.

92. Kelley, W., and Peterson, A., *Difference equations an introduction with applications,* Academic Press, 2000.

93. Khandaker, T., and Raffoul, Y., *Stability properties of linear Volterra difference equations with nonlinear perturbation*, J. Difference Equ. Appl., 10 (2002), pp. 857–874.

94. Kocić, V.L., and Ladas, G., *Global attractivity in nonlinear delay difference equations,* Proceedings of the American Mathematical Society 115 (4), (1992) 1083–1088.

95. Kocić, V.L., and Ladas, G., *Global behavior of nonlinear delay difference equations of higher order with applications,* Kluwer Academic Publishers, Boston 1993.

96. Kolmanovskii, V.B., Castellanos-Velasco, E., and, Torres-Muñoz, J.A., *A survey: stability and boundedness of Volterra difference equations,* Nonlinear Anal., 53 (2003), no. 7–8, pp. 861–928.

97. Krasnoselskii, M.A., *Positive solutions of operator Equations,* Noordhoff, Groningen, (1964).

98. Kublik, C., and Raffoul, Y., *Lyapunov functionals that lead to exponential stability and instability in finite delay Volterra difference equations,* Acta Mathematica Vietnamica October 17, (2014) pp. 77–89.

99. Lakshmikantham, L., and Trigiante, D., *Theory of difference equations: Numerical methods and applications,* Academic Press, New York, 1991.

100. Lauwerier, H., *Mathematical models of epidemics,* Math. Centrum, Amsterdam, 1981.

101. Liao, X.Y., and Cheng, S.S., *Convergent and divergent solutions of a discrete nonautonomous Lotka-Volterra Model,* Tamkang J. Math. 36(2005) 337–344.

102. Liao, X.Y., and Raffoul, Y.N. *Asymptotic behavior of positive solutions of second order nonlinear difference systems,* Nonlinear Studies, 14 (2007), no. 4, pp. 311–318.

103. Liao, X.Y., Li, W.T. and Raffoul, Y.N., *Boundedness in nonlinear functional difference equations via non-negative definite Lyapunov functionals with applications to Volterra discrete systems,* J. Nonlinear studies, 13(1)(2006) 1–14.

104. Liao, X.Y., Zhou, S.F., and Ouyang, Z., *On a stoichiometric two predators on one prey discrete model,* Appl. Math. Lett. 20 (2007) 272–278.

105. Liao, X.Y., Zhu, Y., and Chen, F., *On Asymptotic stability of delay-difference systems,* Appl. Math. Comput. 176 (2006) 759–767.

106. Linh, N., and Phat, V., *Exponential stability of nonlinear time-varying differential equations and applications,* Electronic Journal of Differential Equations, 34 (2001), 1–13.

107. Li, W.T., and Raffoul Y., *Classification and existence of positive solutions of systems of Volterra nonlinear difference equations,* Appl. Math. Comput., 155 (2004), no. 2, pp. 469–478.

108. Liley, D.T.J., and Cadusch, P.J., *A continuum theory of electro-cortical activity,* Neurocomputing 1999; 26–27: 795–800.

109. Lubich, C., *On the stability of linear multistep methods for Volterra integro-differential equations,* IMA J. Numer. Anal., 10 (1983), pp. 439–465.

110. Lyapunov, A.M., *The general problem of the stability of motion,* (Russian) Math. Soc. of Kharkov; English Translation, International Journal of Control, 55 (1992), 531–773.

111. Maroun, M., and Raffoul, Y., *Periodic Solutions in Nonlinear neutral difference equations with Functional Delay,* J. Korean Math. Soc. 42 (2005), No. 2, 255–268.

112. Matsunaga, H., and Hajiri, C.,*Exact stability sets for a linear difference systems with diagonal delay,* J. Math. Anal. 369 (2010) 616–622.

113. Medina, R., *The asymptotic behavior of the solutions of a Volterra difference equations,* Comput. Math. Appl., 181 (1994), no. 1, pp. 19–26.

114. Medina, R., *Solvability of discrete Volterra equations in weighted spaces,* Dynamic Systems and Appl. 5(1996), 407–422.

115. Medina, R., *Stability results for nonlinear difference equations,* Nonlinear Studies, Vol. 6, No. 1, 1999.

116. Medina, R., *Asymptotic equivalence of Volterra difference systems,* Intl. Jou. of Diff. Eqns. and Appl. Vol. 1 No.1(2000), 53–64.

117. Medina, R., *Asymptotic behavior of Volterra difference equations*, Computers and Mathematics with Applications, 41, (2001) 679–687.

118. Merdivenci, F., *Two positive solutions of a boundary value problem for difference equations,* Journal of Difference Equations and Application **1** (1995), 263–270.

119. Mickens, R., *A note on a discretization scheme for Volterra integro-differential equations that preserves stability and boundedness,* Journal of Difference Equations and Application **13**, No.6, (2007), 547–550.

120. Mickens, R., *Difference Equations: Theory and Applications* , 1990, (New York, NY: Chapman and Hall).

121. Mickens, R., *Nonstandard Finite Difference Models of Differential Equations*, 1994, (Singapore: World Scientific).

122. Mickens, R., *A nonstandard finite-difference scheme for the Lotka-Volterra system*, Applied Numerical Mathematics, 45, 2003, 309–314.

123. Miller, R., K., *Nonlinear Volterra Integral Equations*, Benjamin, New York, (1971).

124. Mohler, R., Rajkumar, V., and Zakrzewski, R.,*Nonlinear time-series-based adaptive control applications,* Decision and Control 1991. Proceedings of the 30th IEEE Conference on, pp. 2917–2919 vol.3, 1991.

125. Morshedy, E., *New explicit global asymptotic stability criteria for higher order difference equations,* J. Math. Anal. Appl. Vol. 336, no.1 324 (2007) 262–276.

126. Muroya, Y., *Persistence and global stability in discrete models of Lotka-Volterra type,* J. Math. Anal. Appl. 330 (2007) 24–33.

127. Radin, M., and Raffoul, Y., *Existence and uniqueness of asymptotically constant or periodic solutions in delayed population models,* Journal of Difference Equations and Applications, Volume 20, Issue 5–6, 2014, pp. 706–716.

128. Raffoul, Y., *Boundedness and Periodicity of Volterra Systems of Difference Equations,* Journal of Difference Equations and Applications, 1998, Vol. 4, pp. 381–393.

129. Raffoul, Y., *Periodic solutions for scalar and vector nonlinear difference equations,* Panamer. Math. J., 9(1999), 97–111.

130. Raffoul, Y., *Periodic solutions for nonlinear Volterra difference equations with Infinite delay,* Journal of Nonlinear Differential Equations, 5(1999), 25–33.

131. Raffoul, Y., *T-Periodic solutions and a priori bound* , Mathematical and Computer Modeling, 32(2000), 643–652.

132. Raffoul, Y., *Positive periodic solutions of nonlinear functional difference equations,* Electron. J. Diff. Eqns., 55(2002), 1–8.

133. Raffoul, Y., *General theorems for stability and boundedness for nonlinear functional discrete systems*, J. Math. Analy. Appl., 279 (2003), pp. 639–650.

134. Raffoul, Y., *Stability in neutral nonlinear differential equations with functional delays using fixed point theory,* Mathematical and Computer Modelling, 40(2004), 691–700.

135. Raffoul, Y., *Periodicity in General Delay Nonlinear Difference Equations Using fixed point Theory*, Journal of Difference Equations and Applications, (2004). Vol. 10, pp.1229–1242.

136. Raffoul, Y., *Stability and periodicity in discrete delay equations, J. Math. Anal. Appl.* 324 (2006) 1356–1362.

137. Raffoul, Y., *Periodicity in nonlinears systems with infinite delay, Advances of Dynamical Systems and Applications,* Vol. 3. No. 1 (2008), pp. 185–194.

138. Raffoul, Y., *Inequalities that lead to exponential stability and instability in delay difference equations,* Journal of Inequalities in Pure and Applied Mathematics, Vol. 10, iss.3, art, 70, 2009.

139. Raffoul, Y., *Discrete population models with asymptotically constant or periodic solutions* International Journal of Difference Equations, Volume 6, Number 2, pp. 143–152 (2012).

140. Raffoul, Y., *Stability in functional difference equations using fixed point theory,* Communications of the Korean Mathematical Society 29 (1), (2014), 195–204.

141. Raffoul, Y., *Necessary and sufficient conditions for uniform boundedness In functional difference equations,* EPAM, Volume 2, Issue 2, 2016, Pages 171–180.

142. Raffoul, Y., *Fixed point theory in Volterra summation equations*, preprint.

143. Raffoul, Y., *Total and asymptotic stability in linear Volterra integro-differential equations with nonlinear perturbation*, preprint.

144. Raffoul, Y., *Lyapunov-Razumikhin conditions that leads to stability and boundedness of functional difference equations of Volterra difference type*, preprint.

145. Raffoul, Y., *Uniform asymptotic stability and boundedness in functional finite delays difference equations*, preprint.

146. Raffoul, Y., *Stability in functional difference equations with applications to infinite delay Volterra difference equations*, preprint.

147. Raffoul, Y., Li, W.L., and Liao, X.Y., *Boundedness in nonlinear functional difference equations via non-negative Lyapunov functionals with applications to Volterra discrete systems,* Nonlinear Studies, 13(2006), No. 1, 1–13.

148. Raffoul, Y., Liao, X.Y., and Zhou, S., *On the discrete-time multi-species competition-predation system with several delays,* Appl. Math. Lett. 21 (2008), No. 1, 15, pp.15–22.

149. Raffoul, Y., and Tisdell, C., *Positive periodic solutions of functional discrete systems and with applications to population models,* Advances in Difference Equations Vol.3 (2005), pp. 369–380.

150. Raffoul, Y., and Yankson, E., *Existence of bounded solutions for Almost-Linear Volterra difference equations using fixed point theory and Lyapunov Functionals Nonlinear Studies,* Vol 21, No (2014) pp. 663–674.

151. Raffoul, Y., *Positive periodic solutions of nonlinear functional difference equations,* Electronic Journal of Differential equations, Vol. 2002(2002), No. 55, pp. 1–8.

152. Raffoul, Y., Liao, X. Y., and Shengfan, Z., *On the discrete-time multi-species competition-predation system with several delays,* Appl. Math. Lett. 21 (2008), No. 1, 15, pp.15–22.

153. Robinson, P.A., Rennie, C.J., and Wright, J.J., *Propagation and stability of waves of electrical activity in the cerebral cortex,* Phys Rev E 1997; 56: 826–840.

154. Roger, L., *A Nonstandard discretization method for Lotka-Volterra models that preserves periodic solutions,* Journal of Difference Equations and Applications, (8), Volume 11, 2005, pp 721–733.

155. Saito, Y., Ma, M., and Hara, T., *A necessary and sufficient condition for permanence of a Lotka-Volterra discrete system with delays,* J. Math. Anal. Appl. 256(2001) 162–174.

156. Schaefer, H., *Uber die Method der a priori Scranken, Math Ann.* 129(1955), 45–416.

157. Scheffer, M., *Fish nutrient interplay determines algal biomass: a minimal model,* Oikos 62 (1991), 271–282.

158. Scud, M.F., *Vito Volterra and theoretical ecology,* Theoretical Population Biology 2, 1-23 (1971).

159. Smart, D.R. *Fixed Point Theorems, Cambridge University Press, London,* 1980.

160. Smith, M. J., and Wisten, M.B. *A continuous day-to-day traffic assignment model and the existence of a continuous dynamic user equilibrium,* Annals of Operations Research. (1995). 60 (1): 59–79. doi:10.1007/BF02031940.

161. Song, Y., and Baker, C., *Qualitative behavior of numerical approximations to Volterra integro-differential equations,* Journal of Computational and Applied Mathematics 172 (2004) 101–115.

162. Tang, X. H., and Zou, X., *Global attractivity of nonautonomous Lotka-Volterra competition system without instantaneous negative feedback,* J. Diff. Equ. 192(2003) 502–535.

163. Taniguchi, T.,*Asymptotic behavior of solutions of nonautonomous difference equations,* J. Math. Anal. Appl. 184(2006) 342–347.

164. Yang, P., and Xu, R., *Global attractivity of the periodic Lotka-Volterra system,* J. Math. Anal. Appl. 233(1999) 221–232.

165. Yang, X., *Uniform persistence and periodic solutions for a discrete predator-prey system with delays,* J. Math. Anal. Appl. 316(2006) 161–177.

166. Yankson, E., *Stability in discrete equations with variable delays,* Electron. J. Qual. Theory Differ. Equ. 2009, No. 8, 1–7.

167. Yankson, E., *Stability of Volterra difference delay equations,* Electron. J. Qual. Theory Differ. Equ. 2006, No. 20, 1–14.

168. Yin, W., *Eigenvalue problems for functional differential equations,* Journal of Nonlinear Differential Equations, 3 (1997), 74–82.

169. Wen, X., *Global attractivity of positive solution of multispecies ecological competition-predator delay system (Chinese),* Acta Math. Sinica, 45(1)(2002) 83–92.

170. Wiener, J., *Differential equations with piecewise constant delays,* Trends in Theory and Practice of Nonlinear Differential Equations: Proc. Int. Conf., Arlington/Tex. 1982. Lecture Notes in Pure and Appl. Math. 90, Dekker, New York, 1984, pp. 547–552.

171. Wilson, H.R., and Cowan, J.D., *Excitatory and inhibitory interactions in lo-calized populations of model neurons,* Biophys J 1972; 12: 1–24.

172. Wilson, H.R., and Cowan, J.D., *A mathematical theory of the functional dy-namics of cortical and thalamic nervous tissue,* Kybernetik 1973; 13: 55–80.

173. Wilson, H.R., and Liley, D.T.J., *Simulation of the EEG: dynamic changes in synaptic efficiency, cerebral rhythms, and dissipative and generative activity in cortex,* Biol Cybern 1999; 81: 131–147.

174. Xu, R., Chaplain, M., and Chen, L., *Global asymptotic stability in n-species nonautonomous Lotka-Volterra competitive systems with infinite delays,* Appl. Math. Comput. 130(2002) 295–309.

175. Xu, R., and Chen, L., *Persistence and global stability for a delayed nonau-tonomous predator-prey system without dominating instantaneous negative feedback,* J. Math. Anal. Appl. 262(2001) 50–61.

176. Xu, C., and Li, P., *Dynamics in a discrete predator-prey system with infected prey,* Mathematics Bohemica, Vol. 139 (2014), No. 3, 511–534.

177. Zhang, B., *Asymptotic criteria and integrability properties of the resolvent of Volterra and functional equations,* Funkcialaj Ekvacioj, 40(1997), 355–351.

178. Zhou, Z., and Zou, X., *Stable periodic solutions in a discrete periodic Logistic equation,* Appl. Math. Lett. 16(2003) 165–171.

179. Zhu, J., and Liu, X., *Existence of positive solutions of nonlinear difference equations,* Nonlinear Oscillations, 9 (2006), no. 1, pp. 34–45.

180. Zhang, B., and Vandewalle, S., *General linear methods for Volterra integro-differential equations with memory,* SIAM J. Sci. Comput., 27 (2006), no. 6, pp. 2010–2031.

181. Zhang, S., *Stability of infinite delay difference systems,* Nonlinear Analysis, Method & Applications, (1994) Vol. 22, No. 9, pp. 1121–1129.

182. Zhang, S., and Chen, M.P., *A new Razumikhin theorem for delay difference equations,* Computers Math. Applic. (1998) Vol. 36, No. 10–12, pp. 405–412.

183. Zhang, S., *Stability of neutral delay difference systems,* Computers Math. Ap-plic. (2001) Vol. 42, pp. 291–299.

184. Zhao, M., Xuan, Z., and Li, C., *Dynamics of a discrete-time predator-prey system,* Advances in Difference Equations (2016) 191.

Index

Symbols
l_2-stable, 294, 295

A
A priori bound, 163, 166–170, 173, 182, 196, 200, 201, 203, 318
Ascoli-Arzelà, 97
Asymptotically periodic, 213
Asymptotically stable, 7, 57, 61, 70, 71, 291
Autonomous, 10

B
Banach space, 93, 96, 98–101, 107, 110, 117, 118, 125, 133, 141, 143, 144, 146, 148, 161, 165, 178, 179, 197, 211, 215, 221, 234, 235

C
Cauchy sequence, 96
Classifications of large contraction, 158
Compact, 93, 96, 97, 99, 101, 118, 120, 125, 166, 178–181, 198, 199, 215, 218, 238, 239
Complete metric, 100–102, 112, 133, 206, 209, 212
Cone, 233, 235
Continuous, 10, 34, 55, 90, 118, 122, 124, 125, 127, 165, 233, 291
Contraction, 101, 102, 105, 108, 118, 181, 204
Contraction mapping principle, 53, 93, 96, 122, 135, 141, 143, 182, 204, 226
Coupled system, 213

D
Discretization scheme, 295

E
Equicontinuous, 96, 97, 99, 101, 124, 127

F
Finite delay, 55, 62, 90, 168, 204, 211, 254, 272
Fixed point, 53, 93, 94, 101–103, 107, 114, 115, 118, 121, 122, 130, 132, 133, 140, 145, 147, 149, 161, 163, 165, 166, 178, 182, 196, 204, 206, 209, 212–215, 218, 220, 222, 223, 226, 233–235, 239–241, 263, 312, 316, 318, 319

I
Infinite delay, 142, 171, 196, 197, 199, 208, 214, 225, 234, 253

K
Krasnoselskii, 235

L
Large contraction, 153, 154, 158
Lipschitz, 96, 103, 205, 279
Lyapunov functional, 4, 15, 32, 66, 68–72, 78, 84, 94, 95, 141, 149, 151, 201–203, 271, 274, 282, 284, 287, 290, 294, 298, 303

N
Necessary and sufficient, 1, 48, 50, 62, 72, 132, 133, 137
Neural network, 9, 65
Neutral difference equation, 140, 178, 226, 317
Nonlinear, 20, 53, 95, 281
Norm, 98

P
Periodic solutions, 142, 163, 165–167, 233
Perturbed, 1, 32, 34, 48, 52, 164, 273
Population model, 241, 243, 245
Positive periodic solutions, 233
Predator-prey model, 231, 251

© Springer Nature Switzerland AG 2018
Y. N. Raffoul, *Qualitative Theory of Volterra Difference Equations*,
https://doi.org/10.1007/978-3-319-97190-2

Printed in the United States
By Bookmasters